PROBLEMS ON STATISTICAL MECHANICS

Graduate Student Series in Physics

Other books in the series

Gauge Theories in Particle Physics
I J R AITCHISON and A J G HEY

Introduction to Gauge Field Theory
D BAILIN and A LOVE

Supersymmetric Gauge Field Theory and String Theory
D BAILIN and A LOVE

Mechanics of Deformable Media
A B BHATIA and R N SINGH

Symmetries in Quantum Mechanics
M CHAICHIAN and R HAGEDORN

Hadron Interactions
P D B COLLINS and A D MARTIN

The Physics of Structurally Disordered Matter: An Introduction
N E CUSACK

Basic Ideas and Concepts in Nuclear Physics: An Introductory Approach (2nd edition)
K HEYDE

Collective Effects in Solids and Liquids
N H MARCH and M PARRINELLO

Geometry, Topology and Physics
M NAKAHARA

Supersymmetry, Superfields and Supergravity: An Introduction
P P SRIVASTAVA

Superfluidity and Superconductivity
D R TILLEY and J TILLEY

GRADUATE STUDENT SERIES IN PHYSICS

Series Editor:
Professor Douglas F Brewer, MA, DPhil
Emeritus Professor of Experimental Physics, University of Sussex

---

PROBLEMS ON STATISTICAL MECHANICS

DIEGO A R DALVIT

Universidad de Buenos Aires, Argentina
Los Alamos National Laboratory, USA

JAIME FRASTAI

Buenos Aires, Argentina

IAN D LAWRIE

University of Leeds, England

INSTITUTE OF PHYSICS PUBLISHING
Bristol and Philadelphia

British Library Cataloguing-in-Publication Data

A catalogue record for this book is available from the British Library.

ISBN 0 7503 0520 7 hbk
0 7503 0521 5 pbk

Library of Congress Cataloging-in-Publication Data

Dalvit, Diego A. R.
Problems on statistical mechanics / Diego A. R. Dalvit, Jaime Frastai and Ian D. Lawrie.
p. cm.—(Graduate student series in physics)
Includes index.
ISBN 0 7503 0520 7 (hbk: alk. paper)
ISBN 0 7503 0521 5 (pbk: alk. paper)
1. Statistical mechanics. Problems, exercises, etc.
I. Frastai, Jaime. II. Lawrie, Ian D. III. Title. IV. Series.
QC174.884.D35 1999
530.13′076—dc21 99-16504
CIP

Commissioning Editor: Jim Revill
Editorial Assistant: Victoria Le Billon
Production Editor: Nina Couzin
Production Control: Sarah Plenty and Jenny Troyano
Cover Design: Jeremy Stephens
Marketing Executive: Colin Fenton

Published by Institute of Physics Publishing, wholly owned by The Institute of Physics, London

Institute of Physics Publishing, Dirac House, Temple Back, Bristol BS1 6BE, UK

US Office: Institute of Physics Publishing, The Public Ledger Building, Suite 1035, 150 South Independence Mall West, Philadelphia, PA 19106, USA

Typeset in TeX using the IOP Bookmaker Macros
Printed in the UK by J W Arrowsmith Ltd, Bristol

CONTENTS

PREFACE

This book is intended as a source of material to complement standard textbooks for students following courses on statistical mechanics at an advanced undergraduate or graduate level.

As with most of the sub-disciplines of physics, the insight needed for a thorough understanding of the subject can be gained only through assiduous practice in solving problems. With this in mind, we have attempted to devise a collection of problems that illustrate both the basic principles of statistical mechanics and a varied range of its applications. Applications to real physical systems can be extremely difficult, and have given rise to a vast research literature. Even at the graduate level, the techniques that are needed are often too sophisticated to be dealt with adequately in a book such as this, and considerable simplifications are needed to make calculations reasonably tractable. Therefore, while some of our problems deal with model systems which are reasonable approximations to those found in the real world, others are quite highly idealized, and serve mainly to illustrate basic principles and techniques. By and large, we are not able to approach the current research literature very closely but, where possible, we indicate the directions in which more sophisticated developments are possible.

We assume that our readers already have (or, at least, are in the process of gaining) a basic understanding of the principles of statistical mechanics, such as is offered by standard textbooks and university courses, but we have provided a short summary of the relevant material in chapter 1. The remaining five chapters set out the problems and offer comprehensive solutions to all of them. In some cases, we go somewhat beyond the answers specifically sought in the questions, where we believe that an extended discussion of important ideas will be helpful. Throughout, we try to emphasize efficient and elegant methods of solution, and to bring out the beauty and coherence of the subject as a whole.

We are very grateful to Gabriel B Mindlin for his support and encouragement during the early stages of the writing of this book.

Diego A R Dalvit
Jaime Frastai
Ian D Lawrie
June 1999

1

INTRODUCTION

The problems presented in this volume are intended to illustrate the basic principles and techniques of statistical mechanics as these are normally presented to advanced undergraduate or beginning graduate students. The essential purpose of statistical mechanics is to understand the behaviour of large-scale macroscopic physical systems in terms of the laws of motion that govern their microscopic constituents. A remarkable amount can be understood at the macroscopic level from the purely phenomenological science of thermodynamics, and the insight into this phenomenology that is provided by a statistical treatment of microscopic physics is of prime importance. Accordingly, we begin in chapter 2 by reviewing some of the thermodynamic concepts that form the background to the microscopic theory. Chapter 3 treats the principal ensembles of equilibrium statistical mechanics and their relation to thermodynamic potentials, while chapter 4 investigates the effects that arise specifically from the quantum nature of matter, ignoring most of the complications that are brought about by interactions between particles. Examples of interacting systems are presented in chapter 5, and some of the methods that have been developed to treat macroscopic systems away from true thermal equilibrium are touched on in chapter 6.

Our problems and solutions are offered as an aid to study. They are designed so that readers who have a sound grasp of the material presented in standard textbooks should find them tractable; the formidable technical difficulties that one encounters in dealing with many physically realistic models have prompted the development of highly sophisticated mathematical techniques, but we consider only the simpler methods of approximation to fall within our scope. We would strongly encourage readers to develop their understanding of the subject by tackling the problems as fully as they can before inspecting our solutions. The questions are intended as a stimulus to thought, but not as an examination! In some cases we deliberately leave it to the reader's own good sense to identify assumptions or approximations that may be needed to obtain an answer. On the other hand, we try to give complete solutions and on occasion provide derivations and discussions that are not specifically sought in the questions, where we think that they may be helpful or illuminating.

In the remainder of this chapter, we summarize some essential definitions and results of thermodynamics and statistical mechanics that will be needed to solve the problems. A miscellany of useful mathematical formulae is given in appendix A at the end of the book.

Thermodynamics

Most of the useful results of thermodynamics rest on two phenomenological laws. For any thermodynamic system, the *first law of thermodynamics*, which amounts to a statement of the conservation of energy, can be expressed as

$$\Delta U = \Delta Q - \Delta W$$

where, for any change in the state of the system, ΔU is the change in the internal energy U of the system, ΔQ is the amount of heat absorbed by the system and ΔW is the work done by the system on its surroundings. For an infinitesimal reversible change, we may write $\mathrm{d}U = \mathrm{d}Q - \mathrm{d}W$, where $\mathrm{d}W$ can be expressed in terms of well-defined macroscopic functions of state. For a fluid, with hydrostatic pressure P and volume V, it is $\mathrm{d}W = P\,\mathrm{d}V$; for an elastic band of length l under a tension J it is $\mathrm{d}W = -J\,\mathrm{d}l$. The important case of work done during a change in state of a magnetic material is a potential source of confusion, which is the subject of problem 2.10.

The *second law of thermodynamics* has been stated in several different ways. One of the more useful versions is that of Clausius, to the effect that *no cyclic process is possible, whose only effect is the transfer of heat from a colder body to a hotter body*. Quite remarkably, it is possible to deduce from this statement the existence firstly of an *absolute thermodynamic scale of temperature*, which does not depend on properties of any particular thermometric substance, and secondly of a function of state called the *entropy* S. For an infinitesimal and reversible change between two states of thermal equilibrium at an absolute temperature T, the change in entropy is defined as

$$\mathrm{d}S = \frac{\mathrm{d}Q|_{\mathrm{rev}}}{T}.$$

As a result, the first law can be written completely in terms of functions of state as $\mathrm{d}U = T\,\mathrm{d}S - P\,\mathrm{d}V$ for a fluid system containing a fixed number of particles. If the number N of particles can change, then this statement becomes $\mathrm{d}U = T\,\mathrm{d}S - P\,\mathrm{d}V + \mu\,\mathrm{d}N$, where μ is the chemical potential, and corresponding expressions can be found for other systems. For an irreversible process, it is straightforward to deduce that $\mathrm{d}S > \mathrm{d}Q/T$, and for any change between equilibrium states of an *isolated* system that $\Delta S \geq 0$.

According to information theory, the entropy is associated with the number I of bits of information needed to specify the state of a system according to

$$\Delta S = -\frac{k}{\log_2 \mathrm{e}}\Delta I = -k\,\Delta I \ln 2$$

where k is Boltzmann's constant.

The *third law of thermodynamics* asserts that *the entropy change in any reversible process tends to zero as* $T \rightarrow 0$. This implies that the entropy of a system at $T = 0$ does not depend on the value of any other macroscopic quantity. Since only changes in entropy are defined *a priori*, we are free to specify that $S(T = 0) = 0$.

By writing the differential form of the first law as

$$\mathrm{d}S = \frac{1}{T}\,\mathrm{d}U + \frac{P}{T}\,\mathrm{d}V - \frac{\mu}{T}\,\mathrm{d}N$$

we see that the entropy is naturally regarded as a function $S(U, V, N)$ of a set of 'controlled' variables U, V and N. This means that, in a state of thermal equilibrium for which U, V and N are specified by external constraints, its partial derivatives with respect to these variables yield simple thermodynamic quantities:

$$\frac{1}{T} = \left(\frac{\partial S}{\partial U}\right)_{V,N} \qquad \frac{P}{T} = \left(\frac{\partial S}{\partial V}\right)_{U,N} \qquad \frac{\mu}{T} = -\left(\frac{\partial S}{\partial N}\right)_{U,V}.$$

Moreover, the law of increase in entropy shows that $-S$ is a minimum in the equilibrium state corresponding to these constraints. In this sense, the entropy serves as a 'thermodynamic potential' for a system in which U, V and N are controlled variables. For a system in which some other combination of variables is controlled, the appropriate thermodynamic potential is obtained from a suitable sequence of Legendre transformations. The most commonly used potentials are the internal energy $U(S, V, N)$, the Helmholtz free energy $F(T, V, N)$, the Gibbs free energy $G(T, P, N)$, the enthalpy $H(S, P, N)$ and the grand potential $\Omega(T, V, \mu)$. The definitions of these potentials and the associated differential relations that follow from the first and second laws may be summarized as

$$\begin{aligned}
U &= TS - PV + \mu N & \mathrm{d}U &= T\,\mathrm{d}S - P\,\mathrm{d}V + \mu\,\mathrm{d}N \\
F &= U - TS & \mathrm{d}F &= -S\,\mathrm{d}T - P\,\mathrm{d}V + \mu\,\mathrm{d}N \\
G &= F + PV & \mathrm{d}G &= -S\,\mathrm{d}T + V\,\mathrm{d}P + \mu\,\mathrm{d}N \\
H &= E + PV & \mathrm{d}H &= T\,\mathrm{d}S + V\,\mathrm{d}P + \mu\,\mathrm{d}N \\
\Omega &= -F + \mu N & \mathrm{d}\Omega &= S\,\mathrm{d}T + P\,\mathrm{d}V + N\,\mathrm{d}\mu.
\end{aligned}$$

From these, a variety of useful relations between second derivatives of the potentials, the *Maxwell relations*, can be obtained. Using the Helmholtz free energy, for example, we have

$$P = -\left(\frac{\partial F}{\partial V}\right)_{T,N} \qquad S = -\left(\frac{\partial F}{\partial T}\right)_{V,N}$$

and, for these, the fact that second derivatives of F do not depend on the order of differentiation, that is

$$\left[\frac{\partial}{\partial V}\left(\frac{\partial F}{\partial T}\right)_{V,N}\right]_{T,N} = \left[\frac{\partial}{\partial T}\left(\frac{\partial F}{\partial V}\right)_{T,N}\right]_{V,N}$$

(which is often expressed by saying that $\mathrm{d}F$ is an exact differential) yields the Maxwell relation

$$\left(\frac{\partial S}{\partial V}\right)_{T,N} = \left(\frac{\partial P}{\partial T}\right)_{V,N}.$$

A second set of relations is obtained from the assumption that the entropy S and its natural variables U, V and N are all *extensive*. Thus, if the volume V of our system and the number N of particles that it contains are both increased by the same factor, say λ, in such a way that *intensive* quantities such as temperature, pressure and density remain the same (and hence that the state of any small part of the system is unchanged), then S and U should increase by the same factor. Therefore, the form of the function $S(U, V, N)$ should be such that

$$S(\lambda U, \lambda V, \lambda N) = \lambda S(U, V, N).$$

By differentiating with respect to λ and setting $\lambda = 1$, we obtain

$$U\left(\frac{\partial S}{\partial U}\right)_{V,N} + V\left(\frac{\partial S}{\partial V}\right)_{U,N} + N\left(\frac{\partial S}{\partial N}\right)_{U,V} = S$$

which is equivalent to the *fundamental equation of thermodynamics* or the *Euler equation*

$$U + PV - TS = \mu N.$$

From this equation, it is easily deduced that the chemical potential μ is the Gibbs free energy per particle, $G = \mu N$, and that the grand potential is $\Omega = PV$. By differentiating the fundamental equation

$$\mathrm{d}U + P\,\mathrm{d}V + V\,\mathrm{d}P - \mu\,\mathrm{d}N - N\,\mathrm{d}\mu = T\,\mathrm{d}S + S\,\mathrm{d}T$$

and combining this with the differential form of the first and second laws, we obtain the *Gibbs–Duhem equation*

$$N\,\mathrm{d}\mu = -S\,\mathrm{d}T + V\,\mathrm{d}P.$$

The specific heat of a system is the heat input required to raise its temperature by 1 K, but this amount of heat depends on how much work is done during the process. To be concrete, one normally defines a particular heating process by specifying that some particular variable X is to be held fixed. In that case, for an infinitesimal change, we have

$$C_X \equiv \left(\frac{\mathrm{d}Q}{\mathrm{d}T}\right)_X = T\left(\frac{\partial S}{\partial T}\right)_X.$$

For a fluid, the *principal specific heats*

$$C_V = T\left(\frac{\partial S}{\partial T}\right)_{V,N} = \left(\frac{\partial U}{\partial T}\right)_{V,N}$$

$$C_P = T\left(\frac{\partial S}{\partial T}\right)_{P,N} = \left(\frac{\partial H}{\partial T}\right)_{P,N}$$

are especially useful, and analogous quantities can be defined for other systems. The value of a specific heat naturally depends on the amount of a substance that is involved. Depending on the situation considered, one may be interested in the heat capacity of the entire system, the specific heat per unit volume or the specific hear per unit mass. For all of these macroscopic quantities, we use upper-case symbols (C, C_V, C_P, etc), whose exact meaning should be clear from the context. Occasionally, it is helpful to refer to a specific heat *per particle*, and for these microscopic quantities we use lower case symbols (c, c_V, c_P, etc). Some other useful thermodynamic properties are the coefficient of thermal expansion α, the isothermal compressibility κ_T and the adiabatic compressibility κ_S. They are defined by

$$\alpha \equiv \frac{1}{V}\left(\frac{\partial V}{\partial T}\right)_{P,N} \qquad \kappa_T \equiv -\frac{1}{V}\left(\frac{\partial V}{\partial P}\right)_T \qquad \kappa_S \equiv -\frac{1}{V}\left(\frac{\partial V}{\partial P}\right)_{S,N}.$$

Statistical Ensembles

Equilibrium statistical mechanics is based on the idea that, when we make a controlled measurement on a macroscopic system, microscopic fluctuations are so rapid that the system passes through a sequence of many microscopic states during the finite time needed to make the measurement. Consequently, the measured value of a physical quantity is actually a long-time average of a fluctuating quantity. We assume (or in favourable cases it can be proved) that this long-time average is equivalent to an average over a hypothetical ensemble (a *Gibbs ensemble*) of infinitely many copies of the system, the systems belonging to the ensemble being in different microscopic states consistent with specified values of a small number of macroscopic variables. The statistical weight attached to each microstate is determined by requiring that the probability distribution should be stationary (since we are trying to describe a system in equilibrium) when each system in the ensemble evolves with time according to the microscopic equations of motion. What this stationary distribution is depends on the macroscopic constraints that are applied.

The microcanonical ensemble

The microcanonical ensemble describes a system that is completely isolated from its surroundings—a *closed* system. For a fluid, this means that the number N of particles in the system, the volume V and the total energy E are fixed. The stationary probability distribution is that which assigns equal probabilities to each microstate that is consistent with these constraints. A thermodynamic interpretation is made by identifying the entropy of the closed system as

$$S(E, V, N) = k \ln[\Omega(E, V, N)]$$

where $k = 1.38 \times 10^{-23}$ J K^{-1} is Boltzmann's constant, and $\Omega(E, V, N)$ is, roughly speaking, the number of microstates that are consistent with the total energy being E. (We use the notation E for this fixed energy and U for the average of a fluctuating total energy, or for the internal energy in a purely thermodynamic description of the system.) For a system that has a discrete set of microstates available to it, this definition can be applied directly. For a classical fluid, whose molecules can have positions and momenta within some continuous range, the definition of $\Omega(E, V, N)$ requires a little more care, and this is discussed in problem 3.1.

The canonical ensemble

The canonical ensemble describes a system that has a fixed number N of particles and a fixed volume V but is in thermal contact (and in thermal equilibrium) with a heat bath, with which it can exchange energy. The condition of thermal equilibrium implies that the system has a definite temperature T, which is a property of the heat bath. One finds that the appropriate statistical weight for a microstate of energy E is $\mathcal{P}(E) = Z(T, V, N)^{-1} \mathrm{e}^{-\beta E}$, where $\beta = 1/kT$ and the normalizing factor $Z(T, V, N)$ is called the canonical partition function. When the microstates are discrete (and can be labelled by an index ν), the partition function is

$$Z(T, V, N) = \sum_{\nu} \mathrm{e}^{-\beta E_\nu} = \sum_{E} \Omega(E, V, N)\, \mathrm{e}^{-\beta E}.$$

For a classical fluid of N indistinguishable particles which have coordinates q_i and momenta p_i, we have instead

$$Z(T, V, N) = \frac{1}{h^{3N} N!} \int \mathrm{d}^{3N}p\, \mathrm{d}^{3N}q\, \mathrm{e}^{-\beta \mathcal{H}(p,q)}$$

where $\mathcal{H}(p, q)$ is the Hamiltonian. The factor $1/N!$ enforces the requirement, sometimes called 'correct Boltzmann counting', that two states which differ only by the interchange of two indistinguishable particles are to count as the same state. The quantity h is an arbitrary constant with the dimensions of pq, included to make Z dimensionless. For most purposes the actual value of h is immaterial, but occasionally it is appropriate to identify h as Planck's constant. The connection of this ensemble with thermodynamics is established by identifying the Helmholtz free energy as

$$F(T, V, N) \equiv -kT \ln[Z(T, V, N)].$$

The grand canonical ensemble

The grand canonical ensemble describes an *open* system, whose volume V is fixed, in equilibrium with a reservoir with which it can exchange both energy and particles. A macroscopic state of this system is characterized by its volume, together with definite values of the temperature T and chemical potential μ, both of which are properties of the reservoir. A microstate is now specified by both the energy E and the number N of particles that find themselves instantaneously inside the system, and its statistical weight is $\mathcal{P}(E, N) = \mathcal{Z}(T, V, \mu)^{-1}\, \mathrm{e}^{\beta\mu N}\, \mathrm{e}^{-\beta E}$, where the grand canonical partition function is

$$\mathcal{Z}(T, V, \mu) = \sum_N \mathrm{e}^{\beta\mu N} Z(T, V, N)$$

where $Z(T, V, N)$ is the appropriate canonical partition function, as discussed above. The quantity $z = \mathrm{e}^{\beta\mu}$ is called the *fugacity*. This ensemble has a thermodynamic interpretation which is established by defining the grand potential as

$$\Omega(T, V, \mu) \equiv kT \ln[\mathcal{Z}(T, V, \mu)].$$

The isobaric ensemble

One further ensemble, which is useful for a system of exactly N particles whose pressure is held fixed while its volume is allowed to fluctuate, is the isobaric ensemble. Its partition function, corresponding to a microstate probability $\mathcal{P}(E, V) = \Xi(T, P, N)^{-1}\, \mathrm{e}^{-\beta E - \beta PV}$, is

$$\Xi(T, P, N) = \int \mathrm{d}V\, \mathrm{e}^{-\beta PV}\, Z(T, V, N)$$

and its connection with thermodynamics is through the Gibbs free energy

$$G(T, P, N) \equiv -kT \ln[\Xi(T, P, N)].$$

It is worth observing that the sequence of Legendre transformations which relate the various thermodynamic potentials is reflected by a similar sequence of Laplace transformations which relate the partition functions of the corresponding statistical ensembles.

Given a thermodynamic quantity $\mathcal{A}$, its statistical average over the corresponding ensemble is given roughly by

$$\langle \mathcal{A} \rangle = \sum_\nu \mathcal{P}_\nu \mathcal{A}_\nu$$

where ν labels microstates, although one must of course replace the sum with an appropriately normalized integral in the case of continuous degrees of freedom. With the same caveat, fluctuations in $\mathcal{A}$ can be measured by the variances $(\Delta\mathcal{A})^2$, given by

$$(\Delta\mathcal{A})^2 \equiv \langle (\mathcal{A} - \langle \mathcal{A} \rangle)^2 \rangle = \langle \mathcal{A}^2 \rangle - \langle \mathcal{A} \rangle^2.$$

Table 1.1. Expressions for some useful thermodynamic quantities in terms of the microcanonical, canonical and grand canonical partition functions

	Microcanonical $\Omega(E, V, N)$	Canonical $Z(T, V, N)$	Grand canonical $\mathcal{Z}(T, V, \mu)$
$\frac{S}{k}$	$\ln \Omega$	$\left(\frac{\partial(T \ln Z)}{\partial T}\right)_{V,N}$	$\left(\frac{\partial(T \ln \mathcal{Z})}{\partial T}\right)_{V,\mu}$
F	$E - kT \ln \Omega$	$-kT \ln Z$	$kT\mu^2 \left(\frac{\partial(\mu^{-1} \ln \mathcal{Z})}{\partial \mu}\right)_{T,V}$
U	Fixed $(=E)$	$kT^2\left(\frac{\partial(\ln Z)}{\partial T}\right)_{V,N}$	$-\left(\frac{\partial(\ln \mathcal{Z})}{\partial \beta}\right)_{\beta\mu,V}$
N	Fixed	Fixed	$kT\left(\frac{\partial(\ln \mathcal{Z})}{\partial \mu}\right)_{T,V}$
kT	$\left(\frac{\partial(\ln \Omega)}{\partial E}\right)_{V,N}^{-1}$	Fixed	Fixed
$\frac{\mu}{kT}$	$-\left(\frac{\partial(\ln \Omega)}{\partial N}\right)_{E,V}$	$-\left(\frac{\partial(\ln Z)}{\partial N}\right)_{T,V}$	Fixed
P	$kT\left(\frac{\partial(\ln \Omega)}{\partial V}\right)_{E,N}$	$kT\left(\frac{\partial(\ln Z)}{\partial V}\right)_{T,N}$	$\frac{kT}{V} \ln \mathcal{Z}$
$\frac{C_V}{k}$	$-\beta^2\left(\frac{\partial^2(\ln \Omega)}{\partial E^2}\right)_{V,N}^{-1}$	$\beta^2\left(\frac{\partial^2(\ln Z)}{\partial \beta^2}\right)_{V,N}$	$T\left(\frac{\partial^2(T \ln \mathcal{Z})}{\partial T^2}\right)_{V,\mu}$
$(\Delta N)^2$	0	0	$\left(\frac{\partial^2(\ln \mathcal{Z})}{\partial(\beta\mu)^2}\right)_{\beta,V}$
$(\Delta E)^2$	0	$\left(\frac{\partial^2(\ln Z)}{\partial \beta^2}\right)_{V,N}$	$\left(\frac{\partial^2(\ln \mathcal{Z})}{\partial \beta^2}\right)_{\beta\mu,V}$

An expression for the entropy in terms of the microstate probabilities of the system is

$$S = -k \sum_{\nu} \mathcal{P}_\nu \ln \mathcal{P}_\nu$$

which is known as the Gibbs entropy formula.

Table 1.1 summarizes the ways in which a variety of useful thermodynamic quantities are obtained from the partition functions of the three most commonly used ensembles. The entries for C_V require a little explanation. When heating a gas at constant volume, one can in principle choose to keep either N or μ fixed. In practice, it is normally N that will be fixed, and the first two entries correspond to this situation. In the grand canonical ensemble, this is not a natural procedure and, although a formal expression can be found, it is ungainly and of little practical use. The expression given corresponds to keeping V and μ fixed.

Quantum Statistics

When dealing with quantum-mechanical systems, we need to consider *indistinguishable* particles of two types. Particles of integer spin are said to obey Bose–Einstein statistics and are called bosons; those with half-odd-integer spin obey Fermi–Dirac statistics and are called fermions. Each single-particle state can be occupied by any number of bosons ($n_i = 0, 1, 2, \ldots, \infty$) but, for fermions (which obey the Pauli exclusion principle), a single-particle state can be occupied by at most one particle ($n_i = 0, 1$). In writing the occupation numbers n_i, we use the index i to indicate all the quantum numbers (linear momentum, angular momentum, spin, etc) that are needed to characterize a single-particle state. The distinctive statistics of quantum particles arise from the requirement that multi-particle states be either symmetric (for bosons) or antisymmetric (for fermions) when two indistinguishable particles are interchanged. At high temperatures and low densities, the average occupation number of any single-particle state is much smaller than unity, and the symmetry requirements are of little importance. Under these circumstances, quantum particles behave much like classical particles, and a convenient approximation is provided by the fictitious Maxwell–Boltzmann statistics. Here, both bosons and fermions are treated as classical indistinguishable particles, except that their momenta and other properties are restricted to have the values allowed for single quantum particles (see problems 4.1 and 4.2 for further discussion).

A system of non-interacting particles is most easily treated in the grand canonical ensemble. The grand potential $\Omega = kT \ln \mathcal{Z}$ can be expressed as

$$\Omega = PV = kT \sum_i a^{-1} \ln(1 + a\,\mathrm{e}^{-\beta(\varepsilon_i - \mu)})$$

where ε_i is the energy of the ith single-particle state. Bose–Einstein statistics corresponds to $a = -1$, Fermi–Dirac statistics to $a = 1$ and the limit $a \to 0$ yields the grand potential for Maxwell–Boltzmann particles.

The particles of which ordinary non-relativistic gases are composed cannot normally be created or destroyed, so the number N of such particles is conserved. This is not true, for example, of the photons which constitute black-body radiation, since these can be created and absorbed by the acceleration of charged particles. Under these circumstances, a factor such as $\mathrm{e}^{\beta\mu N}$ cannot appear in a stationary probability distribution, and the chemical potential for photons must be exactly zero. The same applies to quasiparticle excitations that describe collective motion in condensed-matter systems, such as the phonon excitations of a solid (see problem 4.7) or the roton excitations of a superfluid (problem 4.16), which for many purposes can be treated just like a gas of bosons.

The mean occupation number $\langle n_i \rangle$ (which might also be written as $n(\varepsilon_i)$, since it depends only on the single-particle energy ε_i) is given by

$$\langle n_i \rangle = -\frac{1}{\beta}\left(\frac{\partial(\ln \mathcal{Z})}{\partial \varepsilon_i}\right)_{\beta\mu} = \frac{1}{\mathrm{e}^{\beta(\varepsilon_i - \mu)} + a}.$$

For a system of non-interacting particles, the total energy of the system is just the sum of the individual energies of all the particles. The ensemble average of any quantity $\mathcal{A}$ of this kind is given in terms of the mean occupation numbers by

$$\langle \mathcal{A} \rangle = \sum_i \mathcal{A}_i \langle n_i \rangle$$

where $\mathcal{A}_i$ is the value of $\mathcal{A}$ for a single particle in the the ith state. The sum over all states is often well approximated by the integral

$$\int \mathrm{d}\varepsilon \, g(\varepsilon) \mathcal{A}(\varepsilon)$$

where the density of states $g(\varepsilon)$ is the number of single-particle states with energy between ε and $\varepsilon + \mathrm{d}\varepsilon$. Examples of the densities of states for representative systems in a volume V are

$$g(\varepsilon) = \begin{cases} \dfrac{V\varepsilon^2}{\pi^2\hbar^3 c^3} & \text{photons} \\ \dfrac{9N\varepsilon^2}{(\hbar\omega_\mathrm{D})^3} & \text{phonons} \\ \dfrac{(2s+1)Vm^{3/2}\sqrt{\varepsilon}}{\sqrt{2}\pi^2\hbar^3} & \text{non-relativistic particles.} \end{cases}$$

For photons, c is the speed of light. In the case of phonons, we have used the Debye approximation, replacing the true phonon dispersion relation with a linear one, up to a cut-off frequency ω_D, to ensure that the total number of phonon modes is $3N$ for a solid containing N atoms. For non-relativistic particles, the mass of each particle is m, and its spin is s. Making use of this density of states, we can express the mean value $U = \langle E \rangle$ of the total energy and the mean value $\bar{N} = \langle N \rangle$ of the total number of particles as

$$U(T, V, \mu) = \int_0^\infty \mathrm{d}\varepsilon \, g(\varepsilon) n(\varepsilon) \varepsilon \qquad \bar{N}(T, V, \mu) = \int_0^\infty \mathrm{d}\varepsilon \, g(\varepsilon) n(\varepsilon).$$

For all non-relativistic gases, the pressure is related to the internal energy by $PV = \frac{2}{3}U$, while for photons the corresponding relation is $PV = \frac{1}{3}U$. For phonons, on the other hand, we have

$$U_\mathrm{ph}(T, V) = \int_0^{\hbar\omega_\mathrm{D}} \mathrm{d}\varepsilon \, g(\varepsilon) n(\varepsilon) \varepsilon \qquad 3N = \int_0^{\hbar\omega_\mathrm{D}} \mathrm{d}\varepsilon \, g(\varepsilon)$$

where U_ph is *the contribution of phonons* to the internal energy of a solid containing N atoms.

When studying ideal Bose–Einstein gases, one encounters integrals of the type

$$g_n(z) = \frac{1}{\Gamma(n)} \int_0^\infty \frac{u^{n-1}}{z^{-1}\,\mathrm{e}^u - 1} \, \mathrm{d}u \qquad \text{with } 0 \le z = \mathrm{e}^{\beta\mu} \le 1$$

which cannot be evaluated in closed form. In the high-temperature limit, when z is small, one can use the Taylor expansion

$$g_n(z) = \sum_{l=1}^{\infty} \frac{z^l}{l^n} = z + \frac{z^2}{2^n} + \frac{z^3}{3^n} + \cdots.$$

For $z = 1$, we find that $g_n(1)$ diverges for $n \leq 1$, while for $n > 1$ the special cases $g_{3/2}(1) \simeq 2.612$ and $g_{5/2}(1) \simeq 1.341$ are useful. The recursion relation

$$g_{n-1}(z) = z\frac{\partial}{\partial z}[g_n(z)]$$

is sometimes helpful.

The analogous integrals encountered for Fermi–Dirac systems are

$$f_n(z) = \frac{1}{\Gamma(n)} \int_0^\infty \frac{u^{n-1}}{z^{-1}\,\mathrm{e}^u + 1}\,\mathrm{d}u \qquad \text{with } 0 \leq z = \mathrm{e}^{\beta\mu} \leq \infty.$$

The Taylor expansion is given by

$$f_n(z) = \sum_{l=1}^{\infty} (-1)^{(l+1)} \frac{z^l}{l^n} = z - \frac{z^2}{2^n} + \frac{z^3}{3^n} - \cdots.$$

In the low-temperature limit, when z is large, it is convenient to define a new variable $x = \ln z$ and to introduce the integral

$$I_n(x) = \int_0^\infty \frac{u^n}{\mathrm{e}^{u-x} + 1}\,\mathrm{d}u \equiv \Gamma(n+1) f_{n+1}(\mathrm{e}^x)$$

which, for $x \gg 1$, has an asymptotic expansion of the form

$$I_n(x) = \frac{x^{n+1}}{n+1} + \frac{\pi^2}{6} n x^{n-1} + \cdots.$$

These fermionic integrals also satisfy recursion relations, which are

$$f_{n-1} = z\frac{\partial}{\partial z}[f_n(z)] \qquad (n+1) I_n(x) = \frac{\partial}{\partial x}[I_{n+1}(x)].$$

An integral of the form $R(T) = \int_0^\infty \mathrm{d}\varepsilon\, n(\varepsilon)\varphi(\varepsilon)$, where $n(\varepsilon)$ is the Fermi–Dirac distribution, can be estimated in the low-temperature limit $kT \ll \varepsilon_\mathrm{F}$ by the Sommerfeld approximation

$$R(T) \simeq \int_0^{\varepsilon_\mathrm{F}} \varphi(\varepsilon)\,\mathrm{d}\varepsilon + \frac{\pi^2}{6}(kT)^2 \left.\frac{\mathrm{d}\varphi}{\mathrm{d}\varepsilon}\right|_{\varepsilon=\varepsilon_\mathrm{F}} + \frac{7}{360}\pi^4 (kT)^4 \left.\frac{\mathrm{d}^3\varphi}{\mathrm{d}\varepsilon^3}\right|_{\varepsilon=\varepsilon_\mathrm{F}}$$

where the Fermi energy is defined as $\varepsilon_\mathrm{F} = \mu(T = 0)$. For a gas of N free electrons in a volume V, the Fermi energy is

$$\varepsilon_\mathrm{F} = \frac{\hbar^2}{2m}\left(\frac{3\pi^2 N}{V}\right)^{2/3}.$$

Interacting Systems

A dilute gas of N classical particles in a volume V, interacting through a pair potential $\Phi_{ij} = \Phi(|\boldsymbol{r}_i - \boldsymbol{r}_j|)$, has the Hamiltonian

$$\mathcal{H} = \sum_{i=1}^{N} \frac{|\boldsymbol{p}_i|^2}{2m} + \sum_{i<j} \Phi_{ij}.$$

Its equation of state has an expansion in powers of the number density $n = N/V$,

$$P = kT \sum_{j=1}^{\infty} B_j(T) n^j$$

known as the virial expansion. This provides a useful approximation when $\lambda^3 n \ll 1$, where $\lambda = \sqrt{2\pi\hbar^2/mkT}$ is the thermal wavelength. The virial coefficients $B_j(T)$ are given by integrals which can be represented by 'cluster graphs'. The first four of these are

$$B_1 = \frac{1}{V} \text{ ①}$$

$$B_2 = -\frac{1}{2V}$$

$$B_3 = -\frac{1}{3V}$$
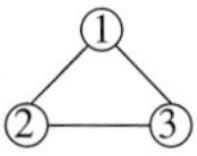

$$B_4 = -\frac{1}{8V}\Bigg(3$$
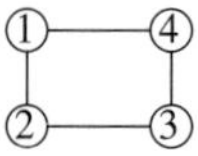
$$+6$$
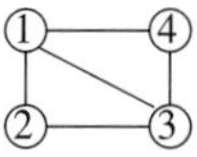
$$+$$
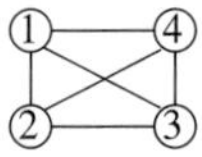
$$\Bigg).$$

Each line joining vertices labelled by i and j represents a factor $e^{-\beta\Phi_{ij}} - 1$ and each vertex i represents a volume integral $\int d^3r_i$. For example,

$$B_1 = \frac{1}{V}\int d^3r_1 = 1$$

$$B_2 = -\frac{1}{2V}\int d^3r_1 \int d^3r_2\,(e^{-\beta\Phi_{1,2}} - 1) = -\tfrac{1}{2}\int d^3r\,(e^{-\beta\Phi(|\boldsymbol{r}|)} - 1).$$

In the theory of phase transitions, a great deal of insight has been gained from the study of the *Ising model*. Originally formulated as an idealized representation of a strongly uniaxial ferromagnet, it is defined by the Hamiltonian

$$\mathcal{H} = -h\sum_{i=1}^{N} S_i - \sum_{i<j} J_{ij} S_i S_j.$$

The 'spin' variables S_i, each taking the values $S_i = \pm 1$, inhabit the sites of a spatial lattice of arbitrary shape and dimension, whose sites are labelled by i.

Each site on a regular lattice has a definite number z of nearest neighbours, called the coordination number. Exchange interactions between the spins on sites i and j are represented by the coupling constants J_{ij} and each spin interacts with an externally applied magnetic field H through its magnetic moment μ, these quantities appearing in the combination $h = \mu H$. If this model is viewed as a description of quantum-mechanical spin-$\frac{1}{2}$ atoms, then the variables S_i stand for $2\hat{s}_z$, where $\hat{s}_z$ is the spin component along the easy axis of the magnet, conventionally taken as the $\hat{z}$ direction. However, since only one spin component appears, there are no non-commuting operators, and it is simpler to regard the S_i as just numbers, which are the eigenvalues of the spin operators.

The net magnetization is defined as

$$M(T, H) = \mu \sum_{i=1}^{N} \langle S_i \rangle = \mu \frac{\partial (\ln Z)}{\partial (\beta h)}$$

where $Z_N(T, h)$ is the canonical partition function. The magnetic susceptibility is

$$\chi(T) = \lim_{H \to 0} \left(\frac{\partial M}{\partial H} \right)$$

and the spin correlation function is

$$G_{ij} = \langle S_i S_j \rangle - \langle S_i \rangle \langle S_j \rangle .$$

There is said to be a *spontaneous magnetization* when $\lim_{H \to 0}[M(T, H)] \neq 0$. For regular lattices of dimension $d \geq 2$, this happens at temperatures below a *critical temperature* (or *Curie temperature*) T_c. Below T_c, the values of widely separated spins are strongly correlated while, above T_c, this correlation decays to zero at large distances; the system is said to undergo an order–disorder transition at the critical temperature. Near T_c, various physical quantities behave in a singular way, typically with a power-law dependence on $T - T_c$, and the corresponding exponents are called *critical exponents*. For example, the magnetic susceptibility diverges as $\chi \sim |T - T_c|^{-\gamma}$, with $\gamma > 0$. The value of γ is found to be universal, in the sense that it does not depend on parameters such as the coupling constants J_{ij}.

For more than two dimensions, an exact calculation of the partition function has not proved possible, and approximation methods, such as the mean-field theory, are used. Within this approximation each spin S_i 'feels' only the influence of an effective magnetic field that takes into account in the average the effect of its neighbours, that is

$$H_{\text{eff}}^{i} = H + \frac{1}{\mu} \sum_{j} J_{ij} \langle S_j \rangle .$$

Thus the mean-field canonical partition function is that of N independent spins, each of which interacts only with the effective field, and is given by

$$Z_N^{\mathrm{MF}}(T, h) = \prod_{i=1}^{N} [2\cosh(\mu\beta H_{\mathrm{eff}}^i)].$$

Given this partition function, the mean spin value $\langle S_i \rangle$ can be calculated in the approximate mean-field ensemble. Since this mean value appears in H_{eff}^i, one obtains a coupled set of self-consistency conditions to be solved for $\langle S_i \rangle$. In the simplest case when J_{ij} has the same value J for each pair of nearest-neighbour spins and is zero otherwise, the mean value $\langle S_i \rangle$ is the same for every spin, and we obtain just one self-consistency condition, which is

$$\langle S \rangle = \tanh(\beta\mu H + \beta J z \langle S \rangle)$$

where z is the coordination number.

Non-Equilibrium Systems

The non-equilibrium evolution of many properties of a gas of N interacting classical particles can be described in terms of the single-particle distribution function $f(\boldsymbol{r}, \boldsymbol{p}, t)$. This function gives the number of particles which, at time t, are contained in the phase-space volume element $\mathrm{d}^3 r\, \mathrm{d}^3 p$ around $(\boldsymbol{r}, \boldsymbol{p})$. It is normalized so that

$$\int f(\boldsymbol{r}, \boldsymbol{p}, t)\, \mathrm{d}^3 r\, \mathrm{d}^3 p = N.$$

This distribution function satisfies the Boltzmann equation

$$\frac{\mathrm{d}f}{\mathrm{d}t} \equiv \frac{\partial f}{\partial t} + \frac{1}{m}\boldsymbol{p} \cdot \boldsymbol{\nabla}_r f + \boldsymbol{F} \cdot \boldsymbol{\nabla}_p f = \left.\frac{\partial f}{\partial t}\right|_{\mathrm{coll}}$$

where $\boldsymbol{F}$ is an external force and the right-hand side accounts for collisions. Integration over the distribution function yields three basic quantities in the form

$$n(\boldsymbol{r}, t) = \int f\, \mathrm{d}^3 p \qquad \text{the particle number density}$$

$$\boldsymbol{j}(\boldsymbol{r}, t) = \int \boldsymbol{v} f\, \mathrm{d}^3 p \qquad \text{the particle current}$$

$$\boldsymbol{j}_\varepsilon(\boldsymbol{r}, t) = \int \varepsilon \boldsymbol{v} f\, \mathrm{d}^3 p \qquad \text{the energy current}$$

where ε is the energy of a single particle and $\boldsymbol{v} = \boldsymbol{p}/m$ is its velocity. Here it is assumed that, although interparticle forces cause scattering, they do not contribute significantly to the energy of the gas.

Assuming that only binary collisions are important and that the momenta of any two particles are uncorrelated in d^3r, one can write the collision term as a difference of gain and loss contributions

$$\left.\frac{\partial f}{\partial t}\right|_{\text{coll}} = \int d^3p_2\, d^3p'_1\, d^3p'_2\, W_{(p,p_2)\to(p'_1,p'_2)}(f'_2 f'_1 - f_2 f)$$

where W, the transition probability per unit time, is given by the two-body scattering process and $\boldsymbol{p}$ is the momentum belonging to the function f that appears on the left-hand side. Inside the collision integral, f'_2 stands for $f(\boldsymbol{r}, \boldsymbol{p}'_2, t)$ and so on.

When the phenomena of interest occur on time scales much longer than the mean collision time τ, one can sometimes use the relaxation time approximation

$$\left.\frac{\partial f}{\partial t}\right|_{\text{coll}} = -\frac{f - f_0}{\tau}$$

where f_0 is the distribution function corresponding to a state of local equilibrium. For a classical gas, one might choose

$$f_0(\boldsymbol{r}, \boldsymbol{p}, t) = \frac{n(\boldsymbol{r}, t)}{[2\pi mkT(\boldsymbol{r}, t)]^{3/2}} \exp\left(-\frac{|\boldsymbol{p} - m\boldsymbol{v}(\boldsymbol{r}, t)|^2}{2mkT(\boldsymbol{r}, t)}\right)$$

where $n(\boldsymbol{r}, t)$, $T(\boldsymbol{r}, t)$ and $\boldsymbol{v}(\boldsymbol{r}, t)$ are the local number density, temperature and velocity, respectively.

Also on long time scales, particle diffusion is described by the diffusion equation

$$\frac{\partial n(\boldsymbol{r}, t)}{\partial t} = D\,\nabla^2 n(\boldsymbol{r}, t) + \mathcal{F}(\boldsymbol{r}, t)$$

where $\mathcal{D} \sim l^2/\tau$ is the diffusion coefficient (l is the mean free path) and $\mathcal{F}(\boldsymbol{r}, t)$ is a source of particles. As will be seen from the problems in chapter 6, this equation can be derived from a variety of microscopic models. The relaxation of a non-uniform temperature distribution $T(\boldsymbol{r}, t)$ in a solid material by the diffusion of heat is described by essentially the same equation, namely

$$C\rho\frac{\partial T(\boldsymbol{r}, t)}{\partial t} = \kappa\,\nabla^2 T(\boldsymbol{r}, t) + Q(\boldsymbol{r}, t)$$

where C is the specific heat per unit mass, ρ is the density, κ is the thermal conductivity and $Q(\boldsymbol{r}, t)$ is a heat source. Obviously, one can define a thermal diffusivity $D = \kappa/C\rho$.

The non-equilibrium evolution of several interesting systems can be modelled as a stochastic process, in which a stochastic variable Z makes random transitions governed by a transition probability $W_{Z\to Z'}(t)$. The process is said to be Markovian if the system has memory only of its immediate past, and in this case the probability density $\mathcal{P}(Z,t)$ evolves according to a *master equation* of the general form

$$\frac{\mathrm{d}\mathcal{P}(Z,t)}{\mathrm{d}t} = \sum_{Z'}[W_{Z'\to Z}\mathcal{P}(Z',t) - W_{Z\to Z'}\mathcal{P}(Z,t)].$$

If Z can assume a continuous range of values, then the probability $\mathcal{P}$ is replaced by a probability density and the sum by an integral. If the transitions occur only at discrete intervals of time, then the differential equation is replaced by a recursion formula.

Finally, the Brownian motion of a particle in a thermal environment can be described phenomenologically by the *Langevin equation*

$$m\ddot{\boldsymbol{r}} + \alpha\dot{\boldsymbol{r}} = \boldsymbol{F}(\boldsymbol{r},t) + \boldsymbol{\eta}(t)$$

where $\boldsymbol{F}(\boldsymbol{r},t)$ is an external force, while the environment gives rise to a random force $\boldsymbol{\eta}(t)$ and a friction coefficient α. In the simplest model of this kind, the random force is taken to have a Gaussian probability distribution and an autocorrelation function

$$\langle\eta_i(t)\eta_j(t')\rangle = A\delta(t-t')$$

where A, the mean square value of the fluctuating force, is a phenomenological parameter. On time scales that are long compared with the relaxation time $\tau = m/\alpha$, the acceleration term $m\ddot{\boldsymbol{r}}$ can often be neglected, and the motion is diffusive. Generalizations of this simple model to a variety of more complicated situations are possible.

2

THERMODYNAMICS

The purpose of this chapter is to review some fundamental concepts of thermodynamics, which form an essential background to the study of statistical mechanics. Problems 2.1–2.3 deal with the first and second laws and the associated concepts of temperature and thermodynamic potentials. A variety of relations between thermodynamic quantities, including the Maxwell relations, can be derived from these basic principles, together with the rules of differential calculus, and the methods for obtaining such relations are illustrated in problems 2.4 and 2.5. Problem 2.6 illustrates the nature of entropy changes in irreversible processes, while problems 2.7–2.9 investigate the equations of state of simple systems and their consequences for various thermodynamic processes. Finally, problem 2.10 deals with the thermodynamic treatment of a magnetic material and problem 2.11 treats the behaviour of relativistic gases in a cosmological setting.

2.1 Questions

• **Problem 2.1** Show that the Helmholtz free energy $F(T, V)$ of a system whose volume is V in contact with a heat bath at temperature T is a minimum at equilibrium. Analogously, show that the Gibbs free energy $G(T, P)$ of a system in contact with a bath at temperature T and pressure P is a minimum at equilibrium.

• **Problem 2.2** Discuss the possibility of a physical system having a negative temperature.

• **Problem 2.3** Compare the decrease in the entropy of a reader's brain during the reading of a book with the increase in entropy due to illumination (by means of an electric light bulb).

• **Problem 2.4** The principal specific heats C_P and C_V of any substance can be expressed in terms of its temperature T, volume V, adiabatic compressibility $\kappa_S \equiv -V^{-1}(\partial V/\partial P)_S$, isothermal compressibility $\kappa_T \equiv -V^{-1}(\partial V/\partial P)_T$ and thermal expansivity $\alpha \equiv V^{-1}(\partial V/\partial T)_P$. Obtain these expressions as follows.

(a) Regarding the entropy S as a function of T and V, show that

$$T\,\mathrm{d}S = C_V\,\mathrm{d}T + \frac{\alpha T}{\kappa_T}\,\mathrm{d}V.$$

(b) Regarding S as a function of T and P, show that

$$T\,\mathrm{d}S = C_P\,\mathrm{d}T - \alpha T V\,\mathrm{d}P.$$

(c) Using these results, prove that

$$C_P - C_V = \frac{TV\alpha^2}{\kappa_T}.$$

(d) Express C_P and C_V in terms of T, V, α, κ_T and κ_S.

• **Problem 2.5** Prove the following useful thermodynamic relations.

(a) $(\partial U/\partial V)_T = T(\partial P/\partial T)_V - P$.

(b) $(\partial N/\partial \mu)_{T,V} = N^2\kappa_T/V$, where μ is the chemical potential and $\kappa_T \equiv -V^{-1}(\partial V/\partial P)_T$ is the isothermal compressibility.

(c) $C_P(\partial T/\partial P)_H = T(\partial V/\partial T)_P - V$, where H denotes enthalpy.

• **Problem 2.6**

(a) A quantity of water, initially at 10 °C, is brought into contact with a heat reservoir at a temperature of 90 °C. What is the entropy change of the entire system when the water reaches the temperature of the bath? Express the answer in terms of the heat capacity C of the water, and assume that C does not depend on temperature.

(b) What is the entropy change of the entire system if we use two reservoirs to heat the water first from 10 to 50 °C, and then from 50 to 90 °C?

(c) Can you suggest how, in principle, water might be heated from an initial temperature T_i to a final temperature T_f with no change in global entropy?

• **Problem 2.7** An ideal classical gas can be defined as one which, with a fixed number of particles and at a constant temperature, satisfies the following conditions:

(i) the internal energy does not depend on the volume;

(ii) the enthalpy does not depend on the pressure.

Using this definition and making appropriate use of thermodynamic potentials, derive the equation of state.

• **Problem 2.8** For an elastic rubber band of length L, at temperature T and under a tension J, it is found experimentally that

$$g(T, L) \equiv \left(\frac{\partial J}{\partial T}\right)_L = \frac{aL}{L_0}\left[1 - \left(\frac{L_0}{L}\right)^3\right]$$

$$f(T, L) \equiv \left(\frac{\partial J}{\partial L}\right)_T = \frac{aT}{L_0}\left[1 + 2\left(\frac{L_0}{L}\right)^3\right]$$

where L_0 is the length of the unstretched band (independent of temperature) and a is a constant.

(a) Obtain the equation of state of this system.

(b) Assume that the heat capacity at constant length of the band is a constant C_L. If the band is stretched, adiabatically and reversibly, from L_0 at an initial temperature T_i to a final length L_f, what is its final temperature T_f?

(c) The band is now released, so that it contracts freely to its natural length L_0. If no heat is exchanged with its surroundings during this contraction, find the changes in its temperature and entropy.

• **Problem 2.9**

(a) Show that, at a given temperature, the specific heat at constant volume C_V of a van der Waals gas with a fixed number of particles N is independent of its volume.

(b) This gas, initially occupying a volume V_i at temperature T_i, undergoes a free adiabatic expansion up to a final volume V_f. Compute the changes in temperature and entropy, assuming that C_V is also independent of temperature.

• **Problem 2.10** Obtain the differential form of the first law of thermodynamics for a rigid magnetic material. Find the thermodynamic potentials for this material and their corresponding Maxwell relations.

• **Problem 2.11** At about 10^{-4} s after the Big Bang, the matter content of the early Universe consisted mainly of electrons and positrons $e^{\pm}$, neutrinos ν and photons γ, all in thermal equilibrium at a temperature of about 10^{12} K. The volume V of a 'comoving' region of space expands with time and, as long as the number of particles in a relativistic gas occupying this region is constant, the temperature of this gas falls, with $T^3 \propto V^{-1}$. As the temperature fell to about 10^{11} K, the rate of interactions between the neutrinos and the other particles became negligible and, thereafter, the evolution of the neutrino gas was completely decoupled from that of the electrons, positrons and photons. At a temperature of about 6×10^9 K (for which kT is equal to the rest energy of an electron), electrons and positrons (of which there were approximately equal numbers) annihilated quite rapidly, producing photons. As a result, the remaining photon gas was at a somewhat higher temperature T_γ than the temperature T_ν of the neutrinos.

Today, the temperature of the photons (the cosmic microwave background) is about 2.7 K. Find the current temperature of the relic neutrinos, assuming that the annihilation process was reversible, and that the Universe at that time was spatially homogeneous. The energy densities of the relativistic gases in question are given by $u_\nu = cT^4$, $u_e = 2cT^4$ and $u_\gamma = \frac{8}{7}cT^4$, where c is a constant. (The factor of 2 accounts for the presence of both electrons and positrons and the factor of $\frac{8}{7}$ for the difference between Bose and Fermi statistics.)

2.2 Answers

• **Problem 2.1** Consider a system whose volume V is fixed, and which is brought into contact with a heat bath at temperature T. As long as the system is not in equilibrium with the bath, it cannot be assigned a definite temperature, so its Helmholtz free energy F is not a well-defined function of temperature and volume. We may, however, attempt to define a non-equilibrium free energy by

$$F = U - TS.$$

Here, we understand by U the internal energy of the *system*, which is well defined even away from equilibrium, and by T the temperature of the *heat bath*, (which will also be the temperature of the system once equilibrium is attained). The non-equilibrium entropy S of the system is not simply a function of U and V but depends on details of the non-equilibrium state. We can appeal to Clausius' statement of the second law to write the inequality

$$\delta S \geq \delta Q/T$$

where δQ is an amount of heat transferred from the heat bath to the system. Since the volume of the system is constant, no work is done while this heat is transferred, so the change in the system's internal energy is $\delta U = \delta Q$. Since the temperature T of the heat bath is fixed, we deduce that

$$\delta F = \delta U - T\,\delta S = \delta Q - T\,\delta S \leq 0.$$

Thus, F always decreases as the system approaches equilibrium and reaches a minimum when equilibrium is attained.

In the case of the Gibbs free energy $G = U + PV - TS$, we must consider a system whose pressure P is held fixed, but whose volume is allowed to vary as it approaches equilibrium. In this case, changes in energy are given by $\delta U = \delta Q - P\,\delta V$. An argument similar to that given above yields

$$\delta G = \delta U + P\,\delta V - T\,\delta S = \delta Q - T\,\delta S \leq 0.$$

Note that F is useful for describing a system whose temperature and volume are subject to external control, whereas G is more useful when the temperature and pressure are externally controlled.

• **Problem 2.2** Consider a system of bounded energy. That is, the system has both a lowest possible energy, which we can take to be $U = 0$, and a greatest possible energy, say U_{max}. For such a system, it is likely that very few microstates (perhaps only one) correspond to $U = 0$, and also that very few correspond to $U = U_{\mathrm{max}}$. Thus, the entropy $S(U)$ is very small both at $U = 0$ and at $U = U_{\mathrm{max}}$, and there is some intermediate energy U^* at which $S(U)$ is a maximum. The inverse temperature $T^{-1} = \partial S/\partial U$ is zero

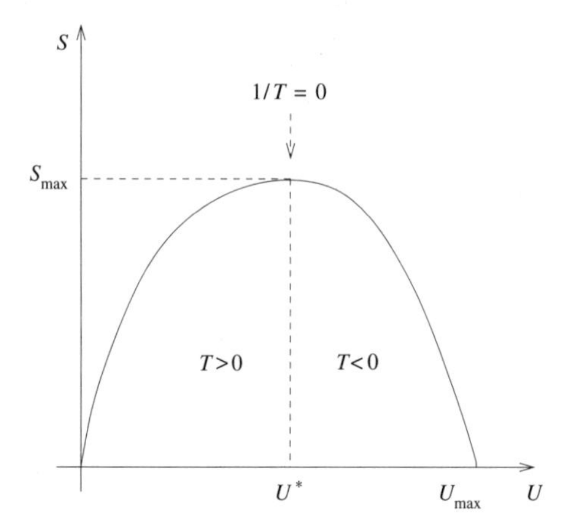

Figure 2.1. Positive- and negative-temperature regions for a system of bounded energy.

at $U = U^*$ and negative for $U > U^*$ (figure 2.1). Note that temperatures $T = +\infty$ and $T = -\infty$ correspond to the same state $T^{-1} = 0$. Moreover, a negative-temperature state has greater energy and is consequently hotter than a positive-temperature state. Clausius' statement of the second law tells us that heat does not flow spontaneously from a cooler body to a hotter body, so the system cannot be heated to an infinite or negative temperature by an inflow of heat, so long as its surroundings are at some finite positive temperature. Heat will always flow out of a negative-temperature system in contact with positive-temperature surroundings.

It is, however, possible to create a negative-temperature state by indirect means. The classic example is that of a collection of magnetic moments of atomic nuclei, which interact very weakly with the remaining degrees of freedom in their host material and can be considered effectively as an isolated system. The energy of a magnetic moment (or *spin* for brevity) is just that due to its orientation in an applied magnetic field. The system has its minimum energy when all the spins are parallel to the field, and its maximum energy when they are all antiparallel to the field. The infinite-temperature state of maximum entropy corresponds to completely random orientations of the spins. At a low positive temperature, the spins will be predominantly parallel to the field. If the direction of the field is now rapidly reversed, the spins find themselves predominantly antiparallel to the field, and thus in a negative-temperature state.

• **Problem 2.3** The change in entropy resulting from the assimilation of an amount ΔI of information (measured in bits) is

$$\Delta S = -\frac{k}{\log_2 \mathrm{e}} \Delta I.$$

A book which uses only the Latin alphabet and punctuation marks requires about 2^5 different characters, so 5 bits are needed to specify any one character. Thus, the information content of a 400-page book, in which each page has 50 lines of 70 characters each is $\Delta I = 5 \times 70 \times 50 \times 400 = 7 \times 10^6$ bits. Then $\Delta S_{\text{brain}} \sim -10^{-16}$ J K^{-1}. On the other hand, an electric bulb emitting a power P of, say, 100 W into its surroundings at a temperature $T = 300$ K causes an entropy increase

$$\Delta S_{\text{bulb}} \sim \frac{Pt}{T} \sim 0.3t \text{ J K}^{-1}$$

where t is the time (in seconds) taken by the reading process. We then have

$$\frac{\Delta S_{\text{brain}}}{\Delta S_{\text{bulb}}} \sim -\frac{10^{-19}}{t^*}$$

where t^* is the time measured in hours. For typical reading times (say a few hours) this ratio is tiny. Of course, it will be even smaller if we take into account entropy production by all the metabolic processes of the reader's body as well as that by the light bulb.

• Problem 2.4

(a) Regarding the entropy as a function of T and V, we have

$$T\,\mathrm{d}S = T\left(\frac{\partial S}{\partial T}\right)_V \mathrm{d}T + T\left(\frac{\partial S}{\partial V}\right)_T \mathrm{d}V.$$

In the first term, we may use the fact that, for a change at constant volume, $T\,\mathrm{d}S = \mathrm{d}Q = C_V\,\mathrm{d}T$ to write $T(\partial S/\partial T)_V = C_V$. To deal with the second term, we note that the thermodynamic potential corresponding to the controlled variables T and V is the Helmholtz free energy $F(T, V)$. The fact that

$$\left[\frac{\partial}{\partial V}\left(\frac{\partial F}{\partial T}\right)_V\right]_T = \left[\frac{\partial}{\partial T}\left(\frac{\partial F}{\partial V}\right)_T\right]_V$$

yields the Maxwell relation

$$\left(\frac{\partial S}{\partial V}\right)_T = \left(\frac{\partial P}{\partial T}\right)_V$$

which, using the identity given in appendix A, can be rewritten as

$$\left(\frac{\partial S}{\partial V}\right)_T = -\frac{(\partial V/\partial T)_P}{(\partial V/\partial P)_T} = \frac{\alpha}{\kappa_T}.$$

We thus obtain the desired relation

$$T\,\mathrm{d}S = C_V\,\mathrm{d}T + \frac{\alpha T}{\kappa_T}\,\mathrm{d}V.$$

(b) In the same way, regarding S as a function of T and P, we have

$$T\,\mathrm{d}S = T\left(\frac{\partial S}{\partial T}\right)_P \mathrm{d}T + T\left(\frac{\partial S}{\partial P}\right)_T \mathrm{d}P$$

and $T(\partial S/\partial T)_P = C_P$. Differentiation of the Gibbs free energy $G(T, P)$ gives the Maxwell relation $(\partial S/\partial P)_T = -(\partial V/\partial T)_P = -\alpha V$, so we obtain our second result

$$T\,\mathrm{d}S = C_P\,\mathrm{d}T - \alpha T V\,\mathrm{d}P.$$

(c) Consider a change in state at constant pressure, $\mathrm{d}P = 0$. From (a) and (b) we obtain

$$T\,\mathrm{d}S = C_V\,\mathrm{d}T + \frac{\alpha T}{\kappa_T}\,\mathrm{d}V = C_P\,\mathrm{d}T.$$

Thus,

$$C_P - C_V = \frac{\alpha T}{\kappa_T}\left(\frac{\partial V}{\partial T}\right)_P = \frac{\alpha^2 T V}{\kappa_T}$$

which is the answer that we wanted.

(d) To obtain expressions for C_P and C_V separately, we need a second equation involving these quantities. Given that the results should involve κ_S, consider setting $\mathrm{d}S = 0$ in the results of (a) and (b):

$$0 = C_V\,\mathrm{d}T|_S + \frac{\alpha T}{\kappa_T}\,\mathrm{d}V|_S$$
$$0 = C_P\,\mathrm{d}T|_S - \alpha T V\,\mathrm{d}P|_S.$$

The ratio of these two equations gives

$$\frac{C_P}{C_V} = -V\kappa_T\left(\frac{\partial P}{\partial V}\right)_S = \frac{\kappa_T}{\kappa_S}.$$

Together with the result of (c), this gives two equations to be solved for C_P and C_V, which finally yields

$$C_P = \frac{\alpha^2 T V}{\kappa_T - \kappa_S} \qquad C_V = \frac{\alpha^2 T V \kappa_S}{\kappa_T(\kappa_T - \kappa_S)}.$$

From these results, and using the stability conditions that C_V, C_P, κ_T and κ_S must be positive, it is easy to see that C_P is always greater than C_V and that κ_T is always greater than κ_S.

• Problem 2.5

(a) For a system with a fixed number of particles, we can apply the fundamental relation $\mathrm{d}U = T\,\mathrm{d}S - P\,\mathrm{d}V$ to a change at fixed temperature to obtain

$$\left(\frac{\partial U}{\partial V}\right)_T = T\left(\frac{\partial S}{\partial V}\right)_T - P.$$

Comparison of the first term with the desired result suggests use of the Maxwell relation $(\partial S/\partial V)_T = (\partial P/\partial T)_V$, and this does indeed yield

$$\left(\frac{\partial U}{\partial V}\right)_T = T\left(\frac{\partial P}{\partial T}\right)_V - P.$$

(b) This relation is most economically obtained from the Gibbs–Duhem equation

$$N\,\mathrm{d}\mu = -S\,\mathrm{d}T + V\,\mathrm{d}P.$$

For a process at constant T and V, we can use this and the identity given in appendix A to obtain

$$N\left(\frac{\partial \mu}{\partial N}\right)_{T,V} = V\left(\frac{\partial P}{\partial N}\right)_{T,V} = -V\frac{(\partial V/\partial N)_{T,P}}{(\partial V/\partial P)_{T,N}} = \frac{(\partial V/\partial N)_{T,P}}{\kappa_T}.$$

Since V and N are extensive quantities, while P and T are intensive, it is clear that the equation of state for any fluid can be written in the form $V = Nf(T, P)$ and hence that $(\partial V/\partial N)_{T,P} = V/N$. (A formal proof of this is

$$\begin{aligned}\left(\frac{\partial V}{\partial N}\right)_{T,P} &= \left[\frac{\partial}{\partial N}\left(\frac{\partial G}{\partial P}\right)_{T,N}\right]_{T,P} \\ &= \left[\frac{\partial}{\partial P}\left(\frac{\partial G}{\partial N}\right)_{T,P}\right]_{T,N} = \left(\frac{\partial \mu}{\partial P}\right)_{T,N} = \frac{V}{N}\end{aligned}$$

where the last step again follows from the Gibbs–Duhem equation.) Consequently, we obtain the desired result

$$\left(\frac{\partial N}{\partial \mu}\right)_{T,V} = \left[\left(\frac{\partial \mu}{\partial N}\right)_{T,V}\right]^{-1} = \frac{N^2\kappa_T}{V}.$$

(c) A useful first step is to express $(\partial T/\partial P)_H$ in terms of derivatives of H, by using the identity given in appendix A to write

$$\left(\frac{\partial T}{\partial P}\right)_H = -\frac{(\partial H/\partial P)_T}{(\partial H/\partial T)_P}.$$

The result then follows from the basic relation $\mathrm{d}H = T\,\mathrm{d}S + V\,\mathrm{d}P$. In the denominator, we have $(\partial H/\partial T)_P = T(\partial S/\partial T)_P = C_P$. For the numerator, using the Maxwell relation $(\partial S/\partial P)_T = -(\partial V/\partial T)_P$, we find that

$$\left(\frac{\partial H}{\partial P}\right)_T = T\left(\frac{\partial S}{\partial P}\right)_T + V = -T\left(\frac{\partial V}{\partial T}\right)_P + V$$

and hence obtain the desired relation

$$\left(\frac{\partial T}{\partial P}\right)_H = C_P^{-1}\left[T\left(\frac{\partial V}{\partial T}\right)_P - V\right].$$

• Problem 2.6

(a) The change in entropy of the water can be found by imagining a reversible process in which the temperature is raised by infinitesimal steps. At each step, the entropy change is

$$\mathrm{d}S_{\mathrm{w}} = \frac{\mathrm{d}Q_{\mathrm{rev}}}{T} = C\frac{\mathrm{d}T}{T}$$

giving a total entropy change

$$\Delta S_{\mathrm{w}} = C\int_{T_{\mathrm{i}}}^{T_{\mathrm{f}}} \frac{\mathrm{d}T}{T} = C\ln\left(\frac{T_{\mathrm{f}}}{T_{\mathrm{i}}}\right).$$

The heat reservoir is at a fixed temperature T_{f}, so its entropy change is just

$$\Delta S_{\mathrm{r}} = -\frac{\Delta Q}{T_{\mathrm{f}}} = -C\frac{T_{\mathrm{f}} - T_{\mathrm{i}}}{T_{\mathrm{f}}}$$

where $\Delta Q = C\,(T_{\mathrm{f}} - T_{\mathrm{i}})$ is the total amount of heat flowing out of the reservoir and into the water. With the specific values $T_{\mathrm{i}} = 283.15$ K and $T_{\mathrm{f}} = 363.15$ K, we find that

$$\Delta S_{\mathrm{total}} = \Delta S_{\mathrm{w}} + \Delta S_{\mathrm{r}} \simeq 0.0285C.$$

(b) Now we repeat the process in two steps, with an intermediate temperature, say, $T_1 = 323.15$ K. The entropy change for the water is, of course, the same as before:

$$\Delta S_{\mathrm{w}} = C\ln\left(\frac{T_1}{T_{\mathrm{i}}}\right) + C\ln\left(\frac{T_{\mathrm{f}}}{T_1}\right) = C\ln\left(\frac{T_{\mathrm{f}}}{T_{\mathrm{i}}}\right).$$

Including the entropy changes for both reservoirs, we obtain

$$\Delta S_{\mathrm{total}} = \Delta S_{\mathrm{w}} + \Delta S_{\mathrm{r1}} + \Delta S_{\mathrm{r2}} \simeq 0.0149C.$$

(c) Clearly, the change in total entropy is smaller when the process is carried out in two steps. This suggests that we consider bringing the water into contact with a sequence of N reservoirs, at temperatures $T_n = T_{\mathrm{i}} + (n/N)(T_{\mathrm{f}} - T_{\mathrm{i}})$. The change in entropy of the water will be $\Delta S_{\mathrm{w}} = C\ln(T_{\mathrm{f}}/T_{\mathrm{i}})$, independent of N. The entropy change for the reservoirs will be

$$\Delta S_{\mathrm{r}} = -C\sum_{n=1}^{N} \frac{T_n - T_{n-1}}{T_n} = -C\sum_{n=1}^{N} \frac{(T_{\mathrm{f}} - T_{\mathrm{i}})/N}{T_{\mathrm{i}} + (n/N)(T_{\mathrm{f}} - T_{\mathrm{i}})}.$$

In the limit of infinitesimal steps, $N \to \infty$, the sum becomes an integral:

$$\lim_{N\to\infty}(\Delta S_{\mathrm{r}}) = -C\int_0^{T_{\mathrm{f}}-T_{\mathrm{i}}} \frac{\mathrm{d}x}{T_{\mathrm{i}} + x} = -C\ln\left(\frac{T_{\mathrm{f}}}{T_{\mathrm{i}}}\right) = -\Delta S_{\mathrm{w}}.$$

So the total entropy change is zero. This, of course, is the fictitious reversible process that we consider to calculate ΔS_{w}; in a reversible process the global change in entropy is zero. It assumes that the set of reservoirs is already available and takes no account of the entropy changes needed to bring these reservoirs to the required temperatures. In practice, any attempt to heat a quantity of water will involve finite irreversible changes and will inevitably cause an increase in global entropy.

• **Problem 2.7** In order to find the equation of state $f(P, V, T) = 0$, we need to find expressions in terms of P, V and T for two of the derivatives $(\partial P/\partial T)_V$, $(\partial V/\partial T)_P$ and $(\partial P/\partial V)_T$ and to integrate them. Clearly, we must apply the given conditions on the internal energy and the enthalpy to the differential relations

$$\mathrm{d}U = T\,\mathrm{d}S - P\,\mathrm{d}V \qquad \mathrm{d}H = T\,\mathrm{d}S + V\,\mathrm{d}P.$$

Using $(\partial U/\partial V)_T = 0$ and $(\partial H/\partial P)_T = 0$, we find that

$$\left(\frac{\partial S}{\partial V}\right)_T = \frac{P}{T} \qquad \left(\frac{\partial S}{\partial P}\right)_T = -\frac{V}{T}.$$

We can eliminate S by using the Maxwell relations. The first equation indicates use of the thermodynamic potential whose natural variables are V and T, namely the Helmholtz free energy $F(T, V)$, whose associated Maxwell relation is $(\partial S/\partial V)_T - (\partial P/\partial T)_V$. Similarly, for the second equation, we need the Maxwell equation associated with the Gibbs free energy $G(T, P)$, which is $(\partial S/\partial P)_T = -(\partial V/\partial T)_P$. In this way, our two equations become

$$\left(\frac{\partial P}{\partial T}\right)_V = \frac{P}{T} \qquad \left(\frac{\partial V}{\partial T}\right)_P = \frac{V}{T}.$$

These differential equations can readily be integrated, giving

$$P = f_1(V)T \qquad V = f_2(P)T$$

where $f_1(\cdot)$ and $f_2(\cdot)$ are two unknown functions. Clearly, however, if both equations are to hold, we must have $f_1(V) = \alpha/V$ and $f_2(P) = \alpha/P$, where α is a single constant of integration. Thus, the equation of state is

$$PV = \alpha T.$$

In all these manipulations, the number N of particles was held fixed, so α may depend on N. Since V is extensive, while P and T are intensive, we see that α must in fact be proportional to N. Finally, therefore, we can write the equation of state as $PV = NkT$ where, of course, the remaining constant k can be identified with Boltzmann's constant.

• Problem 2.8

(a) First of all, we can verify that there exists a well-defined function of state $J(T, L)$, because

$$\left[\frac{\partial}{\partial T}\left(\frac{\partial J}{\partial L}\right)_T\right]_L = \left(\frac{\partial f}{\partial T}\right)_L = \left(\frac{\partial g}{\partial L}\right)_T = \left[\frac{\partial}{\partial L}\left(\frac{\partial J}{\partial T}\right)_L\right]_T.$$

Since $g(T, L)$ in fact depends only on L, we can denote it by $g(L)$. Then, on integrating $(\partial J/\partial T)_L$ at fixed L, we find that

$$J(T, L) = Tg(L) + h_1(L)$$

where $h_1(L)$ is an unknown function. Integration of $(\partial J/\partial L)_T$ at fixed T gives

$$J(T, L) = Tg(L) + h_2(T)$$

since $g'(L) = f(T, L)/T$. Clearly, h_1 and h_2 must both equal the same constant, independent of T and L. In fact, this constant must be zero, since the tension J must vanish when $L = L_0$. Thus, the equation of state is

$$J(T, L) = T\frac{aL}{L_0}\left[1 - \left(\frac{L_0}{L}\right)^3\right].$$

(b) The rubber band obeys thermodynamic relations analogous to those which apply to a fluid. The difference is that the work done on the band when its length increases by an infinitesimal amount $\mathrm{d}L$ is $+J\,\mathrm{d}L$. Thus the first law takes the form

$$\mathrm{d}U = T\,\mathrm{d}S + J\,\mathrm{d}L.$$

Regarding T and L as controlled variables, the appropriate thermodynamic potential is that analogous to the Helmholtz free energy, namely $F(T, L) = U - TS$, for which $\mathrm{d}F = -S\,\mathrm{d}T + J\,\mathrm{d}L$, leading to the Maxwell relation

$$\left(\frac{\partial S}{\partial L}\right)_T = -\left(\frac{\partial J}{\partial T}\right)_L.$$

The fact that $J = Tg(L)$ and hence that $(\partial J/\partial T)_L = J/T$ implies that, as for an ideal gas, the internal energy is a function of T only. This follows from the first law and the above Maxwell relation since, if T is held fixed, we obtain

$$\left(\frac{\partial U}{\partial L}\right)_T = T\left(\frac{\partial S}{\partial L}\right)_T + J = -T\left(\frac{\partial J}{\partial T}\right)_L + J = 0.$$

Thus, for any change in state, we have $\mathrm{d}U = C_L\,\mathrm{d}T$. The first law can now be written as

$$\mathrm{d}S = C_L\frac{\mathrm{d}T}{T} - \frac{J}{T}\,\mathrm{d}L = C_L\frac{\mathrm{d}T}{T} - g(L)\,\mathrm{d}L.$$

This relation is valid for any reversible process. However, since the first term in the last expression involves only temperature and the second involves only length, we can integrate directly to find the entropy difference between the states (T_1, L_1) and (T_2, L_2):

$$\Delta S = C_L \int_{T_1}^{T_2} \frac{\mathrm{d}T}{T} - \int_{L_1}^{L_2} g(L)\,\mathrm{d}L$$
$$= C_L \ln\left(\frac{T_2}{T_1}\right) - aL_0\left[u\left(\frac{L_2}{L_0}\right) - u\left(\frac{L_1}{L_0}\right)\right]$$

where $u(x) = x^2/2 + 1/x$. Since entropy is a function of state, this expression gives correctly the difference in entropy between the initial and the final equilibrium states, whether the actual route taken between these states is reversible or not.

When the band is stretched reversibly and adiabatically from L_0 to L_f, we have $\Delta S = 0$ and so

$$T_\mathrm{f} = T_\mathrm{i} \exp\left\{C_L^{-1} aL_0\left[u\left(\frac{L_\mathrm{f}}{L_0}\right) - u(1)\right]\right\}.$$

Since $u(x)$ is an increasing function for $x > 1$, the temperature of the band has increased.

(c) During the free contraction, no work is done and no heat is exchanged. Therefore, the internal energy and hence the temperature are unchanged. The change in entropy is

$$\Delta S = -aL_0\left[u(1) - u\left(\frac{L_\mathrm{f}}{L_0}\right)\right] > 0.$$

This is quite analogous to the free expansion (Joule effect) of an ideal gas, for which the temperature is also unchanged. Entropy increases since the free contraction is an irreversible process.

• Problem 2.9

(a) To show that C_V is independent of volume, we calculate $(\partial C_V/\partial V)_T$ as follows:

$$\left(\frac{\partial C_V}{\partial V}\right)_T = T\left[\frac{\partial}{\partial V}\left(\frac{\partial S}{\partial T}\right)_V\right]_T$$
$$= T\left[\frac{\partial}{\partial T}\left(\frac{\partial S}{\partial V}\right)_T\right]_V = T\left(\frac{\partial^2 P}{\partial T^2}\right)_V$$

where the last step uses the Maxwell relation $(\partial S/\partial V)_T = (\partial P/\partial T)_V$. From the van der Waals equation of state

$$\left(P + \frac{aN^2}{V^2}\right)\left(\frac{V}{N} - b\right) = kT$$

in which the constants a and b are phenomenological parameters, it is easy to find that $(\partial^2 P/\partial T^2)_V = 0$.

(b) The internal energy U is unchanged in a free adiabatic expansion. However, in contrast with an ideal gas, for which $\mathrm{d}U_{\mathrm{ideal}} = C_V\,\mathrm{d}T$, the internal energy of a van der Waals gas is not a function of T alone. It is helpful to regard U as a function of T and V, expressing infinitesimal changes through the relation derived in problem 2.5(a):

$$\mathrm{d}U = \left(\frac{\partial U}{\partial T}\right)_V \mathrm{d}T + \left(\frac{\partial U}{\partial V}\right)_T \mathrm{d}V = C_V\,\mathrm{d}T + \left[T\left(\frac{\partial P}{\partial T}\right)_V - P\right]\mathrm{d}V.$$

Using the van der Waals equation equation, we find explicitly that

$$\mathrm{d}U = C_V\,\mathrm{d}T + \frac{aN^2}{V^2}\,\mathrm{d}V.$$

Since U is unchanged, we have $\mathrm{d}U = 0$ and

$$T_{\mathrm{f}} - T_{\mathrm{i}} = \int_{T_{\mathrm{i}}}^{T_{\mathrm{f}}} \mathrm{d}T = -\frac{aN^2}{C_V}\int_{V_{\mathrm{i}}}^{V_{\mathrm{f}}} \frac{\mathrm{d}V}{V^2} = -\frac{aN^2}{C_V}\left(\frac{1}{V_{\mathrm{i}}} - \frac{1}{V_{\mathrm{f}}}\right).$$

We see that the temperature falls ($T_{\mathrm{f}} < T_{\mathrm{i}}$) if a is positive, corresponding to an attractive force between two particles in the gas. This reflects the fact that after the expansion the particles are, on average, farther apart, so their potential energy has increased. Since the total internal energy is unchanged, their average kinetic energy and therefore the temperature have decreased. For a gas with no intermolecular forces ($a = 0$), the temperature is unchanged.

Using the van der Waals equation again, we can express infinitesimal changes in entropy as

$$\mathrm{d}S = \frac{1}{T}\,\mathrm{d}U + \frac{P}{T}\,\mathrm{d}V = \frac{C_V}{T}\,\mathrm{d}T + \frac{Nk}{V - Nb}\,\mathrm{d}V.$$

The entropy change for the free expansion could be calculated in two ways. One is to set $\mathrm{d}U = 0$ in the first expression and to evaluate $\Delta S = \int_{\mathrm{i}}^{\mathrm{f}} (P/T)\,\mathrm{d}V$, using the equation of state and our first result to express P and T as functions of V. Since P and T are both positive, entropy increases, as expected for an irreversible change. What is actually calculated here, however, is the total entropy change for a reversible route between the initial and final states where, at each infinitesimal step, heat is added to compensate for the work done. It is slightly quicker to integrate directly the second expression for $\mathrm{d}S$ since this is the sum of a term which depends only on temperature and one which depends only on volume. Thus, the change in entropy is

$$\Delta S = \int_{\mathrm{i}}^{\mathrm{f}} \mathrm{d}S = C_V \ln\left(\frac{T_{\mathrm{f}}}{T_{\mathrm{i}}}\right) + Nk\ln\left(\frac{V_{\mathrm{f}} - Nb}{V_{\mathrm{i}} - Nb}\right).$$

On substituting our previous result for the final temperature T_f, we finally obtain

$$\Delta S = C_V \ln\left[1 - \frac{aN^2}{C_V T_\mathrm{i}}\left(\frac{1}{V_\mathrm{i}} - \frac{1}{V_\mathrm{f}}\right)\right] + Nk \ln\left(\frac{V_\mathrm{f} - Nb}{V_\mathrm{i} - Nb}\right).$$

• **Problem 2.10** The work done in a change in state of a magnetic material can be confusing, so we discuss it in some detail. Taking a coarse-grained view of the material, its state can be specified in terms of a magnetic dipole moment per unit volume, called the *magnetization* $\boldsymbol{M}(\boldsymbol{r})$. Because the equations of electromagnetism are linear, the total magnetic induction $\boldsymbol{B}(\boldsymbol{r})$ at any point inside or outside the material can be expressed as

$$\boldsymbol{B}(\boldsymbol{r}) = \boldsymbol{B}_0(\boldsymbol{r}) + \boldsymbol{B}_\mathrm{M}(\boldsymbol{r}).$$

Here, $\boldsymbol{B}_0(\boldsymbol{r})$ is the field due to some external apparatus (say, a solenoid) and is that which would exist if the material were not present. The second contribution $\boldsymbol{B}_\mathrm{M}(\boldsymbol{r})$ is that produced by the magnetization. The magnetic induction may also be written as $\boldsymbol{B}(\boldsymbol{r}) = \mu_0[\boldsymbol{H}(\boldsymbol{r}) + \boldsymbol{M}(\boldsymbol{r})]$, where $\boldsymbol{H}(\boldsymbol{r})$ is the *magnetic field strength* and (in SI units) $\mu_0 = 4\pi \times 10^{-7}$ H m^{-1} is the permeability of free space. The field strength satisfies $\boldsymbol{\nabla} \times \boldsymbol{H} = \boldsymbol{j}_0$, where $\boldsymbol{j}_0$ is the current density due to the external apparatus, but it is important to note that $\boldsymbol{B}_0$ is not necessarily equal (or even parallel) to $\mu_0\boldsymbol{H}$.

Consider now a small dipole, which may be modelled as a loop of area a carrying a current i. We can define a vector $\boldsymbol{a}$ whose magnitude is a and whose direction is normal to the plane of the loop, its sense being determined by the direction of the current i according to the right-hand screw rule. In terms of this vector, the dipole moment of the loop is $\boldsymbol{m} = \boldsymbol{a}i$. Its dipole moment is $\boldsymbol{m} = \boldsymbol{a}i$. We suppose that at the position of the dipole there is a magnetic induction $\boldsymbol{B}$ produced by a circuit external to the dipole carrying a current I. The flux produced by this circuit and linked with the dipole is $\boldsymbol{B} \cdot \boldsymbol{a}$, so the mutual inductance between the dipole and the external circuit is $L = \boldsymbol{B} \cdot \boldsymbol{a}/I$. Therefore, the magnetic flux produced by the dipole and linked with the external circuit is $\phi = Li = \boldsymbol{B} \cdot \boldsymbol{a}i/I = \boldsymbol{B} \cdot \boldsymbol{m}/I$. If the dipole moment $\boldsymbol{m}$ changes with time, an EMF of magnitude $|\mathrm{d}\phi/\mathrm{d}t| = |I^{-1}\boldsymbol{B} \cdot \mathrm{d}\boldsymbol{m}/\mathrm{d}t|$ is induced in the external circuit, and the rate at which the dipole does work on this circuit is

$$\frac{\mathrm{d}W}{\mathrm{d}t} = -I\frac{\mathrm{d}\phi}{\mathrm{d}t} = -\boldsymbol{B} \cdot \frac{\mathrm{d}\boldsymbol{m}}{\mathrm{d}t}.$$

The change in $\boldsymbol{m}$ may be due either to a reorientation of the dipole or to a change in i. Note that, even if this change is induced by a change in $\boldsymbol{B}$, the ratio $\boldsymbol{B}/I$ is constant, and the instantaneous rate of working does not depend directly on $\mathrm{d}\boldsymbol{B}/\mathrm{d}t$. In a short time interval $\mathrm{d}t$, therefore, the work done by the dipole on the external circuit is $\mathrm{d}W = -\boldsymbol{B} \cdot \mathrm{d}\boldsymbol{m}$. It is easy to check (again, because of the linearity of the equations) that the same result is obtained if $\boldsymbol{B}$ arises from several external circuits (some of which might be other small dipoles) carrying different currents.

The magnetic material may be regarded as consisting of a large number of small dipoles, the dipole moment associated with a volume element d^3r being $\boldsymbol{m} = \boldsymbol{M}(\boldsymbol{r})\,\mathrm{d}^3r$. For each dipole, the work done in a change in state is

$$\mathrm{d}W = -\boldsymbol{B}(\boldsymbol{r}) \cdot \mathrm{d}\boldsymbol{M}(\boldsymbol{r}) = -[\boldsymbol{B}_0(\boldsymbol{r}) + \boldsymbol{B}_\mathrm{M}(\boldsymbol{r})] \cdot \mathrm{d}\boldsymbol{M}(\boldsymbol{r}).$$

However, the contribution from $\boldsymbol{B}_\mathrm{M}$ is work done on other dipoles in the material and does not change the overall internal energy of the material. Consequently, the work done by the material on the external apparatus is

$$\mathrm{d}W = -\int \mathrm{d}^3r\, \boldsymbol{B}_0(\boldsymbol{r}) \cdot \mathrm{d}\boldsymbol{M}(\boldsymbol{r}).$$

For simplicity, we now consider the special case in which $\boldsymbol{B}_0(\boldsymbol{r})$ and $\boldsymbol{M}(\boldsymbol{r})$ are spatially uniform and parallel to each other. Let $B_0 = |\boldsymbol{B}_0|$ be the magnitude of the applied field, and $M = |\int \mathrm{d}^3r\, \boldsymbol{M}(\boldsymbol{r})|$ the total magnetic moment of the sample. Then the work done in a change of state is just $\mathrm{d}W = -B_0\,\mathrm{d}M$ and, for a reversible change, the first law is

$$\mathrm{d}U = \mathrm{d}Q - \mathrm{d}W = T\,\mathrm{d}S + B_0\,\mathrm{d}M.$$

From the equality of partial derivatives of $U(S, M)$, we obtain the Maxwell relation

$$\left(\frac{\partial T}{\partial M}\right)_S = \left(\frac{\partial B_0}{\partial S}\right)_M .$$

By analogy with a P–V–T system, we can perform a sequence of Legendre transformations and define

(i) a Helmholtz free energy $F(T, M) = U - TS$ for which

$$\mathrm{d}F = -S\,\mathrm{d}T + B_0\,\mathrm{d}M$$

with the associated Maxwell relation $(\partial S/\partial M)_T = -(\partial B_0/\partial T)_M$,

(ii) a Gibbs free energy $G(T, B_0) = F - B_0 M$ for which

$$\mathrm{d}G = -S\,\mathrm{d}T - M\,\mathrm{d}B_0$$

with the associated Maxwell relation $(\partial S/\partial B_0)_T = (\partial M/\partial T)_{B_0}$ and

(iii) an enthalpy $H(S, B_0) = G + TS$ for which

$$\mathrm{d}H = T\,\mathrm{d}S - M\,\mathrm{d}B_0$$

with the associated Maxwell relation $(\partial T/\partial B_0)_S = -(\partial M/\partial S)_{B_0}$.

• **Problem 2.11** At an instant just before the annihilation of $e^{\pm}$, neutrinos and photons were at the same temperature, say, T. We consider a region of space whose volume at that time was V. A short time later, when the annihilation was complete, the volume had increased, say, to V'. We call the photon temperature at this later time T'_γ and the neutrino temperature T'_ν. Since then, both temperatures have fallen by the same factor, so their ratio now is $T_\nu^{\rm now}/T_\gamma^{\rm now} = T'_\nu/T'_\gamma$, and we need to calculate this ratio.

The neutrinos were unaffected by the annihilation, so we have

$$VT^3 = V'T'^3_\nu .$$

If the Universe were homogeneous, there would be no temperature gradient and therefore no heat flow. Consequently, if the annihilation process was reversible (which is a matter of assumption), then the total entropy of the region considered remained constant. It will therefore be useful to find the entropies of the gases in terms of their volume and temperature. From the first law, the total energy U and entropy S of a gas in a volume V are related by

$$\left(\frac{\partial U}{\partial T}\right)_V = T\left(\frac{\partial S}{\partial T}\right)_V$$

and, if $U = cT^4V$, we can integrate this relation to obtain $S = \frac{4}{3}cT^3V$. Our previous result shows that the entropy of the neutrinos is unchanged, and we can now find out what is implied by the assumption that the entropy of the electron–photon gas is also unchanged. Just before the annihilation, the entropy in the volume V was

$$S = S_e + S_\gamma = \tfrac{4}{3} \times 2cT^3V + \tfrac{4}{3} \times \tfrac{8}{7}cT^3V = \tfrac{88}{21}cT^3V.$$

After the annihilation, the entropy of the photons at a temperature T'_γ in a volume V' was

$$S' = \tfrac{4}{3} \times \tfrac{8}{7}cT'^3_\gamma V' = \tfrac{32}{21}cT'^3_\gamma V'.$$

Equating S and S' and then using our first result, we find that

$$T'^3_\gamma = \tfrac{11}{4}\frac{V}{V'}T^3 = \tfrac{11}{4}T'^3_\nu .$$

Since the ratio of volumes V/V' cancels out, it does not matter exactly which two instants of time we choose before and after the annihilation. Indeed, we could have assumed that the annihilation was instantaneous (and so $V' = V$) and obtained the same result, but this assumption is not entirely realistic. Thus, the ratio T'_ν/T'_γ is equal to $(\frac{4}{11})^{1/3}$ and the present neutrino temperature is

$$T_\nu^{\rm now} = (\tfrac{4}{11})^{1/3} \times 2.7\ \mathrm{K} \simeq 1.9\ \mathrm{K}.$$

3

STATISTICAL ENSEMBLES

This chapter focuses on the ensembles of equilibrium statistical mechanics and their relationships with thermodynamics. Problems 3.1–3.8 deal with the microcanonical ensemble, treating systems with both discrete and continuously distributed microstates. Although the microcanonical ensemble is inconvenient to use and can often be avoided in practice, it plays an important role in the foundations of statistical mechanics and affords a number of useful insights. The canonical ensemble is explored for a variety of systems in problems 3.9–3.22, of which the last five concentrate on the molecular degrees of freedom (translations, rotations and vibrations) that contribute to the specific heats of gases. The use of the grand canonical ensemble is studied in problems 3.23–3.26, while the (possibly less familiar) isobaric ensemble is illustrated in problem 3.27. The remaining problems 3.28–3.30 explore the relationships between the three principal ensembles.

3.1 Questions

• **Problem 3.1** Find the entropy $S(E, V, N)$ of an ideal gas of N classical monatomic particles, with a fixed total energy E, contained in a d-dimensional box of volume V. Deduce the equation of state of this gas, assuming that N is very large.

• **Problem 3.2** Consider an ideal gas of N quantum-mechanical monatomic particles, with a fixed total energy E, contained in a d-dimensional hypercubic box, of side L. For the case when E is much greater than the ground-state energy, obtain an approximate expression for the entropy $S(E, V, N)$, and compare your expression with the entropy of a classical gas. In this approximation, and when N is large, what is the equation of state of the quantum gas? What is the probability of finding a particle with momentum $\boldsymbol{p}$ in this gas?

• **Problem 3.3** Compute the volume in phase space occupied by a three-dimensional ultrarelativistic classical gas of N particles and fixed energy E. Find the entropy, the temperature, the equation of state and the specific heat ratio $\gamma = C_P/C_V$.

• **Problem 3.4** For a collection of N three-dimensional classical harmonic oscillators of frequency ω and fixed total energy E, compute the entropy S

and the temperature T. Discuss whether the oscillators should be treated as distinguishable or indistinguishable.

• **Problem 3.5** For a collection of N three-dimensional quantum harmonic oscillators of frequency ω and total energy E, compute the entropy S and the temperature T.

• **Problem 3.6** Using the microcanonical ensemble, compute the Helmholtz free energy $F(T, N)$ as a function of temperature for a system of N identical but distinguishable particles, each of which has two energy levels. Explore the limits $T \to 0$ and $T \to \infty$ of the energy, the entropy and the occupation numbers.

Show that the maximum (minimum) entropy corresponds to minimum (maximum) information on the system. How many bits of information are lost if the system evolves from an initial state of zero temperature to a final state of infinite temperature? What is the highest temperature at which this system can exist?

• **Problem 3.7** Consider a system of N identical but distinguishable particles, each of which has two energy levels with energy 0 or $\varepsilon > 0$. The upper energy level has a g-fold degeneracy while the lower level is non-degenerate. The total energy of the system is E.

(a) Using the microcanonical ensemble, find the occupation numbers n_+ and n_0 in terms of the temperature of the system (where n_+ corresponds to the upper level and n_0 to the lower one).

(b) Consider the case $g = 2$. If the system has energy $E = 0.75N\varepsilon$ and is brought into contact with a bath at constant temperature $T = 500$ K, in what direction does heat flow?

• **Problem 3.8** A system of N three-level particles has a Hamiltonian of the form

$$\mathcal{H} = -h \sum_{i=1}^{N} S_i \qquad S_i = 1,\ 0,\ -1$$

where h is a positive constant. (This might represent the energy levels of spin-1 particles in an applied magnetic field.) If n_S is the average number of particles in the state S ($S = 1, 0, -1$), use the microcanonical ensemble to find the ratio n_{-1}/n_1 in terms of the temperature in the limit $N \to \infty$. Hence find the Helmholtz free energy $F(T, N)$. Check your answer for $F(T, N)$ by using the canonical ensemble (which is much easier).

Identify the limits in which the information on the state of the system is maximum and minimum and find the entropy in these cases.

• **Problem 3.9** Consider a system of N non-interacting quantum-mechanical harmonic oscillators in three dimensions. Compute the canonical partition function of the system $Z(T, N)$. Verify that the same answer is obtained by regarding the system as consisting of either

(a) $3N$ one-dimensional oscillators or

(b) N three-dimensional oscillators.

• **Problem 3.10** Compute the average energy and the heat capacity of a classical system of N non-identical particles in d spatial dimensions, that has a Hamiltonian of the form

$$\mathcal{H} = \sum_{i=1}^{N} A_i |\boldsymbol{p}_i|^s + B_i |\boldsymbol{q}_i|^t .$$

The parameters A_i and B_i characterize individual particles, while s and t are positive integers, and the system is maintained at a fixed temperature T. As a special case, obtain the average energy and heat capacity for N three-dimensional harmonic oscillators.

• **Problem 3.11** Find approximate expressions for the partition function $Z(T, L)$ for a single quantum-mechanical particle in an infinite one-dimensional square well of width L in the limits of high and low temperatures. Obtain the heat capacity c_L and the equation of state $f(T, P, L) = 0$ in these limits.

• **Problem 3.12** A gas of N classical point particles lives in a two-dimensional surface. Find the internal energy $U(T, A)$, where A is the surface area, and the equation of state when the surface is

(a) a sphere and

(b) a torus.

• **Problem 3.13** Consider a classical system whose Hamiltonian can be expressed as $\mathcal{H} = \mathcal{H}_0 + \lambda \mathcal{H}_1$, where $\lambda \ll 1$. Show that the expansion of the Helmholtz free energy in powers of λ has the form

$$F = F_0 + \lambda \langle \mathcal{H}_1 \rangle_0 + \cdots$$

where F_0 and $\langle \cdots \rangle_0$ denote the free energy and an expectation value calculated with $\lambda = 0$, and find the next term in this series. Within this expansion, find the internal energy $U = \langle \mathcal{H} \rangle$ correct to first order in λ.

• **Problem 3.14** A lattice in a d-dimensional space has N sites, each occupied by an atom whose magnetic moment is $\boldsymbol{\mu}$ and is in contact with a heat reservoir at fixed temperature T. The atoms do not interact with each other, but they do interact with an applied magnetic field $\boldsymbol{H} = H(\boldsymbol{r})\hat{\boldsymbol{z}}$.

(a) Express the canonical partition function of this system in terms of a suitable product of integrals over the angles θ_i between the magnetic moments and the $\hat{z}$ direction.

(b) In the case when $d = 3$, find the average magnetization M and the susceptibility per lattice site χ.

• **Problem 3.15** Consider an ideal classical diatomic gas whose molecules have an electric dipole moment $\boldsymbol{\mu}$. The system is contained in a box of volume V, with a uniform applied electric field $\boldsymbol{E}$. Ignoring interactions between molecules,

(a) find the electric polarization $\boldsymbol{P}$ and

(b) calculate the dielectric constant of the gas in the low-field limit $|\boldsymbol{\mu}\cdot\boldsymbol{E}| \ll kT$.

• **Problem 3.16** A 'lattice gas' consists of a lattice of N sites, each of which can be empty, in which case its energy is zero, or occupied by one particle, in which case its energy is ε. Each particle has a magnetic moment of magnitude μ which, in the presence of an applied magnetic field H, can adopt two orientations (parallel or antiparallel to the field).

(a) Find the canonical partition function for this system.

(b) Evaluate the average energy and the magnetization of the system.

• **Problem 3.17** A lattice in one dimension has N sites and is at temperature T. At each site there is an atom which can be in either of two energy states: $E_i = \pm\varepsilon$. When L consecutive atoms are in the $+\varepsilon$ state, we say that they form a cluster of length L (provided that the atoms adjacent to the ends of the cluster are in the $-\varepsilon$ state). In the limit $N \to \infty$,

(a) compute the probability $\mathcal{P}_L$ that a given site belongs to a cluster of length L and

(b) calculate the mean length of a cluster $\langle L\rangle_T$ and determine its low- and high-temperature limits.

• **Problem 3.18** At a given temperature, the difference between the specific heats of a diatomic ideal gas and a monatomic gas is due in part to the rotational energy of the diatomic molecules. A rigid quantum rotator has energy levels $E_{\text{rot}}(l)$ with degeneracy $g(l)$ given by

$$E_{\text{rot}}(l) = l(l+1)\frac{\hbar^2}{2I} \qquad g(l) = 2l+1 \qquad l = 0, 1, 2, \ldots$$

where I is the moment of inertia.

(a) Find the canonical partition function of a gas of N non-interacting diatomic molecules.

(b) Evaluate the specific heat of this gas at high temperatures and at the lowest temperatures at which the rotational motion of the molecules makes a significant contribution.

• **Problem 3.19** When a diatomic molecule vibrates, its moment of inertia depends to a small extent on the vibrational state. Consequently, the rotational and vibrational motions are not completely independent. Under suitable conditions, the spectrum of vibrational and rotational energies can be approximated as

$$E_{n,l} = \hbar\omega(n + \tfrac{1}{2}) + \frac{\hbar^2}{2I}l(l+1) + \alpha l(l+1)(n + \tfrac{1}{2})$$

where the first two terms correspond to vibrational and rotational motion respectively, and the last term is a small correction that arises from the interdependence of vibrations and rotations. The various molecular constants satisfy $\hbar\omega \gg \hbar^2/2I \gg \alpha$. For a gas of such molecules, compute the internal energy for temperatures in the range $\hbar\omega > kT \gg \hbar^2/2I$.

• **Problem 3.20** A methane molecule CH_4 consists of four hydrogen atoms, situated at the corners of a regular tetrahedron, with a carbon atom situated at the centre of mass of this tetrahedron. Treating this molecule as a rigid classical object (which can rotate but does not vibrate), calculate the canonical partition function and the specific heat at constant volume of methane gas, in terms of temperature T, the mass m of a hydrogen atom, the mass M of the whole molecule and the length a of each C–H bond.

• **Problem 3.21** For a gas of diatomic molecules which are composed of indistinguishable atoms of spin s,

(a) obtain the partition function due to rotations Z_{rot} and

(b) find the rotational contribution to the specific heat C_V at constant volume.

• **Problem 3.22** The potential energy between the atoms of an hydrogen molecule can be modelled by means of the Morse potential

$$V(r) = V_0 \left[\exp\left(\frac{-2(r - r_0)}{a} \right) - 2\exp\left(\frac{-(r - r_0)}{a} \right) \right]$$

where $V_0 = 7 \times 10^{-12}$ erg, $r_0 = 8 \times 10^{-9}$ cm and $a = 5 \times 10^{-9}$ cm.

(a) Find the lowest angular frequency of rotational motion and the frequency of small-amplitude vibrations.

(b) Estimate the temperatures $T_{\rm rot}$ and $T_{\rm vib}$ at which rotations and vibrations respectively begin to contribute significantly to the internal energy.

• **Problem 3.23** Using the grand canonical ensemble, evaluate the chemical potential $\mu(T, P)$ for an ultrarelativistic gas contained in a box of volume V.

• **Problem 3.24** A fluid of particles with a repulsive interparticle potential can be modelled as a 'lattice gas' in the following way. Consider the container to be divided into N cells, each of volume v, comparable with the volume of a particle. An unoccupied cell and a cell occupied by one particle have zero energy. A cell occupied by two particles has an energy of ε, and no cell may be occupied by more than two particles. Use the grand canonical ensemble to find the average energy per cell, the concentration c of particles (c is the total number of particles divided by N) and the pressure P in terms of temperature and chemical potential. Find approximate expressions for the average energy per cell and for the pressure in terms of T and c in the limits that c is very small and that c is close to its maximum value.

• **Problem 3.25** Consider a box containing an ideal classical gas at pressure P and temperature T. The walls of the box have N_0 absorbing sites, each of which can absorb one molecule of the gas. Let $-\varepsilon$ be the energy of an absorbed molecule.

(a) Find the fugacity $z_{\rm g} = {\rm e}^{\beta\mu}$ of the gas in terms of temperature and pressure.

(b) Find the mean number of absorbed molecules $\langle N \rangle$ and investigate its low- and high-pressure limits.

• **Problem 3.26** A classical ideal gas is contained in a box of fixed volume V whose walls have N_0 absorbing sites. Each of these sites can absorb at most two particles, the energy of each absorbed particle being $-\varepsilon$. The total number of particles N is fixed and greater than $2N_0$. Use the grand canonical ensemble to obtain the equation of state of the gas in the presence of the absorbing walls and to find the average number of absorbed particles in the limits $T \to 0$ and $T \to \infty$.

• **Problem 3.27** A long elastic molecule can be modelled as a linear chain of N links. The state of each link is characterized by two quantum numbers l and n. The length of a link is either $l = a$ or $l = b$. The vibrational state of a link is modelled as a harmonic oscillator whose angular frequency is ω_a for a link of length a and ω_b for a link of length b. Thus, the energy of a link is

$$E(n, l) = \begin{cases} (n + \frac{1}{2})\hbar\omega_a & \text{for } l = a \\ (n + \frac{1}{2})\hbar\omega_b & \text{for } l = b \end{cases} \qquad n = 0, 1, 2, \ldots, \infty.$$

If the chain is immersed in a bath at temperature T and held under a tension F, find the mean energy and mean length of the chain and analyse their low- and high-temperature limits.

• **Problem 3.28** The energy of a system described by the canonical ensemble fluctuates, because the system is able to exchange energy with a heat bath, and the size of these fluctuations can be characterized by the variance $(\Delta_C E)^2 = \langle E^2 \rangle_C - \langle E \rangle_C^2$, where the subscript C denotes the canonical ensemble. For a system described by the grand canonical ensemble, both energy and particle number fluctuate, and variances $(\Delta_G E)^2$ and $(\Delta_G N)^2$ can be defined in a corresponding manner. One might suppose that particles exchanged with a reservoir would carry energy and, for this reason, that $(\Delta_G E)^2 > (\Delta_C E)^2$.

Express $(\Delta_G E)^2$ and $(\Delta_C E)^2$ in terms of derivatives of appropriate thermodynamic functions. In the thermodynamic limit, assume that $\langle E \rangle_C = \langle E \rangle_G = U$ and $\langle N \rangle_G = N$ and use thermodynamic reasoning to show that

$$(\Delta_G E)^2 - (\Delta_C E)^2 = (\Delta_G N)^2 \left[\left(\frac{\partial U}{\partial N} \right)_{T,V} \right]^2 .$$

• **Problem 3.29** Consider a system of non-interacting, identical but distinguishable particles. Using both the canonical and the grand canonical ensembles, find the partition function and the thermodynamic functions $U(T, V, N)$, $S(T, V, N)$ and $F(T, V, N)$, where T is the temperature, V the volume and N the number of particles, in terms of the single-particle partition function $Z_1(T, V)$. Verify that $U_G = U_C$, where the subscripts C and G denote the canonical and grand canonical ensembles respectively. If s and f are the entropy and the Helmholtz free energy per particle respectively, show that, when N is large, $(s_G - s_C)/k = -(f_G - f_C)/kT \simeq (\ln N)/N$.

• **Problem 3.30** Consider a system of identical but distinguishable particles, each of which has two states, with energies ε and $-\varepsilon$ available to it. Use the microcanonical, canonical and grand canonical ensembles to calculate the mean entropy per particle as a function of the mean energy per particle in the limit of a very large system. Verify that all three ensembles yield identical results in this limit.

3.2 Answers

• **Problem 3.1** The gas in question is a closed, thermally isolated system, described by the microcanonical ensemble. Its entropy is defined, roughly, by $S = k \ln \Omega$, where Ω is the number of microstates corresponding to the fixed energy E. However, the coordinates q_i and momenta p_i ($i = 1, \ldots, dN$) of the particles in a classical gas can assume a continuous range of values, and some care is needed to obtain a well-defined entropy. To this end, we define $\Omega(E, V, N; \Delta E)$ as the volume of a thin shell in phase space corresponding to a narrow range of energies between E and $E + \Delta E$. The precise definition of entropy is then

$$S(E, V, N; \Delta E) = k \ln \left(\frac{\Omega(E, V, N; \Delta E)}{h^{dN} N!} \right).$$

In this expression, h is an arbitrary constant with the dimensions of (length $\times$ momentum), included to make the argument of the logarithm dimensionless. (It is sometimes convenient to identify h as Planck's constant, to facilitate a comparison between classical and quantum gases.) Since ΔE is small, we have $\Omega(E, V, N; \Delta E) = \Delta E \, \Sigma(E, V, N)$ where $\Sigma(E, V, N)$ is a measure of the area of a constant energy surface. Consequently, both h and ΔE contribute to S only an additive constant, which has no physical significance. The factor $N!$ enforces the condition (sometimes referred to as 'correct Boltzmann counting') that states which differ only by one of the $N!$ permutations of the N identical particles are to count as the same state.

To calculate $\Omega(E, V, N; \Delta E)$, we express it as $\Omega = \Delta E \, \partial \mathcal{V}/\partial E$, where $\mathcal{V}$ is the volume of phase space corresponding to energies less than or equal to E. The Hamiltonian of the whole system is

$$\mathcal{H} = \sum_{i=1}^{dN} \frac{p_i^2}{2m}$$

for particles of mass m, so we obtain

$$\mathcal{V}(E, V, N) = \int_{\mathcal{H} \leq E} \prod_{i=1}^{dN} \mathrm{d}p_i \, \mathrm{d}q_i = V^N \int_{\sum p^2 < 2mE} \mathrm{d}^{dN} p$$

where the factor V^N is the result of integrating over the coordinates. The momentum integral is precisely the volume of a dN-dimensional sphere with radius $R = \sqrt{2mE}$ given in appendix A. We therefore have

$$\mathcal{V}(E, V, N) = \frac{V^N (2\pi m E)^{dN/2}}{(dN/2)\Gamma(dN/2)}$$

and the entropy can be written as

$$S(E, V, N; \Delta E) = k\left\{N \ln\left(\frac{V(2\pi m E)^{d/2}}{h^d}\right) - \ln\left[\Gamma\left(\frac{dN}{2}\right)\right] - \ln(N!) + \ln\left(\frac{\Delta E}{E}\right)\right\}.$$

For large values of N, we can make use of Stirling's approximation $\ln(N!) \simeq N \ln N - N$ to get

$$S(E, V, N) \simeq Nk\left\{\ln\left[\frac{V}{N}\left(\frac{4\pi m E}{dNh^2}\right)^{d/2}\right] + \frac{d+2}{2}\right\}.$$

Note that, in this approximation, the entropy is extensive, since S/N depends on E, V and N only in the combinations V/N and E/N.

To obtain the equation of state, we use the first law of thermodynamics to write

$$P = T\left(\frac{\partial S}{\partial V}\right)_E = T\frac{Nk}{V}$$

which gives the familiar result $PV = NkT$. We can also find the temperature as a function of E, V and N, by using

$$\frac{1}{T} = \left(\frac{\partial S}{\partial E}\right)_V = \frac{dNk}{2E}.$$

This can be rearranged to give the usual relation $E = (d/2)NkT$ for the energy of a monatomic classical ideal gas.

• **Problem 3.2** A free particle of mass m confined to a hypercubic box has energy eigenfunctions

$$\psi_{\boldsymbol{k}}(\boldsymbol{x}) = \left(\frac{2}{L}\right)^{d/2} \prod_{i=1}^{d} \sin(k_i x_i)$$

where, in order for the wavefunction to vanish both at $x_i = 0$ and at $x_i = L$, the components of the wavevector are restricted to the values $k_i = n_i\pi/L$, where each n_i is a positive integer. The energy of this state is $E_n = \hbar^2 k^2/2m = (\pi^2\hbar^2/2mL^2)\sum_i n_i^2$. An energy eigenfunction for the N-particle system is a sum of products of N of these single-particle functions which for bosons is symmetric, or for fermions is antisymmetric, under the interchange of any pair of particles. The total energy is

$$E = \sum_{i=1}^{N} \frac{\hbar^2 |\boldsymbol{k}_i|^2}{2m} = \sum_{j=1}^{dN} \frac{\hbar^2}{2m} k_j^2 = \epsilon \sum_{j=1}^{dN} n_j^2$$

where $\epsilon = \pi^2\hbar^2/2mL^2$. The entropy is $S(E, V, N) = k\ln[\Omega(E, V, N)]$, where $\Omega(E, V, N)$ is the number of distinct states with energy E. An exact expression for this number is difficult to obtain. Clearly, we are allowed to choose for E only values which are integer multiples of ϵ. In the case of bosons, Ω is related to the number of sets of integers $\{n_j\}$ whose squares add to E/ϵ. In the case of fermions, only those sets are allowed which are consistent with all the particles (which obey the Pauli exclusion principle) being in different quantum states. A good approximate answer can be obtained when E is large, however. If $\epsilon/E \ll 1$, then E can be treated as a continuous variable and, if E is much greater than the ground-state energy, then many states are available, only a small fraction of which are forbidden (for fermions) by the exclusion principle.

The dN-dimensional space of the integers $\{n_j\}$ can be divided into cells of unit volume, each containing exactly one state. If we allow for a small uncertainty in E, then the number of states with this energy is, to a good approximation, the number of cells intersected by a sphere of radius $\sqrt{E/\epsilon}$, or the area of this sphere, which is $2\pi^{dN/2}(E/\epsilon)^{dN/2-1}/\Gamma(dN/2)$. More precisely, we wish to include only that segment of the sphere for which all the n_j are positive, which is a fraction $1/2^{dN}$ of the whole sphere, and we must also divide by the number $N!$ of permutations of the N particles, which do not count as distinct states. In this way, we obtain

$$\begin{aligned}\Omega(E, V, N) &= \frac{2\pi^{dN/2}}{2^{dN}N!\Gamma(dN/2)}\left(\frac{E}{\epsilon}\right)^{dN/2-1} \\ &= \frac{V^N}{N!\Gamma(dN/2)}\left(\frac{2\pi mE}{h^2}\right)^{dN/2}\frac{2\epsilon}{E}\end{aligned}$$

where $V = L^d$ and $h = 2\pi\hbar$. Note that, in contrast with the energy surface in phase space considered in the previous problem, the surface area needed here is a dimensionless quantity. For particles of spin s, there will be an additional factor $(2s+1)^N$ corresponding to the number of distinct spin configurations. For the entropy of a gas of spin-0 particles, we have

$$S(E, V, N) = k\left\{N\ln\left[V\left(\frac{2\pi mE}{h^2}\right)^{d/2}\right] - \ln\left[\Gamma\left(\frac{dN}{2}\right)\right] - \ln(N!) + \ln\left(\frac{2\epsilon}{E}\right)\right\}.$$

Apart from the last term, which is unimportant when N is large, this is identical with the expression obtained in the previous problem for a classical gas, but here h really is Planck's constant. Obviously, when N is large, the equation of state is the same as that of a classical gas: $P = T(\partial S/\partial V)_E = NkT/V$ or $PV = NkT$. The idea of determining the momentum of a specific particle in a quantum-mechanical gas is somewhat ill defined, since the particles are in principle indistinguishable. It is nevertheless reasonable to ask what fraction of the particles have, on average, a momentum in a given range and this can loosely be interpreted to mean that the probability of a particle selected at random having its momentum in this range. With this caveat, let us first ask for the probability that a randomly selected particle has energy $E_p = |\boldsymbol{p}|^2/2m$, while the remaining energy $E - E_p$ is distributed in an unspecified way among the $N - 1$ other particles. This probability is

$$\mathcal{P}(E_p) = \frac{\Omega(E_p, V, 1)\Omega(E - E_p, V, N - 1)}{N\Omega(E, V, N)}.$$

The numerator is the number of states available to a pair of systems, one consisting of a single particle with energy E_p and the other of $N - 1$ particles with total energy $E - E_p$. The factor N^{-1} accounts for the fact that the N ways of abstracting one particle lead to indistinguishable states, and $\mathcal{P}(E_p)$ is the ratio of the resulting number of distinct states to the number of states available to the whole N-particle system. Using the expression for Ω obtained above, we find that

$$\mathcal{P}(E_p) = \frac{\Gamma(dN/2)}{\Gamma[d(N-1)/2]\Gamma(d/2)} \frac{\epsilon}{E} \left(\frac{E_p}{E}\right)^{d/2-1} \left(1 - \frac{E_p}{E}\right)^{d(N-1)/2-1}.$$

To appreciate the meaning of this result, recall that the energy is quantized in units of ϵ, but that our approximation for Ω is valid for large energies, which can be treated as continuously distributed. Thus, to verify that $\sum_{E_p} \mathcal{P}(E_p) = 1$, we replace $\sum_{E_p}$ by $\epsilon^{-1} \int_0^E \mathrm{d}E_p$ and easily verify that the resulting integral is equal to unity, by using the standard result

$$\int_0^1 \mathrm{d}x\, x^{\alpha-1}(1-x)^{\beta-1} = \frac{\Gamma(\alpha)\Gamma(\beta)}{\Gamma(\alpha+\beta)}.$$

Also, in this approximation, the system is effectively isotropic. It is straightforward to convert $\mathcal{P}(E_p)$ into a probability distribution for momentum using

$$\int_{E_p \le E} \mathrm{d}^d p \propto \int_0^{\sqrt{2mE}} p^{d-1}\, \mathrm{d}p \propto \int_0^E E_p^{d/2-1}\, \mathrm{d}E_p.$$

The factor $E_p^{d/2-1}$ in $\mathcal{P}(E_p)$ corresponds to this momentum-space volume element. The omitted constants are not hard to find but are not needed, since we already know that the probability is correctly normalized. We thus have

$$\mathcal{P}(\boldsymbol{p})\, \mathrm{d}^d p = \mathcal{N}(E) \left(1 - \frac{E_p}{E}\right)^{d(N-1)/2-1} \mathrm{d}^d p$$

where $\mathcal{N}(E)$ is a normalizing factor such that $\int \mathcal{P}(\boldsymbol{p})\, \mathrm{d}^d p = 1$. It is instructive to evaluate this momentum-space probability density in the *thermodynamic limit* that $N \to \infty$ with E/N and V/N fixed. In fact, as was shown in problem 3.1, we have $E/N = (d/2)kT$; so, using the fact that $\lim_{N\to\infty}[(1 - x/N)^N] = \mathrm{e}^{-x}$, we find that

$$\mathcal{P}(\boldsymbol{p})\, \mathrm{d}^d p = \frac{\mathrm{e}^{-E_p/kT}\, \mathrm{d}^d p}{\int \mathrm{e}^{-E_p/kT}\, \mathrm{d}^d p}.$$

This expression is valid for high temperatures, where the quantum gas behaves in the same way as a classical gas and is, of course, the expression that we would immediately write down using the canonical ensemble for a system in thermal contact with a heat bath at temperature T. This illustrates the fact that the two ensembles are equivalent (for systems without long-range forces) in the thermodynamic limit.

• **Problem 3.3** As explained in detail for problem 3.1, it is necessary to allow for a small uncertainty ΔE in the energy of the gas. Then the phase space volume occupied by the gas is given by $\Omega(E, V, N; \Delta E) = \Delta E\, \partial\mathcal{V}(E, V, N)/\partial E$, where $\mathcal{V}$ is the volume of a region D within which the energy is less than or equal to E. For highly relativistic particles, their rest energy can be neglected, and the total energy of the gas is

$$\mathcal{H} = c\sum_{i=1}^{N} p_i$$

where $p_i = |\boldsymbol{p}_i|$, so D is the region $\sum_i p_i \leq E/c$ and

$$\mathcal{V} = \int_D \prod_{i=1}^{N} \mathrm{d}^3 q_i\, \mathrm{d}^3 p_i .$$

Since the positions and the directions of the momenta of the particles are unconstrained, we can write for each particle $\int \mathrm{d}^3 q\, \mathrm{d}^3 p = 4\pi V \int p^2\, \mathrm{d}p$, so

$$\mathcal{V} = (4\pi V)^N \int_D \prod_{i=1}^{N} \int p_i^2\, \mathrm{d}p_i \equiv (4\pi V)^N I_N(E).$$

To evaluate the integral $I_N(E)$, we note that, if the momentum of one particle is fixed to the value p, then the integral for the remaining particles is $I_{N-1}(E - pc)$. We therefore have

$$I_N(E) = \int_0^{E/c} \mathrm{d}p\, p^2 I_{N-1}(E - pc).$$

Then dimensional analysis tells us that $I_N(E) = C_N (E/c)^{3N}$, and substituting this in the above equation yields a recursion relation for the number C_N:

$$\begin{aligned}
C_N &= \left(\frac{c}{E}\right)^{3N} \int_0^{E/c} \mathrm{d}p\, p^2 C_{N-1} \left(\frac{E}{c} - p\right)^{3N-3} \\
&= \int_0^1 \mathrm{d}x\, x^2 (1 - x)^{3N-3} C_{N-1} \\
&= \frac{2(3N-3)!}{(3N)!} C_{N-1} .
\end{aligned}$$

Clearly, we have $C_1 = \int_0^1 x^2\, \mathrm{d}x = \frac{1}{3}$ and the solution to the recursion relation is $C_N = 2^N/(3N)!$, giving

$$\mathcal{V} = \frac{1}{(3N)!}\left(\frac{8\pi V E^3}{c^3}\right)^N \qquad \Omega = \frac{1}{(3N-1)!}\left(\frac{8\pi V E^3}{c^3}\right)^N \frac{\Delta E}{E}.$$

When N is large, we use Stirling's approximation to evaluate the entropy as

$$S(E, V, N) = k \ln\left(\frac{\Omega}{h^{3N} N!}\right) \simeq Nk\left[\ln\left(\frac{8\pi V E^3}{27 h^3 c^3 N^4}\right) + 4\right].$$

It is now easy to compute the temperature and pressure:

$$\frac{1}{T} = \left(\frac{\partial S}{\partial E}\right)_{V,N} = \frac{3Nk}{E} \qquad P = T\left(\frac{\partial S}{\partial V}\right)_{E,N} = \frac{NkT}{V}.$$

The equation of state is thus the same as for a non-relativistic gas: $PV = NkT$. Making use of these results, we can write

$$S(T, V, N) = Nk \ln\left(\frac{c_1 T^3 V}{N}\right) \qquad S(T, P, N) = Nk \ln\left(\frac{c_2 T^4}{P}\right)$$

where c_1 and c_2 are constants. The specific heats are then

$$C_V = T\left(\frac{\partial S}{\partial T}\right)_{V,N} = 3Nk \qquad C_P = T\left(\frac{\partial S}{\partial T}\right)_{P,N} = 4Nk$$

and we finally obtain $\gamma = C_P/C_V = \frac{4}{3}$, compared with $\gamma = \frac{5}{3}$ for a non-relativistic monatomic gas.

• **Problem 3.4** The Hamiltonian of the set of N classical harmonic oscillators is

$$\mathcal{H} = \sum_{i=1}^{3N}\left(\frac{p_i^2}{2m} + \frac{m\omega^2}{2} q_i^2\right).$$

We need to find the volume in phase space given by

$$\mathcal{V}(E, N) = \int_{\mathcal{H}<E} \prod_{i=1}^{3N} \mathrm{d}p_i\, \mathrm{d}q_i$$

which can be computed by means of the substitutions

$$p_i = \sqrt{2m}\, x_i \qquad i = 1, \ldots, 3N$$

$$q_i = \sqrt{\frac{2}{m\omega^2}}\, x_{3N+i} \qquad i = 1, \ldots, 3N.$$

In terms of these variables, the energy constraint is $E = \sum_{i=1}^{6N} x_i^2$ and the volume is

$$\mathcal{V}(E, N) = \left(\frac{2}{\omega}\right)^{3N} \int_D \prod_{i=1}^{6N} \mathrm{d}x_i = \left(\frac{2}{\omega}\right)^{3N} \frac{\pi^{3N}}{3N\Gamma(3N)} E^{3N}.$$

Here, D is the region in which the constraint is satisfied, and the remaining integral is the volume of a $6N$-dimensional sphere with radius $R = \sqrt{E}$, which is given in appendix A. To compute the entropy (see problem 3.1 for details) we need the volume $\Omega(E, N; \Delta E)$ of the energy shell of thickness ΔE, which is

$$\Omega(E, N; \Delta E) = \Delta E \frac{\partial \mathcal{V}(E, N)}{\partial E} = \left(\frac{2}{\omega}\right)^{3N} \frac{\pi^{3N}}{\Gamma(3N)} E^{3N} \frac{\Delta E}{E}.$$

For large N, we use Stirling's approximation to obtain

$$S(E, N) = k \ln\left(\frac{\Omega}{h^{3N}}\right) \simeq 3Nk\left[\ln\left(\frac{2\pi E}{3h\omega N}\right) + 1\right].$$

In obtaining this expression, we have taken the oscillators to be *distinguishable* and have not inserted a factor of $1/N!$ in the first logarithm (see the discussion below). The temperature is given by

$$\frac{1}{T} = \left(\frac{\partial S}{\partial E}\right)_N = \frac{3Nk}{E}.$$

Thus, the average energy per oscillator is $3kT$, in accordance with the equipartition theorem.

Whether the oscillators should be treated as distinguishable or indistinguishable depends on the physical situation that we have in mind. We have treated them as distinguishable (although they are identical), with the result that the entropy is extensive: S/N is a function just of E/N. This would be appropriate if, for example, we consider the oscillators as modelling atoms in a lattice, assuming that each atom oscillates independently about its own lattice site. We can then distinguish the atoms by the sites to which they are attached. On the other hand, we might consider the oscillators as a gas of N particles, all inhabiting a single potential well. In that case, any position in space could be occupied by any of the particles, and we should treat them as indistinguishable. The entropy would differ from that calculated above by an amount $\Delta S = -Nk(\ln N - 1)$ and would not be extensive. Since the entropy difference depends only on N, quantities such as the temperature and specific heat are the same in either case. The question of distinguishability is important only when we consider changing the number of particles. In the case of a lattice, this could be done only by having two lattices with the same average energy per particle ($E_1/N_1 = E_2/N_2$). We are then at liberty to regard these either as two separate systems or as one combined system, and our choice should have no physical consequences. This is

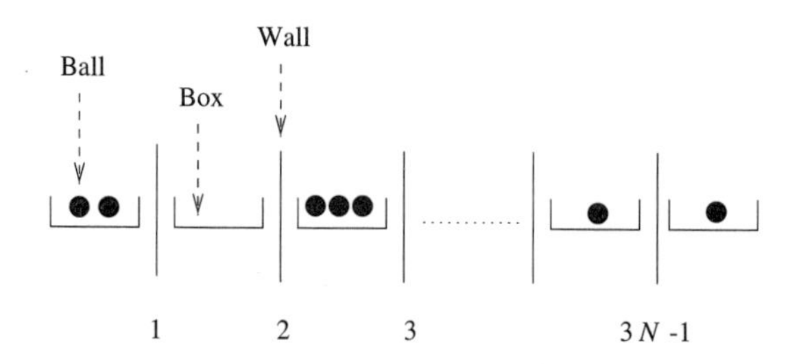

Figure 3.1. Particular arrangement of M balls into $3N$ boxes, {bbwwbbbw...}, for counting the number of sets of non-negative integers which add to M.

indeed so: since the entropy is extensive, we obtain a total entropy $S = S_1 + S_2$ in either case. By contrast, the number of particles in a gas can be changed by bringing the gas into contact with a particle reservoir, and the rate of particle exchange will depend on the number of particles per unit volume in the gas (and possibly on other factors). There is therefore a real physical difference between having N potential wells each containing one particle, and having a single well containing N particles. This difference is reflected in the non-extensive part of the entropy ΔS.

• **Problem 3.5** In order to solve this problem using the microcanonical ensemble we have to count all possible states of the entire system with its total energy fixed. The energy of the system is

$$E = \sum_{i=1}^{3N} E_i = \sum_{i=1}^{3N} (n_i + \tfrac{1}{2})\hbar\omega$$

where n_i is the number of quanta of energy in the ith oscillator ($n_i = 0, 1, \ldots, \infty$). We can write this as

$$\sum_{i=1}^{3N} n_i = \frac{E}{\hbar\omega} - \frac{3N}{2} \equiv M.$$

The number of possible states is the number of sets of non-negative integers $\{n_i\}$ which add to M. This combinatorial problem can be solved by considering the $3N$ oscillators as boxes, into which we place M indistinguishable balls, representing the quanta of energy. To count the number of ways in which this can be done, we treat the boxes (oscillators) as distinguishable, and imagine them to be arranged in a row and separated by $3N - 1$ walls, as illustrated in figure 3.1. A given state of the balls in the boxes can be viewed as a sequence of balls and walls. For example, the sequence {bbwwbbbw...} shown in the figure corresponds to two balls in box 1, no ball in box 2, three balls in box 3, and so on. If the balls and walls were all distinguishable, the total number of arrangements of these $M + 3N - 1$ objects would be $(M + 3N - 1)!$. Since the balls are indistinguishable from each other, and so are the walls, we divide by the $M!$ permutations of the balls and the $(3N - 1)!$ permutations of the walls to obtain the number of distinct arrangements. Hence, the number of distinct possible states is

$$\Omega(E, N) = \frac{(M + 3N - 1)!}{M!(3N - 1)!} = \frac{(E/\hbar\omega + 3N/2 - 1)!}{(E/\hbar\omega - 3N/2)!(3N - 1)!}.$$

The entropy is $S(E, N) = k\ln[\Omega(E, N)]$. In the thermodynamic limit, that $E, N \to \infty$ with E/N fixed, we use Stirling's approximation to obtain

$$S(E, N) \simeq Nk\left[\left(\frac{E}{N\hbar\omega} + \frac{3}{2}\right)\ln\left(\frac{E}{N\hbar\omega} + \frac{3}{2}\right) - \left(\frac{E}{N\hbar\omega} - \frac{3}{2}\right)\ln\left(\frac{E}{N\hbar\omega} - \frac{3}{2}\right)\right].$$

Note that, as in problem 3.4, the entropy is extensive in the thermodynamic limit, provided that the oscillators are distinguishable.

The temperature is now given by

$$\frac{1}{T} = \left(\frac{\partial S}{\partial E}\right)_N = \frac{k}{\hbar\omega}\ln\left(\frac{E/N\hbar\omega + \frac{3}{2}}{E/N\hbar\omega - \frac{3}{2}}\right).$$

This can easily be solved for E in terms of T and N, and we get

$$E(T, N) = \tfrac{3}{2}N\hbar\omega\coth\left(\frac{\hbar\omega}{2kT}\right)$$

which is the result that would be obtained from the canonical ensemble. At high temperatures, $kT \gg \hbar\omega$ or, equivalently, when $E \gg 3N\hbar\omega$, we recover the classical result of problem 3.4, namely $E \simeq 3NkT$.

• **Problem 3.6** Let us suppose that each particle can have an energy $\pm\varepsilon$, and that n_+ particles have energy $+\varepsilon$. Obviously, the number of particles with energy $-\varepsilon$ is $n_- = N - n_+$ and, if the total energy is E, then $n_\pm = (N \pm E/\varepsilon)/2$. The number of microstates with energy E is just the number of ways that n_+ (or n_-) particles can be chosen from N, namely

$$\Omega(E, N) = \binom{N}{n_+} = \frac{N!}{n_+!(N - n_+)!}.$$

As usual, we obtain a compact form for the entropy in the thermodynamic limit. In this case, we take N, n_+ and n_- all to be large. Using Stirling's approximation and the definition

$$x = \frac{n_+}{N} = \frac{1}{2}\left(1 + \frac{E}{N\varepsilon}\right)$$

we find that

$$S(E, N) = k \ln \Omega \simeq -Nk[x \ln x + (1-x)\ln(1-x)]$$

and note that this is extensive, provided that the particles are distinguishable (cf problems 3.4 and 3.5). The temperature is given by

$$\frac{1}{T} = \left(\frac{\partial S}{\partial E}\right)_N = \frac{1}{2N\varepsilon}\left(\frac{\partial S}{\partial x}\right)_N = \frac{k}{2\varepsilon}\ln\left(\frac{1-x}{x}\right)$$

and this relation can be inverted to yield

$$x = (\mathrm{e}^{2\varepsilon/kT} + 1)^{-1} \qquad E(T, N) = -N\varepsilon \tanh\left(\frac{\varepsilon}{kT}\right).$$

It is now a simple matter to compute the Helmholtz free energy:

$$F(T, N) = E(T, N) - TS(T, N) = -N\varepsilon - NkT\ln(1 + \mathrm{e}^{-2\varepsilon/kT}).$$

Of course, this result would also be obtained from the canonical ensemble:

$$\mathrm{e}^{-F/kT} = \prod_{i=1}^{N} \sum_{E_i=\pm\varepsilon} \mathrm{e}^{-E_i/kT} = \mathrm{e}^{N\varepsilon/kT}(1 + \mathrm{e}^{-2\varepsilon/kT})^N.$$

In the limit $T \to 0$, we find that $x \to 0$, so $n_+(0, N) = 0$, $n_-(0, N) = N$, $E(0, N) = -N\varepsilon$ and $S(0, N) = 0$. Clearly, all the particles have their lowest energy and the entropy vanishes because there is only one state with this minimum energy. Since x must lie between zero and one, S cannot be negative, so $S = 0$ is its minimum value. We need N bits of information to specify this state.

In the limit $T \to \infty$, we get $x \to \frac{1}{2}$, so $n_+(\infty, N) = n_-(\infty, N) = N/2$, $E(\infty, N) = 0$ and $S(\infty, N) = Nk\ln 2$. Intuitively, it is fairly clear that this state, in which the energy levels are equally occupied, has the greatest possible degree of disorder. Indeed, it is easy to check that S attains its maximum value when $x = \frac{1}{2}$. Since every particle has equal probabilities of being in either of its two energy levels, the information content of this state is zero. Consequently, N bits of information are lost if the system is heated from $T = 0$ to $T = \infty$. This can be confirmed using the relation between information and entropy:

$$\Delta I = \frac{S(0, N) - S(\infty, N)}{k\ln 2} = N.$$

The maximum energy that the system can hold is, of course, $E_{\max} = +N\varepsilon$, which corresponds to $x = 1$ or to $T = 0$. As discussed in problem 2.2, the range of temperatures available to the system is best understood in terms of the parameter $\beta = 1/kT$. The lowest possible temperature $T = 0$ corresponds to the largest possible value of β, namely $\beta = +\infty$, where $E = -N\varepsilon$ and $x = 0$.

As E is increased, x increases smoothly through zero. As E approaches $E_{\max}$, x approaches its maximum value of unity, while β approaches its minimum value of $-\infty$. This corresponds to T *increasing* towards zero through negative values.

• Problem 3.7

(a) The number of states with total energy $E = n_+\varepsilon$ is

$$\Omega(E, N) = \binom{N}{n_+} g^{n_+} = \frac{N!\, g^{n_+}}{n_+!(N - n_+)!}$$

where the binomial coefficient counts the number of ways of choosing n_+ distinguishable particles to occupy the upper level, and g^{n_+} is the number of ways of assigning these particles to the g degenerate states. For large N, we use Stirling's approximation to compute the entropy as

$$S(E, N) = k \ln \Omega(E, N) \simeq -Nk[x \ln x + (1 - x) \ln(1 - x) - x \ln g]$$

where $x = n_+/N = E/N\varepsilon$, and the temperature is given by

$$\beta \equiv \frac{1}{kT} = \frac{1}{k}\left(\frac{\partial S}{\partial E}\right)_N = \frac{1}{N\varepsilon k}\left(\frac{\partial S}{\partial x}\right)_N = \frac{1}{\varepsilon} \ln\left(\frac{g(1 - x)}{x}\right).$$

Solving this relation for x, we find that

$$n_+ = Nx = N\frac{g\,\mathrm{e}^{-\beta\varepsilon}}{1 + g\,\mathrm{e}^{-\beta\varepsilon}} \qquad n_0 = N - n_+ = N\frac{1}{1 + g\,\mathrm{e}^{-\beta\varepsilon}}.$$

(b) If $E = 0.75N\varepsilon$, then $x = 0.75$ and, with $g = 2$, we get $\beta\varepsilon = \ln(\frac{2}{3}) = -0.405$. Thus, the temperature is negative. Indeed, the temperature is negative whenever $E > 2N\varepsilon/3$ (or $n_+ > 2N/3$). As discussed in problems 2.2 and 3.6, a system with negative temperature has an energy greater than that for which the entropy is a maximum (figure 3.2) and is hotter than a system at any positive temperature.

According to the second law of thermodynamics, heat should therefore flow from the negative-temperature system into a heat bath at any positive temperature. It is interesting to check this conclusion from the point of view of entropy changes. If an amount ΔE of energy flows from the system into the heat bath, then the total entropy change is

$$\begin{aligned}\Delta S = \Delta S|_{\text{bath}} + \Delta S|_{\text{system}} &= \left(\left.\frac{\partial S}{\partial E}\right|_{\text{bath}} - \left.\frac{\partial S}{\partial E}\right|_{\text{system}}\right)\Delta E \\ &= \left(\frac{1}{T_{\text{bath}}} - \frac{1}{T_{\text{system}}}\right)\Delta E.\end{aligned}$$

In order for ΔS to be positive, ΔE must be positive if $1/T_{\text{system}}$ is smaller than $1/T_{\text{bath}}$ and vice versa. Clearly, for the purpose of calculating this sign, a

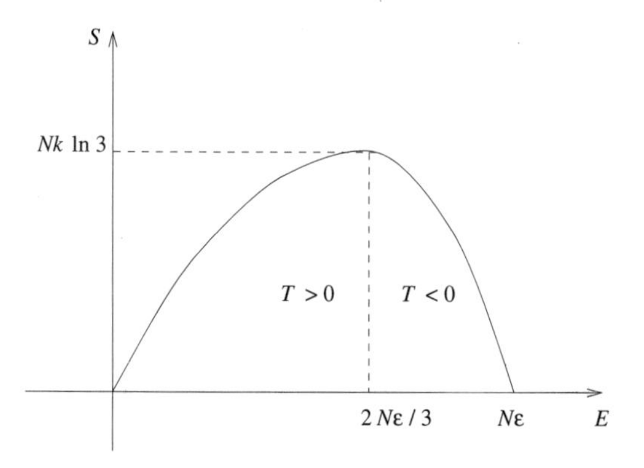

Figure 3.2. Entropy versus energy for a two-level system with a degenerate level whose degeneracy is $g = 2$.

negative value of $1/T$ counts as smaller than a positive value. Thus, from this point of view also, we see that a negative temperature is higher than a positive temperature and, in the case at hand, heat flows from the negative-energy system into the positive-energy bath.

• **Problem 3.8** Using the microcanonical ensemble, we have to count the number of ways in which the N particles can be distributed amongst the three energy levels, subject to the two constraints

$$N = n_1 + n_{-1} + n_0 \qquad E = -h(n_1 - n_{-1}).$$

Clearly, if n_0 specific particles have $S = 0$, then the remaining $N - n_0$ particles can be treated as having two energy levels (cf problems 3.6 and 3.7). Thus the number of microstates is

$$\Omega(E, N) = \sum_{n_0=0 \text{ or } 1}^{N-|E|/h} \binom{N}{n_0} \binom{N - n_0}{n_1} \equiv \sum_{n_0=0 \text{ or } 1}^{N-|E|/h} \Phi(n_0, E, N).$$

The first factor is the number of ways of choosing n_0 particles out of N, while the second is the number of ways of choosing $n_1 = (N - n_0 - E/h)/2$ of the remaining $N - n_0$ particles to occupy the lower level $S = 1$. The limits for n_0 can be found by rearranging the constraints to read $n_0 = N - E/h - 2n_1$ and $n_{-1} = E/h + n_1$. From the first of these, we see that the smallest value of n_0 is zero if $N - E/h$ is even and one if $N - E/h$ is odd. The largest value of n_0 corresponds to the smallest value of n_1. From the second constraint, we see that this is $n_1 = 0$ if $E \geq 0$ and $n_1 = -E/h$ if $E < 0$. Then the first constraint gives the largest value of n_0 as $N - |E|/h$.

The remaining sum over n_0 is difficult to evaluate in closed form. In the limit $N \to \infty$, however, it can be approximated by retaining only a single term, corresponding to the value of n_0 for which $\Phi(n_0, E, N)$ is a maximum. This can be seen from the following argument (a version of the central limit theorem). Define $\varepsilon = E/Nh$, $x = n_0/N$, $y = n_1/N = (1 - \varepsilon - x)/2$ and $z = n_{-1}/N = (1 + \varepsilon - x)/2$. Then, using Stirling's approximation in the form $N! \simeq N^N \mathrm{e}^{-N}$, etc, and approximating the sum by an integral, we get

$$\Phi(n_0, E, N) = \frac{N!}{n_0!\,n_1!\,n_{-1}!} \simeq \left(\frac{1}{x^x y^y z^z}\right)^N \equiv f(x)^N$$

$$\Omega(E, N) \simeq N \int_0^{1-|\varepsilon|} \mathrm{d}x\, f(x)^N.$$

If $f(x)$ is a maximum when $x = \bar{x}$, we write $x = \bar{x} + N^{-1/2}\xi$ and use a Taylor expansion together with the result $\lim_{N\to\infty}[(1 + \alpha/N)^N] = \mathrm{e}^{\alpha}$ to obtain

$$f(x)^N \simeq [f(\bar{x}) + (2N)^{-1} f''(\bar{x})\xi^2]^N \simeq f(\bar{x})^N\, \mathrm{e}^{-\sigma\xi^2}$$

and

$$\Omega(E, N) \simeq N^{1/2} f(\bar{x})^N A \qquad A = \int_{-\infty}^{\infty} \mathrm{d}\xi\, \mathrm{e}^{-\sigma\xi^2}.$$

In this expression, $\sigma = -f''(\bar{x})/2f(\bar{x})$ and the limits of the integral, which are proportional to $\sqrt{N}$ have been taken to $\pm\infty$. Evidently, the function $f(x)^N$ has a sharp peak at $x = \bar{x}$, with a width proportional to $N^{-1/2}$. In this way, we obtain the large-N limit of the entropy as

$$\begin{aligned} S(E, N) &= k \ln[\Omega(E, N)] \\ &\simeq Nk\left(\ln[f(\bar{x})] + \frac{1}{2N}\ln N + \frac{1}{N}\ln A + \cdots\right) \\ &\simeq Nk \ln[f(\bar{x})]. \end{aligned}$$

To find $\bar{x}$, it is easiest to maximize the quantity $\ln f = -x\ln x - y\ln y - z\ln z$:

$$\frac{\partial(\ln f)}{\partial x} = -(\ln x + 1) - (\ln y + 1)\frac{\partial y}{\partial x} - (\ln z + 1)\frac{\partial z}{\partial x} = \tfrac{1}{2}\ln\left(\frac{yz}{x^2}\right) = 0$$

giving $\bar{x}$ as the positive solution of the equation $\bar{x}^2 = y(\bar{x})z(\bar{x})$, which is

$$\bar{x} = \tfrac{1}{3}(\sqrt{4 - 3\varepsilon^2} - 1).$$

We next find the temperature, given by $1/kT = \beta = k^{-1}(\partial S/\partial E)_N = h^{-1}\,\mathrm{d}\{\ln[f(\bar{x})]\}/\mathrm{d}\varepsilon$. Since $[\partial(\ln f)/\partial x]_{x=\bar{x}} = 0$, we need to take account only of the ε dependence of $y(\bar{x})$ and $z(\bar{x})$, keeping $\bar{x}$ fixed. Thus we have

$$\beta = \frac{1}{h}\left(-\frac{1}{2}\frac{\partial \ln f}{\partial y} + \frac{1}{2}\frac{\partial \ln f}{\partial z}\right)_{x=\bar{x}} = -\frac{1}{2h}\ln\left(\frac{z(\bar{x})}{y(\bar{x})}\right)$$

and the ratio of occupation numbers is

$$\frac{n_{-1}}{n_1} = \frac{z(\bar{x})}{y(\bar{x})} = \mathrm{e}^{-2\beta h}.$$

Now let $r = \mathrm{e}^{-\beta h}$. From the constraint equation $(n_0 + n_1 + n_{-1})/N = \bar{x} + y(\bar{x}) + z(\bar{x}) = 1$, and our two results $\bar{x}^2 = y(\bar{x})z(\bar{x})$ and $z(\bar{x})/y(\bar{x}) = r^2$, we find that

$$\bar{x} = (r + 1 + r^{-1})^{-1} \qquad y(\bar{x}) = \bar{x}/r \qquad z(\bar{x}) = \bar{x}r.$$

Using these expressions, we obtain

$$S = Nk\ln[f(\bar{x})] = -Nk[\ln\bar{x} + \bar{x}(r - r^{-1})\ln r]$$
$$E = Nh\varepsilon = Nh[z(\bar{x}) - y(\bar{x})] = Nh\bar{x}(r - r^{-1})$$
$$T = \frac{1}{\beta k} = -\frac{h}{k\ln r}$$

which can be assembled to give the Helmholtz free energy

$$F(T, N) = E - TS = NkT\ln\bar{x} = -NkT\ln(\mathrm{e}^{-\beta h} + 1 + \mathrm{e}^{\beta h}).$$

Using the canonical ensemble, we easily obtain the partition function

$$Z(T, N) = \prod_{i=1}^{N}\sum_{S_i} \mathrm{e}^{-\beta h S_i} = (\mathrm{e}^{-\beta h} + 1 + \mathrm{e}^{\beta h})^N$$

and the Helmholtz free energy $F(T, N) = -kT\ln[Z(T, N)]$ clearly agrees with our microcanonical result.

We expect information to be a minimum in the limit $T \to \infty$. Indeed, we then have $r = 1$ and $\bar{x} = y(\bar{x}) = z(\bar{x}) = \frac{1}{3}$, or $n_0 = n_1 = n_{-1} = N/3$, so each particle has equal probabilities of occupying all three energy levels. In this state, the entropy is $S = Nk\ln 3$. The information content is a maximum when we know exactly the state of every particle. This happens when the energy has its minimum value $E = -Nh$ or $\beta \to +\infty$, so that $r = 0$ and $n_1 = Ny(\bar{x}) = N$, while $n_0 = n_{-1} = 0$, and all the particles occupy the lowest level. It also happens when the energy has its maximum value $E = +Nh$, corresponding to a negative temperature, $\beta \to -\infty$, where $r \to \infty$, giving $n_{-1} = Nz(\bar{x}) = N$, and $n_0 = n_1 = 0$. In either of these states, we have $S = 0$.

• **Problem 3.9** The Hamiltonian for a three-dimensional oscillator is simply the sum of Hamiltonians for three one-dimensional oscillators, so we are free to regard the system as a collection either of $3N$ one-dimensional oscillators or of N three-dimensional oscillators. In either case, since the oscillators do not interact, the partition function $Z(T, N)$ is the product of the partition functions for the individual oscillators:

$$Z(T, N) = \prod_{i=1}^{M} Z_i(T) = \prod_{i=1}^{M} \left(\sum_{n_i=0}^{\infty} g(n_i) \, \mathrm{e}^{-\beta E_i(n_i)} \right)$$

where M is either $3N$ or N and, for the ith oscillator, $g(n_i)$ is the degeneracy of the energy level labelled by n_i.

(a) For a one-dimensional oscillator of angular frequency ω, the energy levels are $E_i(n_i) = (n_i + \frac{1}{2})\hbar\omega$ and are non-degenerate, so $g(n_i) = 1$. Therefore,

$$Z(T, N) = \prod_{i=1}^{3N} \left(\sum_{n_i=0}^{\infty} \mathrm{e}^{-\beta\hbar\omega(n_i+1/2)} \right) = \left(\mathrm{e}^{-\beta\hbar\omega/2} \sum_{n=0}^{\infty} \mathrm{e}^{-\beta\hbar\omega n} \right)^{3N}.$$

The sum is easily evaluated by recalling the binomial expansion

$$L(q) \equiv \sum_{n=0}^{\infty} q^n = \frac{1}{1-q}$$

which is valid for $q < 1$, and (with $q = \mathrm{e}^{-\beta\hbar\omega}$) we obtain the answer

$$Z(T, N) = \left(\frac{\mathrm{e}^{-\beta\hbar\omega/2}}{1 - \mathrm{e}^{-\beta\hbar\omega}} \right)^{3N}.$$

(b) The energy levels of an isotropic three-dimensional oscillator are $E_i(n_i) = (n_i + 3/2)\hbar\omega$. The degeneracy of one of these levels is the number of sets of three non-negative integers that add to n_i. This can be found, for example, by the method suggested in the solution to problem 3.5 and is given by

$$g(n_i) = \frac{(n_i + 2)!}{2! n_i!} = \tfrac{1}{2}n_i^2 + \tfrac{3}{2}n_i + 1.$$

The partition function for one such oscillator is therefore

$$Z_i = \mathrm{e}^{-3\beta\hbar\omega/2} \sum_{n_i=0}^{\infty} (\tfrac{1}{2}n_i^2 + \tfrac{3}{2}n_i + 1) \, \mathrm{e}^{-\beta\hbar\omega n_i}.$$

The sums required here can be evaluated using the previous formula, together with

$$\sum_{n=0}^{\infty} n q^n = q \frac{\mathrm{d}}{\mathrm{d}q}[L(q)] = \frac{q}{(1-q)^2}$$

$$\sum_{n=0}^{\infty} n^2 q^n = \left(q \frac{\mathrm{d}}{\mathrm{d}q} \right)^2 [L(q)] = \frac{q(1+q)}{(1-q)^3}.$$

The total partition function is then found to be

$$Z(T,N) = \left[\mathrm{e}^{-3\beta\hbar\omega/2} \left(\frac{1}{2}\frac{q(1+q)}{(1-q)^3} + \frac{3}{2}\frac{q}{(1-q)^2} + \frac{1}{1-q} \right) \right]^N$$

again with $q = \mathrm{e}^{-\beta\hbar\omega}$, and after a little arrangement we recover our first result.

• **Problem 3.10** The canonical partition function $Z(T, N)$ of the entire system is

$$Z = \int \mathrm{d}^{dN} p \, \mathrm{d}^{dN} q \, \mathrm{e}^{-\beta\mathcal{H}(q,p)}$$

where the regions of integration are unbounded, both for momenta and for spatial coordinates. Explicitly, we have

$$Z = \int \prod_{i=1}^{N} (\mathrm{d}p_{i1} \cdots \mathrm{d}p_{id})(\mathrm{d}q_{i1} \cdots \mathrm{d}q_{id})$$
$$\times \exp\left\{ -\beta\left[A_i (p_{i1}^2 + \cdots + p_{id}^2)^{s/2} + B_i (q_{i1}^2 + \cdots + q_{id}^2)^{t/2} \right] \right\}.$$

Clearly, this is a product of $2N$ integrals, each of which has the general form

$$I(\lambda, r, d) = \int \mathrm{d}^d x \exp\left[-\lambda \left(x_1^2 + \cdots + x_d^2 \right)^{r/2} \right].$$

Making a change in integration variable $u = x\lambda^{1/r}$, we find that

$$I(\lambda, r, d) = c(r, d)\lambda^{-d/r}$$

where $c(r, d) = I(1, r, d)$. This integral can be computed (it is given by $c(r, d) = \pi^{d/2}(d/r)!/(d/2)!$) but is not needed for our purposes. For the N momentum integrals we have $\lambda = \beta A_i$ and $r = s$, while for the N coordinate integrals we have $\lambda = \beta B_i$ and $r = t$. Thus, the partition function is

$$Z(T, N) = [c(s, d)c(t, d)]^N \left(\prod_{i=1}^{N} A_i^{-d/s} B_i^{-d/t} \right) \beta^{-N(d/s+d/t)}.$$

From this we easily obtain the internal energy and heat capacity:

$$U(T, N) = -\frac{\partial}{\partial\beta}\{\ln[Z(T, N)]\} = Nd\left(\frac{1}{s} + \frac{1}{t}\right)kT$$
$$C = \frac{\partial U(T, N)}{\partial T} = Nd\left(\frac{1}{s} + \frac{1}{t}\right)k.$$

These results are a generalization of the equipartition theorem, and we note that U and C depend on the powers of momenta and coordinates appearing in the Hamiltonian, but not on the coefficients A_i and B_i. For harmonic oscillators in three dimensions, we set $d = 3$ and $s = t = 2$ to obtain $U = 3NkT$ and $C = 3Nk$. These are, of course, independent of the masses and frequencies of the oscillators, which need not be the same for each oscillator.

• **Problem 3.11** The quantization of a particle in a one-dimensional square well gives rise to a spectrum of allowed energies given by

$$E(n) = \frac{n^2\hbar^2\pi^2}{2mL^2} \qquad n = 1, 2, 3, \ldots$$

and so the canonical partition function has the form

$$Z(T, L) = \sum_{n=1}^{\infty} \mathrm{e}^{-n^2\theta/T} \qquad \theta = \frac{\hbar^2\pi^2}{2mL^2k}.$$

In the low-temperature limit, it is useful to express $Z(T, L)$ in terms of the variable $x = \mathrm{e}^{-\theta/T} = \mathrm{e}^{-\beta k\theta}$ (with $\beta = 1/kT$, as usual). Since x is small when $T \ll \theta$, we need only the first few terms of the power series expansion:

$$\begin{aligned} Z(T, L) &= \mathrm{e}^{-\theta/T} + \mathrm{e}^{-4\theta/T} + \mathrm{e}^{-9\theta/T} + \mathrm{e}^{-16\theta/T} + \cdots \\ &= x(1 + x^3 + x^8 + x^{15} + \cdots) \end{aligned}$$

which gives

$$\ln Z = \ln x + x^3 + \cdots.$$

Derivatives with respect to β and L of a quantity which depends only on x are

$$\frac{\partial}{\partial\beta} = -k\theta x \frac{\partial}{\partial x} \qquad \frac{\partial}{\partial L} = \frac{2\theta}{LT} x \frac{\partial}{\partial x}.$$

We first find the average energy

$$U = -\frac{\partial(\ln Z)}{\partial\beta} = k\theta x \frac{\partial(\ln Z)}{\partial x} = k\theta(1 + 3x^3 + \cdots).$$

Then the heat capacity is

$$c_L = -k\beta^2 \frac{\partial U}{\partial\beta} = 9k\left(\frac{\theta}{T}\right)^2 (x^3 + \cdots) \simeq 9k\left(\frac{\theta}{T}\right)^2 \mathrm{e}^{-3\theta/T}.$$

Note that c_L vanishes as $T \to 0$ because of the exponential factor. This illustrates the third law of thermodynamics and is due to the existence of a non-zero energy gap $\Delta E = 3k\theta$ between the ground state and the first excited state.

To obtain the equation of state, we need only calculate the (one-dimensional) pressure as a function of T and L. We find that

$$P = -\frac{\partial F(T, L)}{\partial L} = kT \frac{\partial(\ln Z)}{\partial L} = \frac{2kT}{L}\left(\frac{\theta}{T}\right)(1 + 3x^2 + \cdots).$$

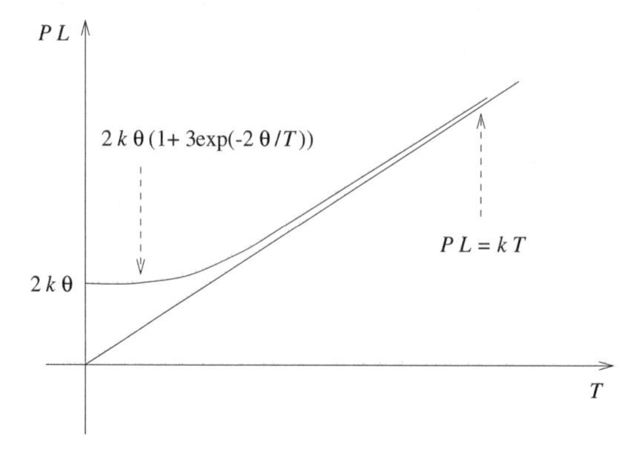

Figure 3.3. Equation of state for a one-dimensional gas. The straight line represents the classical behaviour.

Note, however, that, since Z is actually a function only of the single variable x, there is a simple exact relation between pressure and average energy, which is $P = 2U/L$.

In the high-temperature limit $T \gg \theta$, the energy levels which contribute significantly to Z are closely spaced, and we can approximate the sum by an integral. Using $p = n\pi\hbar/L$ and $\mathrm{d}n = (L/\pi\hbar)\,\mathrm{d}p$, we get

$$Z(T, L) = \frac{L}{\pi\hbar}\int_0^\infty \mathrm{e}^{-p^2/2mkT}\,\mathrm{d}p = L\left(\frac{mkT}{2\pi\hbar^2}\right)^{1/2} = \beta^{-1/2}L\left(\frac{m}{2\pi\hbar^2}\right)^{1/2}.$$

It is now a simple matter to compute

$$U = -\frac{\partial}{\partial\beta}(\ln Z) = \tfrac{1}{2}kT$$

$$c_L = \left(\frac{\partial U}{\partial T}\right)_L = \tfrac{1}{2}k$$

$$P = \left(\frac{\partial}{\partial L}(kT\ln Z)\right)_T = \frac{kT}{L}.$$

As might have been expected, these are precisely the results for a one-dimensional classical gas. The complete equation of state, interpolating between the low- and high-temperature limits, is sketched in figure 3.3.

• Problem 3.12

(a) The position of a point particle on the surface of a sphere of radius r can be specified in spherical polar coordinates as

$$\boldsymbol{r} = r\sin\theta\cos\phi\,\hat{\boldsymbol{x}} + r\sin\theta\sin\phi\,\hat{\boldsymbol{y}} + r\cos\theta\,\hat{\boldsymbol{z}}$$

with $0 \leq \theta \leq \pi$ and $0 \leq \phi \leq 2\pi$. To find the correct phase space variables, we first consider the Lagrangian for a single particle:

$$L = \tfrac{1}{2}m|\dot{\boldsymbol{r}}|^2 = \tfrac{1}{2}mr^2(\dot{\theta}^2 + \sin^2\theta\,\dot{\phi}^2).$$

The momenta conjugate to θ and ϕ are $p_\theta = \partial L/\partial\dot{\theta} = mr^2\dot{\theta}$ and $p_\phi = \partial L/\partial\dot{\phi} = mr^2\sin^2\theta\,\dot{\phi}$, and the Hamiltonian for a single particle then reads

$$\mathcal{H}_1(\theta, \phi, p_\theta, p_\phi) = \frac{1}{2mr^2}p_\theta^2 + \frac{1}{2mr^2\sin^2\theta}p_\phi^2.$$

Note that r is now a geometrical constant and not a dynamical coordinate. The single-particle partition function Z_1 is given by

$$\begin{aligned} Z_1 &= \int \exp\left[-\left(\frac{1}{2mr^2kT}p_\theta^2 + \frac{1}{2mr^2kT\sin^2\theta}p_\phi^2\right)\right] \mathrm{d}p_\theta\,\mathrm{d}p_\phi\,\mathrm{d}\theta\,\mathrm{d}\phi \\ &= \int (2\pi mr^2kT)^{1/2}(2\pi mr^2kT\sin^2\theta)^{1/2}\,\mathrm{d}\theta\,\mathrm{d}\phi \\ &= (2\pi mr^2kT)2\pi\int_0^\pi |\sin\theta|\,\mathrm{d}\theta \\ &= 2\pi mAkT \end{aligned}$$

where $A = 4\pi r^2$ is the area of the sphere. The total partition function is $Z(T, A) = Z_1^N$, and we obtain the internal energy and equation of state as

$$\begin{aligned} U &= kT^2\frac{\partial(\ln Z)}{\partial T} = NkT \\ P &= \frac{\partial}{\partial A}(kT\ln Z) = \frac{NkT}{A}. \end{aligned}$$

(b) The position of a particle on a torus of minor radius a and major radius b ($b > a$) can be specified in terms of angles α and β (each of which ranges from 0 to 2π) as

$$\boldsymbol{r} = (b + a\cos\alpha)\cos\beta\,\hat{\boldsymbol{x}} + (b + a\cos\alpha)\sin\beta\,\hat{\boldsymbol{y}} + a\sin\alpha\,\hat{\boldsymbol{z}}.$$

Proceeding as before, we find for the Hamiltonian of a single particle that

$$\mathcal{H}_1(\alpha, \beta, p_\alpha, p_\beta) = \frac{1}{2ma^2}p_\alpha^2 + \frac{1}{2m(b + a\cos\alpha)^2}p_\beta^2$$

and for the single-particle partition function that

$$\begin{aligned} Z_1 &= \int_0^{2\pi}\int_0^{2\pi} (2\pi ma^2kT)^{1/2}[2\pi m(b + a\cos\alpha)^2kT]^{1/2}\,\mathrm{d}\alpha\,\mathrm{d}\beta \\ &= 2\pi mAkT \end{aligned}$$

where $A = 4\pi^2ab$ is the area of the torus. This is identical with the result obtained above for a sphere, so the average energy and equation of state are also the same.

• **Problem 3.13** In the canonical ensemble with $\lambda = 0$, the expectation value of a quantity X is

$$\langle X \rangle_0 = Z_0^{-1} \sum \mathrm{e}^{-\beta \mathcal{H}_0} X \qquad Z_0 = \sum \mathrm{e}^{-\beta \mathcal{H}_0}.$$

So the full partition function is

$$\begin{aligned} Z &= \sum \mathrm{e}^{-\beta \mathcal{H}_0} \mathrm{e}^{-\lambda \beta \mathcal{H}_1} \\ &= Z_0 \langle \mathrm{e}^{-\lambda \beta \mathcal{H}_1} \rangle_0 \\ &= Z_0 (1 - \lambda \beta \langle \mathcal{H}_1 \rangle_0 + \tfrac{1}{2} \lambda^2 \beta^2 \langle \mathcal{H}_1^2 \rangle_0 + \cdots). \end{aligned}$$

The Helmholtz free energy is therefore given by

$$\begin{aligned} F &= -kT \ln Z \\ &= -kT \ln Z_0 - kT \ln(1 - \lambda \beta \langle \mathcal{H}_1 \rangle_0 + \tfrac{1}{2} \lambda^2 \beta^2 \langle \mathcal{H}_1^2 \rangle_0 + \cdots) \\ &= F_0 + \lambda \langle \mathcal{H}_1 \rangle_0 - \tfrac{1}{2} \lambda^2 \beta \langle \mathcal{H}_1^2 \rangle_0 + \tfrac{1}{2} \lambda^2 \beta \langle \mathcal{H}_1 \rangle_0^2 + \cdots \\ &= F_0 + \lambda \langle \mathcal{H}_1 \rangle_0 - \tfrac{1}{2} \lambda^2 \beta \langle (\mathcal{H}_1 - \langle \mathcal{H}_1 \rangle_0)^2 \rangle_0 + \cdots. \end{aligned}$$

The combination of expectation values appearing at order λ^n is the *cumulant average* of $\mathcal{H}_1^n$.

The internal energy is, of course, $U = -\partial(\ln Z)/\partial \beta$. In carrying out the differentiation, we must be careful to account for the β dependence of the expectation values as well as the explicit factors of β. To evaluate the first two terms, we use $\partial(\ln Z_0)/\partial \beta = -\langle \mathcal{H}_0 \rangle_0$ and

$$\begin{aligned} \frac{\partial}{\partial \beta} (\langle \mathcal{H}_1 \rangle_0) &= \frac{\partial}{\partial \beta} \left(Z_0^{-1} \sum \mathrm{e}^{-\beta \mathcal{H}_0} \mathcal{H}_1 \right) \\ &= -Z_0^{-1} \frac{\partial Z_0}{\partial \beta} Z_0^{-1} \sum \mathrm{e}^{-\beta \mathcal{H}_0} \mathcal{H}_1 - Z_0^{-1} \sum \mathrm{e}^{-\beta \mathcal{H}_0} \mathcal{H}_0 \mathcal{H}_1 \\ &= \langle \mathcal{H}_0 \rangle_0 \langle \mathcal{H}_1 \rangle_0 - \langle \mathcal{H}_0 \mathcal{H}_1 \rangle_0 \\ &= -\langle (\mathcal{H}_0 - \langle \mathcal{H}_0 \rangle_0)(\mathcal{H}_1 - \langle \mathcal{H}_1 \rangle_0) \rangle_0 \end{aligned}$$

to obtain

$$U = \langle \mathcal{H}_0 \rangle_0 + \lambda \langle \mathcal{H}_1 \rangle_0 - \lambda \beta \langle (\mathcal{H}_0 - \langle \mathcal{H}_0 \rangle_0)(\mathcal{H}_1 - \langle \mathcal{H}_1 \rangle_0) \rangle_0 + \mathrm{O}(\lambda^2).$$

• **Problem 3.14**

(a) Since the magnetic moments do not interact with each other, the partition function is a product of single-site partition functions:

$$Z(T, \boldsymbol{H}, N) = \prod_{i=1}^{N} \int \mathrm{d}\Omega_i \, \mathrm{e}^{-\beta E_i(\mu_i)}$$

where $\mathrm{d}\Omega_i$ is an element of solid angle into which the ith moment may point. The magnetic field at the position $\boldsymbol{r}_i$ of the ith site is $H_i\hat{\boldsymbol{z}} = H(\boldsymbol{r}_i)\hat{\boldsymbol{z}}$ and, if the moment at this site makes an angle θ_i ($0 \leq \theta_i \leq \pi$) with the field, then its energy is $E_i(\boldsymbol{\mu}_i) = -\mu H_i \cos\theta_i$, where $\mu = |\boldsymbol{\mu}_i|$ is the magnitude of the magnetic moment of each atom. Since the integrand in the above expression depends only on θ_i, the remaining angular integrations can be performed, with the result

$$\int \mathrm{d}\Omega\, f(\theta) = \int_0^\pi \mathrm{d}\theta\, S_{d-1}(\sin\theta)\, f(\theta)$$

where $S_n(\sin\theta) = 2\pi^{n/2}(\sin\theta)^{n-1}/\Gamma(n/2)$ is the surface area of an n-dimensional sphere of radius $\sin\theta$. We therefore obtain

$$Z(T, H, N) = \left(\frac{2\pi^{(d-1)/2}}{\Gamma[(d-1)/2]}\right)^N \prod_{i=1}^N \int_0^\pi \mathrm{d}\theta_i\, (\sin\theta_i)^{d-2}\, \mathrm{e}^{\beta\mu H_i \cos\theta_i}.$$

(b) In particular, setting $d = 3$ we get

$$Z(T, H, N) = (2\pi)^N \prod_{i=1}^N \int_0^\pi \mathrm{d}\theta\, \sin\theta\, \mathrm{e}^{\beta H_i \mu \cos\theta} = (4\pi)^N \prod_{i=1}^N \frac{\sinh(\beta\mu H_i)}{\beta\mu H_i}.$$

The average magnetization of the ith site is $|\langle\boldsymbol{\mu}_i\rangle| = \partial(\ln Z)/\partial(\beta H_i)$ and for the whole lattice we get

$$M = \sum_{i=1}^N \frac{\partial}{\partial(\beta H_i)}(\ln Z) = \mu \sum_{i=1}^N L(\beta\mu H_i)$$

where $L(x) = \coth x - 1/x$ is the Langevin function. The magnetic susceptibility per site is

$$\chi = \lim_{H_i \to 0}\left(\frac{\partial M}{\partial H_i}\right) = \frac{\mu^2}{3kT}$$

which is Curie's law for a paramagnetic material.

• Problem 3.15

(a) When interactions between molecules are ignored, the total Hamiltonian is the sum of Hamiltonians for individual molecules, each of which has the form $\mathcal{H} = |\boldsymbol{p}|^2/2m - \boldsymbol{\mu}\cdot\boldsymbol{E}$. In order to find the polarization of the gas (which is the average dipole moment per unit volume), we first find the canonical partition function $Z(T, \boldsymbol{E}, V, N)$, and then use

$$P_\alpha = V^{-1}\sum_{i=1}^N \langle\mu_{i\alpha}\rangle = V^{-1}\frac{\partial(\ln Z)}{\partial(\beta E_\alpha)}$$

for each component ($\alpha = x, y, z$). Since the total Hamiltonian is a sum of kinetic and electric terms for each of the N particles, the partition function can be written in the form

$$Z(T, \boldsymbol{E}, V, N) = \frac{1}{N!}[Z_1^{\text{kin}}(T, V)]^N [Z_1^{\text{elect}}(T, \boldsymbol{E})]^N$$

where $1/N!$ accounts for the 'correct Boltzmann counting' of indistinguishable particles. Clearly, we have $P_\alpha = NV^{-1}\partial(\ln Z_1^{\text{elect}})/\partial(\beta E_\alpha)$, so the kinetic terms can be ignored. To evaluate Z_1^{elect}, we take $\boldsymbol{E} = (0, 0, E_z)$ and specify the orientation of $\boldsymbol{\mu}$ using polar angles θ and ϕ:

$$Z_1^{\text{elect}}(T, \boldsymbol{E}) = \int_0^{2\pi} \mathrm{d}\phi \int_0^{\pi} \mathrm{d}\theta \, \sin\theta \, \mathrm{e}^{\beta\mu E_z \cos\theta} = 4\pi \frac{\sinh(\beta\mu E_z)}{\beta\mu E_z}$$

where $\mu = |\boldsymbol{\mu}|$. The only non-zero component of the polarization is

$$P_z = \frac{N}{V}\frac{\partial Z_1^{\text{elect}}}{\partial(\beta E_z)} = \frac{N}{V}\mu\left(\coth(\beta\mu E_z) - \frac{1}{\beta\mu E_z}\right).$$

The function in large parentheses is, of course, the Langevin function which also appears in the analogous magnetic problem (see problem 3.14).

(b) The dielectric constant ε of the gas is defined by $\boldsymbol{D} = \varepsilon\boldsymbol{E}$, where (in suitable units) $\boldsymbol{D} \equiv \boldsymbol{E} + 4\pi\boldsymbol{P}$ is the electric displacement. We therefore have $\varepsilon = 1 + 4\pi P_z/E_z$. In general, ε depends on E but, in the low-field limit $\beta\mu E \ll 1$ (which applies to most practical situations), the polarization is linear in E and we have

$$\frac{P_z}{E_z} \simeq \frac{N}{V}\frac{\mu^2}{3kT}$$

giving

$$\varepsilon = 1 + 4\pi \frac{N}{V}\frac{\mu^2}{3kT}.$$

• Problem 3.16

(a) The canonical partition function for the N sites in this lattice can be expressed as the product of partition functions for individual sites, $Z(T, N) = Z_1(T)^N$, because these sites are independent. The energy of each site can be expressed as

$$\mathcal{H}_1 = n(\varepsilon - \mu H s)$$

where $n = 0$ if the site is unoccupied and $n = 1$ if it is occupied, while $s = \pm 1$, depending on the orientation of the magnetic moment. So the single-site partition function is

$$Z_1 = \sum_{n=0,1}\sum_{s=\pm 1} \mathrm{e}^{-\beta n(\varepsilon - \mu H s)} = 1 + \mathrm{e}^{-\beta\varepsilon}(\mathrm{e}^{\beta\mu H} + \mathrm{e}^{-\beta\mu H}).$$

Note that the quantity $Z(T, N)$ calculated here is indeed the *canonical* partition function for a lattice of N sites, the state of each site being specified by the variables n and s. However, if the system is regarded as a model of a gas, then the number of gas molecules, $N_{\text{gas}} = \sum_{i=1}^{N} n_i$, is not fixed but varies between 1 and N. From this point of view, $Z(T, N)$ is the *grand canonical* partition function for the gas, provided that the energy ε is taken to include the chemical potential.

(b) The average energy and magnetization are given by appropriate derivatives of $Z(T, N)$:

$$U = -\left(\frac{\partial\{\ln[Z(T,N)]\}}{\partial\beta}\right)_H = 2N\frac{\varepsilon\cosh(\beta\mu H) - \mu H\sinh(\beta\mu H)}{\mathrm{e}^{\beta\varepsilon} + 2\cosh(\beta\mu H)}$$

$$M = \left(\frac{\partial\{\ln[Z(T,N)]\}}{\partial(\beta H)}\right)_\beta = 2N\mu\frac{\sinh(\beta\mu H)}{\mathrm{e}^{\beta\varepsilon} + 2\cosh(\beta\mu H)}.$$

• **Problem 3.17**

(a) Let us call $\mathcal{P}_+$ and $\mathcal{P}_-$ the probabilities of given sites having energies $+\varepsilon$ and $-\varepsilon$ respectively. In the absence of interactions between sites, they are given by the canonical ensemble as

$$\mathcal{P}_+ = \frac{\mathrm{e}^{-\beta\varepsilon}}{\mathrm{e}^{-\beta\varepsilon} + \mathrm{e}^{\beta\varepsilon}} \qquad \mathcal{P}_- = \frac{\mathrm{e}^{\beta\varepsilon}}{\mathrm{e}^{-\beta\varepsilon} + \mathrm{e}^{\beta\varepsilon}}.$$

A given site belongs to a cluster of length L, if (i) it has energy $+\varepsilon$, (ii) for some n, with $0 \le n \le L-1$, there are n sites to its left and $L-n-1$ sites to its right which also have energy $+\varepsilon$ and (iii) the sites $n+1$ places to its left and $(L-n)$ places to its right have energy $-\varepsilon$. The probability for this situation is $\mathcal{P}_+^L\mathcal{P}_-^2$. Thus, the total probability that the site belongs to a cluster of length L is

$$\mathcal{P}_L = \begin{cases} L\mathcal{P}_+^L\mathcal{P}_-^2 & L \ge 1 \\ \mathcal{P}_- & L = 0. \end{cases}$$

For $L \ge 1$, the factor L accounts for the L possible values of n, while a cluster of length $L = 0$ corresponds to the given site having energy $-\varepsilon$, for which the probability is of course $\mathcal{P}_-$. We have ignored the possibility that the given site is within a distance L of the ends of the whole lattice, which would change the probabilities. For any finite L, this is legitimate in the limit $N \to \infty$. Note that

$$\sum_{L=1}^{\infty} L\mathcal{P}_+^L = \mathcal{P}_+\frac{\partial}{\partial\mathcal{P}_+}\left(\sum_{L=0}^{\infty}\mathcal{P}_+^L\right) = \mathcal{P}_+\frac{\partial}{\partial\mathcal{P}_+}\left(\frac{1}{1-\mathcal{P}_+}\right) = \frac{\mathcal{P}_+}{(1-\mathcal{P}_+)^2} = \frac{\mathcal{P}_+}{\mathcal{P}_-^2}$$

and therefore

$$\sum_{L=0}^{\infty}\mathcal{P}_L = \mathcal{P}_- + \mathcal{P}_-^2\sum_{L=1}^{\infty}L\mathcal{P}_+^L = \mathcal{P}_- + \mathcal{P}_+ = 1.$$

Consequently, with the above interpretation of a cluster of length zero, we may say that, in some particular state of the whole lattice, any arbitrarily chosen site belongs to a cluster of some length.

(b) The average length of a cluster could be defined in several ways, which are not necessarily equivalent. We adopt the following definition, for which the answer can be computed straightforwardly. Choose a particular site on the lattice. For each state of the system, it belongs, as discussed above, to a cluster of some length L, and we average this length over all states of the system, with the weighting prescribed by the canonical ensemble. In the limit $N \to \infty$, this average length is independent of which site we choose. With this definition, we have

$$\begin{aligned}\langle L\rangle &= \sum_{L=0}^{\infty} L\,\mathcal{P}_L = \mathcal{P}_-^2 \sum_{L=1}^{\infty} L^2\,\mathcal{P}_+^L = \mathcal{P}_-^2\mathcal{P}_+\frac{\partial}{\partial\mathcal{P}_+}\left(\sum_{L=1}^{\infty} L\mathcal{P}_+^L\right)\\ &= \mathcal{P}_-^2\mathcal{P}_+\frac{\partial}{\partial\mathcal{P}_+}\left(\frac{\mathcal{P}_+}{(1-\mathcal{P}_+)^2}\right) = \frac{\mathcal{P}_+(1+\mathcal{P}_+)}{\mathcal{P}_-}.\end{aligned}$$

In terms of temperature (with $\beta = 1/kT$), this is

$$\langle L\rangle_T = \mathrm{e}^{-2\beta\varepsilon}\,\frac{\mathrm{e}^{\beta\varepsilon} + 2\mathrm{e}^{-\beta\varepsilon}}{\mathrm{e}^{\beta\varepsilon} + \mathrm{e}^{-\beta\varepsilon}}.$$

At $T = 0$, all sites of the lattice should be in the lower energy level, so every cluster has length zero, and we find that indeed $\langle L\rangle_0 = 0$. As $T \to \infty$, $\mathcal{P}_+$ and $\mathcal{P}_-$ both approach $\frac{1}{2}$, and $\langle L\rangle_T$ approaches its maximum value, $\langle L\rangle_\infty = \frac{3}{2}$.

• Problem 3.18

(a) Since the molecules do not interact, we can express the canonical partition function as $Z(T, V, N) = Z_1^N/N!$, where Z_1 is the partition function for a single molecule. The quantum Hamiltonian for each molecule is the sum of a translational term and a rotational term and can be diagonalized in the basis of the Hilbert space formed by the simultaneous eigenstates $|\boldsymbol{p}, l\rangle$ of momentum and angular momentum. We then have

$$\langle \boldsymbol{p}, l|H|\boldsymbol{p}, l\rangle = \frac{|\boldsymbol{p}|^2}{2m} + l(l+1)\frac{\hbar^2}{2I}$$

and the partition function Z_1 can be written as the product of translational and rotational contributions, namely

$$Z_{\text{trans}} = \frac{V}{h^3}\int \mathrm{d}^3p\, \mathrm{e}^{-\beta|\boldsymbol{p}|^2/2m} = \frac{V}{\lambda^3}$$

$$Z_{\text{rot}} = \sum_{l=0}^{\infty} g(l)\,\mathrm{e}^{-\beta l(l+1)\hbar^2/2I}$$

where the thermal de Broglie wavelength is $\lambda = (h^2/2\pi mkT)^{1/2}$ and the degeneracy is $g(l) = 2l + 1$. The translational part Z_{trans} has been approximated by its classical version because, in all practical situations, the energy-level spacing is much smaller than kT, and the sum over these levels can be replaced by an integral (see problem 3.11). Thus we obtain

$$Z(T, V, N) = \frac{1}{N!}\left(\frac{V}{\lambda^3}\sum_{l=0}^{\infty} g(l)\,\mathrm{e}^{-\beta l(l+1)\hbar^2/2I}\right)^N.$$

The remaining sum can be evaluated analytically only in the limits of high or low temperatures.

(b) At sufficiently high temperatures, we might expect that the spacing of rotational energy levels would become sufficiently small compared with kT that the above sum could be replaced with an integral, corrections to this approximation becoming significant as the temperature is reduced. This expectation can be made precise by means of the Euler–Maclaurin formula (given in appendix A) which is

$$\sum_{k=0}^{\infty} f(k) = \int_0^{\infty} f(k)\,\mathrm{d}k + \tfrac{1}{2}f(0) - \tfrac{1}{12}f'(0) + \tfrac{1}{720}f'''(0) - \cdots.$$

In this form, the formula is valid for smooth functions $f(k)$ which vanish, and all of whose derivatives vanish, as $k \to \infty$, as is often the case in applications to statistical mechanics. Of course, this formula is useful only in circumstances where all but a few terms of the series can be neglected. In the present case, let us define $\theta = \hbar^2/2Ik$ and $f(l) = (2l+1)\,\mathrm{e}^{-l(l+1)\theta/T}$. The Euler–Maclaurin series can be converted into a power series expansion of Z_{rot} in powers of θ/T, of which we need keep only a few terms when $T \gg \theta$. However, these two series do not agree term by term. Indeed, let us expand $f(l)$ in powers of θ/T:

$$f(l) = \sum_{n=0}^{\infty} (2l+1)\frac{[-l(l+1)]^n}{n!}\left(\frac{\theta}{T}\right)^n = (2l+1) - (2l^3 + 3l^2 + l)\frac{\theta}{T} + \cdots.$$

We see that the coefficient of $(\theta/T)^n$ is a polynomial of degree $2n+1$ in l, and will contribute to all terms of the Euler–Maclaurin series up to $f^{(2n+1)}(0)$. To obtain our answer correct to order θ/T, we need the terms up to $f'''(0)$. In this way, we find that

$$\begin{aligned} Z_{\text{rot}} &= \int_0^{\infty} f(l)\,\mathrm{d}l + \tfrac{1}{2}f(0) - \tfrac{1}{12}f'(0) + \tfrac{1}{720}f'''(0) - \cdots \\ &= \frac{T}{\theta} + \frac{1}{3} + \frac{1}{15}\frac{\theta}{T} + \mathrm{O}\left[\left(\frac{\theta}{T}\right)^2\right]. \end{aligned}$$

The internal energy and specific heat are then given by

$$U = -N\left(\frac{\partial(\ln Z_{\text{trans}})}{\partial\beta} + \frac{\partial(\ln Z_{\text{rot}})}{\partial\beta}\right)$$
$$= \frac{5}{2}NkT\left\{1 - \frac{2}{15}\frac{\theta}{T} - \frac{2}{225}\left(\frac{\theta}{T}\right)^2 + \mathrm{O}\left[\left(\frac{\theta}{T}\right)^3\right]\right\}$$
$$C_V = \left(\frac{\partial U}{\partial T}\right)_V = \frac{5}{2}Nk\left\{1 + \frac{2}{225}\left(\frac{\theta}{T}\right)^2 + \mathrm{O}\left[\left(\frac{\theta}{T}\right)^3\right]\right\}.$$

Of course, the high-temperature limit $C_V \simeq 5Nk/2$ agrees with the classical equipartition theorem.

At low temperatures $T \ll \theta$, successive terms of the sum over l decrease rapidly, and we need to keep only a few of them. Keeping just the first two terms, we get

$$Z_{\text{rot}} = 1 + 3\,\mathrm{e}^{-2\theta/T} + \cdots$$

and

$$C_V = \tfrac{3}{2}Nk + 12Nk\left(\frac{\theta}{T}\right)^2(\mathrm{e}^{-2\theta/T} + \cdots).$$

The second term is the rotational contribution. It vanishes as $T \to 0$, in accordance with the third law of thermodynamics, and this is a quantum-mechanical effect (cf problem 3.11). The first term is due to translational motion, and agrees with the classical equipartition theorem when only translational degrees of freedom are taken into account. It does not vanish at $T = 0$, although the fully quantum-mechanical result needed at extremely (and impracticably) low temperatures does vanish (problem 3.11).

Readers who are not familiar with the Euler–Maclaurin formula may like to obtain it for themselves. The following suggests a simple method (which, however, does not prove that the series actually converges to the desired result). First, express the integral as

$$\int_0^\infty f(x)\,\mathrm{d}x = \sum_{k=0}^{\infty}\int_0^1 f(k+x)\,\mathrm{d}x$$
$$= \sum_{k=0}^{\infty}\int_0^1\left(f(k) + xf'(k) + \frac{x^2 f''(k)}{2} + \cdots\right)\mathrm{d}x$$
$$= \sum_{k=0}^{\infty}\left(f(k) + \frac{f'(k)}{2} + \frac{f''(k)}{6} + \cdots\right).$$

Now assume that

$$\sum_{k=0}^{\infty} f(k) = \int_0^\infty f(x)\,\mathrm{d}x + a_0 f(0) + a_1 f'(0) + a_2 f''(0) + \cdots$$

and that the sums of $f'(k)$, $f''(k)$, etc, can be expressed in the same way, with the same coefficients a_i. Substitute these expressions into the first result. The integral of $f(x)$ cancels. Assuming that $f(\infty) = f'(\infty) = \cdots = 0$, the integral of $f'(x)$ is $-f(0)$, the integral of $f''(x)$ is $-f'(0)$ and so on. By setting the coefficients of $f(0)$, $f'(0)$, etc, equal to zero, a sequence of equations is obtained which can be solved for the constants a_i.

• **Problem 3.19** Let us define the characteristic temperatures $\theta_1 = \hbar^2/2Ik$ and $\theta_2 = \hbar\omega/k$. The partition function for a single molecule is

$$Z_1 = Z_{\text{trans}} \sum_{l=0}^{\infty} \sum_{n=0}^{\infty} (2l+1)\, \mathrm{e}^{-l(l+1)\theta_1/T}\, \mathrm{e}^{-(n+1/2)\theta_2/T}\, \mathrm{e}^{-(n+1/2)l(l+1)\alpha/kT}$$

where (for molecules of mass m in a container of volume V) $Z_{\text{trans}} = V(mkT/2\pi\hbar^2)^{3/2}$ corresponds to the translational motion of the centre of mass of the molecule. For temperatures $T \gg \theta_1$, we can replace the sum over l by an integral. We then obtain

$$\begin{aligned}
Z_1 &= Z_{\text{trans}} \sum_{n=0}^{\infty} \int_0^{\infty} \mathrm{d}l\, (2l+1)\, \mathrm{e}^{-l(l+1)[\theta_1/T+(n+1/2)\alpha/kT]}\, \mathrm{e}^{-(n+1/2)\theta_2/T} \\
&= Z_{\text{trans}} \sum_{n=0}^{\infty} \left(\frac{\theta_1}{T} + \frac{\alpha}{kT}(n+\tfrac{1}{2}) \right)^{-1} \mathrm{e}^{-(n+1/2)\theta_2/T} \\
&= Z_{\text{trans}} \frac{T}{\theta_1} \left(1 + \frac{\alpha}{k\theta_1} \frac{\partial}{\partial(\theta_2/T)} + \cdots \right) Z_{\text{vib}} \\
&= Z_{\text{trans}} \frac{T}{\theta_1} Z_{\text{vib}} \left(1 - \frac{\alpha \varepsilon_{\text{vib}}}{k^2\theta_1\theta_2} + \cdots \right)
\end{aligned}$$

where

$$Z_{\text{vib}} = \sum_{n=0}^{\infty} \mathrm{e}^{-(n+1/2)\theta_2/T} = \mathrm{e}^{-\theta_2/2T}(1 - \mathrm{e}^{-\theta_2/T})^{-1}$$

is the vibrational partition function, and

$$\varepsilon_{\text{vib}} = -\frac{\partial(\ln Z_{\text{vib}})}{\partial\beta} = \frac{\hbar\omega}{2} \coth\left(\frac{\hbar\omega}{2kT}\right)$$

is the average of the purely vibrational energy. Here, we have used the approximation that $\alpha/k\theta_1$ is small, while $\varepsilon_{\text{vib}}/k\theta_2$ is not large in the temperature range considered. Finally, for a gas of N molecules, we obtain the internal energy as

$$U = -N \frac{\partial(\ln Z_1)}{\partial\beta} = \tfrac{5}{2}NkT + \frac{N\hbar\omega}{2} \coth\left(\frac{\hbar\omega}{2kT}\right) - \frac{N\alpha I\omega}{2\hbar} \operatorname{cosech}^2\left(\frac{\hbar\omega}{2kT}\right) + \cdots.$$

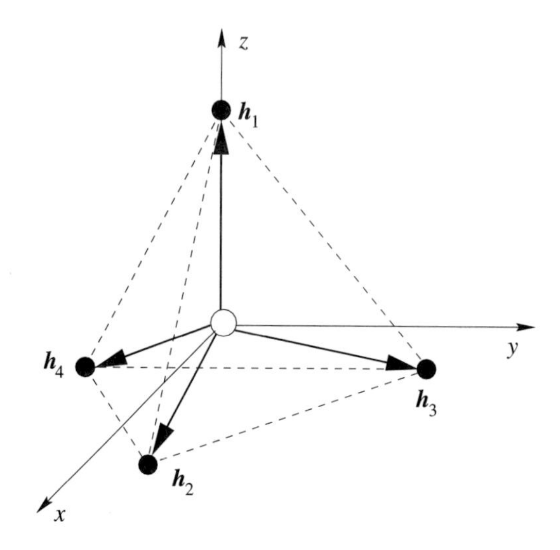

Figure 3.4. A particular orientation of the methane molecule. The four filled spheres on the corners are the hydrogen atoms and the one in the centre is the carbon atom.

For this range of temperatures, translations and rotations contribute classically, giving rise to the first term, while the second is the quantum-mechanical contribution of vibrations. The last term is, of course, the leading correction due to interaction between the vibrational and rotational degrees of freedom.

• **Problem 3.20** The specific heat of methane, which has three translational and three rotational degrees of freedom is, of course, given immediately by the classical equipartition theorem as $C_V = 3Nk$. However, it is interesting to obtain this result by calculating the partition function explicitly. For a gas of N molecules, this partition function is $Z = Z_1^N/N!$, where Z_1 is the partition function for a single molecule. The function Z_1 is the product of a translational part associated with motion of the centre of mass,

$$Z_{\text{trans}} = h^{-3} \int \mathrm{d}^3q\, \mathrm{d}^3p\, \mathrm{e}^{-|p|^2/2M} = V\left(\frac{2\pi MkT}{h^2}\right)^{3/2}$$

and a rotational part Z_{rot}. To calculate Z_{rot}, we first need an expression for the rotational energy, which can be given in terms of Euler angles θ, ϕ and ψ, with $0 \leq \theta < \pi$, $0 \leq \phi < 2\pi$ and $0 \leq \psi < 2\pi$. There is one particular orientation of the molecule, shown in figure 3.4, in which the positions of the four hydrogen atoms are given in Cartesian coordinates by the four vectors

$$\boldsymbol{h}_1 = \begin{pmatrix} 0 \\ 0 \\ a \end{pmatrix} \qquad \boldsymbol{h}_2 = \begin{pmatrix} 2\sqrt{2}a/3 \\ 0 \\ -a/3 \end{pmatrix}$$

$$\boldsymbol{h}_3 = \begin{pmatrix} -\sqrt{2}a/3 \\ \sqrt{2/3}a \\ -a/3 \end{pmatrix} \qquad \boldsymbol{h}_4 = \begin{pmatrix} -\sqrt{2}a/3 \\ -\sqrt{2/3}a \\ -a/3 \end{pmatrix}.$$

A general orientation can be reached by the following sequence of rotations: (i) a rotation through ψ about the z axis; (ii) a rotation through θ about the y axis; (iii) a rotation through ϕ about the z axis. In this orientation, the position of the ith hydrogen atom is $\boldsymbol{r}_i(\theta, \phi, \psi) = \mathbf{R}(\theta, \phi, \psi)\boldsymbol{h}_i$, where $\mathbf{R}(\theta, \phi, \psi)$ is the product of three rotation matrices

$$\mathbf{R}(\theta, \phi, \psi) = \begin{pmatrix} \cos\phi & \sin\phi & 0 \\ -\sin\phi & \cos\phi & 0 \\ 0 & 0 & 1 \end{pmatrix}$$
$$\times \begin{pmatrix} \cos\theta & 0 & -\sin\theta \\ 0 & 1 & 0 \\ \sin\theta & 0 & \cos\theta \end{pmatrix} \begin{pmatrix} \cos\psi & \sin\psi & 0 \\ -\sin\psi & \cos\psi & 0 \\ 0 & 0 & 1 \end{pmatrix}.$$

We can then obtain the Lagrangian (which is equal to the kinetic energy) as

$$L = \frac{m}{2}\sum_i \dot{\boldsymbol{r}}_i \cdot \dot{\boldsymbol{r}}_i = \frac{m}{2}\sum_i \boldsymbol{h}_i^{\mathrm{T}}\dot{\mathbf{R}}^{\mathrm{T}}\dot{\mathbf{R}}\boldsymbol{h}_i = \frac{m}{2}\frac{4}{3}a^2 \,\mathrm{Tr}(\dot{\mathbf{R}}^{\mathrm{T}}\dot{\mathbf{R}})$$
$$= \tfrac{1}{2}I(\dot{\theta}^2 + \dot{\phi}^2 + \dot{\psi}^2 + 2\dot{\phi}\dot{\psi}\cos\theta)$$

where $I = 8ma^2/3$ is the moment of inertia of the molecule about an axis of symmetry. From this Lagrangian, we obtain the Hamiltonian

$$\mathcal{H} = \frac{1}{2I}\left(p_\theta^2 + \frac{1}{\sin^2\theta}(p_\phi^2 + p_\psi^2 - 2p_\phi p_\psi \cos\theta)\right)$$

and the rotational partition function is

$$Z_{\mathrm{rot}} = \frac{1}{h^3}\int \mathrm{d}p_\theta\, \mathrm{d}p_\phi\, \mathrm{d}p_\psi \int \mathrm{d}\theta\, \mathrm{d}\phi\, \mathrm{d}\psi\, \mathrm{e}^{-\beta\mathcal{H}}.$$

To evaluate the momentum integral, we write $\beta\mathcal{H} = \sum_{ij} p_i A_{ij} p_j$ and use the result

$$\int \prod_i \mathrm{d}p_i \exp\left(-\sum_{ij} p_i A_{ij} p_j\right) = \pi^{3/2}(\det \mathbf{A})^{-1/2}$$

to obtain

$$Z_{\mathrm{rot}} = \int \mathrm{d}\theta\, \mathrm{d}\phi\, \mathrm{d}\psi \left(\frac{2\pi IkT}{h^2}\right)^{3/2} \sin\theta = 8\pi^2 \left(\frac{2\pi IkT}{h^2}\right)^{3/2}.$$

Thus, the total partition function is

$$Z = \frac{1}{N!}(Z_{\text{trans}} Z_{\text{rot}})^N = \frac{(8\pi^2 V)^N}{N!}\left(\frac{4\pi^2 M I k^2 T^2}{h^4}\right)^{3N/2}$$

giving the internal energy as $U = -\partial(\ln Z)/\partial\beta = 3NkT$ and the specific heat at constant volume as $C_V = (\partial U/\partial T)_V = 3Nk$.

• Problem 3.21

(a) By treating a diatomic molecule as a quantum-mechanical rigid rotor, one obtains a spectrum of energies $E_{\text{rot}} = l(l+1)\hbar^2/2I$, where I is the moment of inertia, with $l = 0, 1, 2, 3, \ldots, \infty$, the lth level having a degeneracy $g(l) = 2l + 1$. For molecules whose constituent atoms are indistinguishable, the wavefunction must be symmetric under interchange of the two particles if these have integer spin s, or antisymmetric if s is half an odd integer (for brevity, we shall simply say 'half-integer'). A useful basis for the Hilbert space of these wavefunctions consists of the products of orbital and spin functions of the form $\Psi = \phi_{lm}\chi_{\text{spin}}$. When the two atoms are interchanged, the orbital function ϕ_{lm} changes by a factor $(-1)^l$, so it is symmetric if l is even and antisymmetric if l is odd. We therefore need a basis for the spin states such that χ_{spin} is also either symmetric or antisymmetric. One basis for the spin states consists of the $(2s+1)^2$ states $\chi_\nu(1)\chi_{\nu'}(2)$, where ν is the spin component for atom 1 along a chosen quantization axis and ν' is the corresponding quantity for atom 2. A little thought shows that we can construct from these states a basis consisting of $(s+1)(2s+1)$ symmetric states of the form

$$[\chi_\nu(1)\chi_{\nu'}(2) + \chi_{\nu'}(1)\chi_\nu(2)]/\sqrt{2}$$

and $s(2s+1)$ antisymmetric states of the form

$$[\chi_\nu(1)\chi_{\nu'}(2) - \chi_{\nu'}(1)\chi_\nu(2)]/\sqrt{2}$$

which together span the same $(2s+1)^2$-dimensional space.

We now come to a tricky point. In a real diatomic molecule, the spin-s particles are the nuclei, and interactions which cause transitions between different nuclear spin states are very weak. Certainly, their contribution to the energy levels is negligible, but there is a further question as to whether transitions occur rapidly enough for true thermal equilibrium to be maintained. Suppose that they do, and that s is a half-integer (for example $s = \frac{1}{2}$ for the well-studied case of hydrogen). We have to sum over states which are antisymmetric overall, so the partition function for one molecule is

$$Z_{\text{rot}} = s(2s+1)z_{\text{even}} + (s+1)(2s+1)z_{\text{odd}}$$

where the prefactors are the degeneracies of the appropriate spin states, while

$$z_{\text{even}} = \sum_{l \text{ even}} (2l+1)\,\mathrm{e}^{-l(l+1)\theta/T} = \sum_{k=0}^{\infty}(4k+1)\,\mathrm{e}^{-k(2k+1)(2\theta/T)}$$

is the sum over symmetric orbital states ($l = 2k$), with degeneracy $(2l+1)$ and

$$z_{\mathrm{odd}} = \sum_{l \text{ odd}} (2l+1)\,\mathrm{e}^{-l(l+1)\theta/T} = \sum_{k=0}^{\infty} (4k+3)\,\mathrm{e}^{-(2k+1)(k+1)(2\theta/T)}$$

is the sum over antisymmetric orbital states ($l = 2k+1$). The characteristic temperature associated with the energy-level spacing is $\theta = \hbar^2/2Ik$. The remaining sums cannot be evaluated in closed form, but the partition function, and hence the specific heat, can be estimated numerically, with results that disagree markedly with experimental observations of hydrogen.

To obtain results which do agree with observations, we must assume that transitions occur often enough for the nuclei in a stored sample to come into equilibrium over a long period of time, but that they occur negligibly often over the period of time taken to measure the specific heat. Since the interaction energy is negligible, the equilibrium ratio of numbers of molecules in symmetric and antisymmetric spin states is determined purely by the degeneracies: a fraction $F_{\mathrm{S}} = (s+1)(2s+1)/(2s+1)^2 = (s+1)/(2s+1)$ are in symmetric spin states, while a fraction $F_{\mathrm{A}} = s/(2s+1)$ are in antisymmetric spin states. These two sets of molecules can be treated as a mixture of two different gases. The rotational partition function for the whole gas of N molecules is then

$$Z_{N\,\mathrm{rot}} = \begin{cases} (z_{\mathrm{even}})^{F_{\mathrm{S}}N}(z_{\mathrm{odd}})^{F_{\mathrm{A}}N} & \text{for integer values of } s \\ (z_{\mathrm{even}})^{F_{\mathrm{A}}N}(z_{\mathrm{odd}})^{F_{\mathrm{S}}N} & \text{for half-integer values of } s. \end{cases}$$

(b) For the two components of the gas, let us define $c_V^{\mathrm{even}} = (\partial u_{\mathrm{even}}/\partial T)_V$, with $u_{\mathrm{even}} = -\partial(\ln z_{\mathrm{even}})/\partial\beta$, and c_V^{odd} in the corresponding way. It is then a simple matter to find

$$C_V^{\mathrm{rot}} = \begin{cases} N(F_{\mathrm{S}}c_V^{\mathrm{even}} + F_{\mathrm{A}}c_V^{\mathrm{odd}}) & \text{for integer values of } s \\ N(F_{\mathrm{A}}c_V^{\mathrm{even}} + F_{\mathrm{S}}c_V^{\mathrm{odd}}) & \text{for half-integer values of } s. \end{cases}$$

For $T \gg \theta$ or $T \ll \theta$, approximate expressions for c_V^{even} and c_V^{odd} can be obtained by the methods indicated in problem 3.18. At high temperatures, in particular, the sums can be replaced by integrals and we find that $c_V^{\mathrm{even}} \simeq c_V^{\mathrm{odd}} \simeq k$ and $C_V^{\mathrm{rot}} \simeq Nk$, which is the classical result, independent of the nuclear spins. Nevertheless, the relative intensities of lines in the emission spectrum of an incandescent gas will reveal the presence of the appropriate fractions of molecules in even and odd rotational states.

• Problem 3.22

(a) A hydrogen molecule can rotate about an axis perpendicular to the line joining its two atoms and can also undergo radial oscillations. For a rigid quantum-mechanical rotor, the energy levels are $E_{\mathrm{rot}} = \hbar^2 l(l+1)/2I$, where l is an integer and I is the moment of inertia. The lowest non-zero rotational energy is obtained for $l = 1$, when $E_{\mathrm{rot}} = \hbar^2/I$ and, comparing this with the classical energy $E_{\rm rot} = I\omega_{\rm rot}^2/2$, we can identify a corresponding frequency $\omega_{\rm rot} = \sqrt{2}\hbar/I$, in terms of which we have $E_{\rm rot} = \hbar\omega_{\rm rot}/\sqrt{2}$. The equilibrium separation of the atoms corresponds to the minimum of the interatomic potential which, in the case of the Morse potential, is r_0. The moment of inertia is then $I = mr_0^2/2$, where $m \simeq 1.67 \times 10^{-24}$ g is the mass of a hydrogen atom (roughly that of a proton). Thus we find that

$$\omega_{\rm rot} = \frac{2\sqrt{2}\hbar}{mr_0^2} \simeq 2.8 \times 10^{13}\ \text{rad s}^{-1}.$$

To investigate small-amplitude radial oscillations, we expand the potential to quadratic order about its minimum: $V(r) = -V_0 + (V_0/a^2)(r - r_0)^2 + \cdots$. In this approximation, the radial motion is that of a simple harmonic oscillator whose mass is the reduced mass $\mu = m/2$ and whose angular frequency is given by $V_0/a^2 = \mu\omega_{\rm vib}^2/2$, or

$$\omega_{\rm vib} = \sqrt{\frac{4V_0}{ma^2}} \simeq 8.2 \times 10^{14}\ \text{rad s}^{-1}.$$

Of course, the vibrational energy levels are given by $E_{\rm vib} = (n + \frac{1}{2})\hbar\omega_{\rm vib}$, for integer values of n.

(b) Roughly speaking, rotations will contribute significantly to the internal energy when kT is greater than the minimum energy $\hbar\omega_{\rm rot}/\sqrt{2}$ needed to excite the rotational motion. Thus $T_{\rm rot} = \hbar\omega_{\rm rot}/\sqrt{2}k \simeq 150$ K. Similarly, the lowest temperature at which vibrations contribute is given roughly by $T_{\rm vib} = \hbar\omega_{\rm vib}/k \simeq 6300$ K. At room temperature, for example, we see that rotational states are significantly excited. Excitation of the vibrational states, however, requires temperatures higher than those needed to dissociate the molecules.

• **Problem 3.23** For a gas of non-interacting particles, the grand canonical partition function is

$$\mathcal{Z}(T, V, \mu) = \sum_{N=0}^{\infty} \frac{1}{N!}\,\mathrm{e}^{\beta\mu N}\, Z_1^N(T, V) = \exp[\mathrm{e}^{\beta\mu}\, Z_1(T, V)]$$

where Z_1 is the single-particle canonical partition function. Since the energy of an ultrarelativistic particle of momentum $\boldsymbol{p}$ is $\varepsilon = c|\boldsymbol{p}|$, we have

$$Z_1 = \frac{1}{h^3}\int \mathrm{d}^3q\, \mathrm{d}^3p\, \mathrm{e}^{-\beta c|\boldsymbol{p}|} = \frac{4\pi V}{h^3}\int_0^\infty \mathrm{d}p\, p^2\, \mathrm{e}^{-\beta c p} = \frac{8\pi V}{(\beta c h)^3}.$$

Using the relation with thermodynamics $PV = kT \ln \mathcal{Z}$ we finally get

$$\mu(T, P) = kT \ln\left(\frac{h^3c^3P}{8\pi(kT)^4}\right).$$

• **Problem 3.24** For a fixed number M of particles, the canonical partition function is

$$Z(T,V,M)=\sum_{n_1,n_2}\binom{N}{n_1}\binom{N-n_1}{n_2}\mathrm{e}^{-\beta\varepsilon n_2}$$

where n_1 and n_2 are the numbers of cells occupied respectively by one or two particles, and the total volume is $V = Nv$. The sum is constrained so that $n_1+2n_2 = M$ and the degeneracy of the energy level $E = n_2\varepsilon$ is the product of two binomial coefficients, the first giving the number of ways in which the n_1 singly occupied cells may be chosen from N and the second giving the number of ways in which the n_2 doubly occupied cells may be chosen from the remaining $N - n_1$. This constrained sum can be evaluated approximately when N, n_1 and n_2 are very large (see, for example, problem 3.8) but the problem is much easier to treat using the grand canonical ensemble. The grand partition function is

$$\begin{aligned}
\mathcal{Z}(T,V,\mu) &= \sum_{M=0}^{2N}\mathrm{e}^{\beta\mu M}Z(T,V,M)\\
&= \sum_{n_1,n_2}\mathrm{e}^{\beta\mu(n_1+2n_2)}\binom{N}{n_1}\binom{N-n_1}{n_2}\mathrm{e}^{-\beta\varepsilon n_2}\\
&= \sum_{n_1=0}^{N}\binom{N}{n_1}\mathrm{e}^{\beta\mu n_1}\sum_{n_2=0}^{N-n_1}\binom{N-n_1}{n_2}\mathrm{e}^{-\beta(\varepsilon-2\mu)n_2}\\
&= (1+z+\mathrm{e}^{-\beta\varepsilon}z^2)^N.
\end{aligned}$$

The sum over M converts the original constrained sum over n_1 and n_2 into one which is unconstrained in the range $0 \le n_1 + n_2 \le N$. The sum over n_2 is of the form

$$\sum_{n=0}^{N}\binom{N}{n}x^n = (1+x)^N$$

and, after carrying out this sum, the remaining sum over n_1 is of the same form. The result is conveniently expressed in terms of the fugacity $z = \mathrm{e}^{\beta\mu}$. Writing $\mathcal{Z}_1(T,\mu) = 1 + z + \mathrm{e}^{-\beta\varepsilon}z^2$, we calculate the average energy per cell, the concentration and the pressure as

$$\begin{aligned}
\frac{E}{N} &= -N^{-1}\left(\frac{\partial(\ln\mathcal{Z})}{\partial\beta}\right)_{\beta\mu} = \varepsilon\frac{\mathrm{e}^{-\beta\varepsilon}z^2}{\mathcal{Z}_1(T,\mu)}\\
c &= N^{-1}\left(\frac{\partial(\ln\mathcal{Z})}{\partial(\beta\mu)}\right)_{\beta} = \frac{z+2\,\mathrm{e}^{-\beta\varepsilon}z^2}{\mathcal{Z}_1(T,\mu)}\\
P &= \left(\frac{\partial(kT\ln\mathcal{Z})}{\partial V}\right)_{\beta\mu} = \frac{1}{v}\left(\frac{\partial(kT\ln\mathcal{Z})}{\partial N}\right)_{\beta\mu} = \frac{kT}{v}\ln[\mathcal{Z}_1(T,\mu)].
\end{aligned}$$

The expression for c can be solved to obtain z (which cannot be negative) as a function of c and T:

$$z = \frac{c - 1 + \sqrt{(1-c)^2 + 4c(2-c)\,\mathrm{e}^{-\beta\varepsilon}}}{2(2-c)\,\mathrm{e}^{-\beta\varepsilon}}$$

and hence E/N and P can also be expressed in terms of c and T. When c is very small, we have $z \simeq c$ and

$$\frac{E}{N} \simeq \varepsilon c^2\,\mathrm{e}^{-\beta\varepsilon} \qquad P \simeq \frac{c}{v}kT.$$

Since c is the average number of particles in a cell of volume v, c/v is just the number of particles per unit volume; so the result for P reproduces the equation of state for a classical ideal gas. Since our model does not take into account the kinetic energy of the particles, the result for E/N does not contain the usual contribution of $3kT/2$ but gives the leading contribution from intermolecular forces. The factor of c^2 can be understood in terms of a pairwise interaction. When c is close to its maximum value of 2, we have $z \simeq \mathrm{e}^{\beta\varepsilon}\,\delta^{-1}$, where $\delta = 2 - c$, and

$$\frac{E}{N} \simeq \varepsilon(1-\delta) \qquad P \simeq -\frac{2kT}{v}\ln\delta.$$

Clearly, as $c \to 2$ or $\delta \to 0$, the maximum possible number of particles are being fitted into the container. The energy per cell approaches its maximum value of ε, and the pressure becomes infinite because, when the particles are closely packed, the fluid becomes incompressible.

• Problem 3.25

(a) The grand canonical partition function of an ideal classical gas is

$$\mathcal{Z}(T, V, \mu) = \sum_{N=0}^{\infty} \frac{1}{N!} z^N Z_1^N = \mathrm{e}^{zZ_1}$$

where $z = \mathrm{e}^{\beta\mu}$ is the fugacity and $Z_1 = V(2\pi mkT/h^2)^{3/2}$ is the canonical partition function of one particle in the gas. According to the fundamental (or the Euler) equation of thermodynamics, the grand potential $\Omega(T, V, \mu) = -U + TS + \mu N$ is equal to PV. It is related to the grand canonical partition function by $\Omega(T, V, \mu) = kT\ln[\mathcal{Z}(T, V, \mu)]$, so the pressure is

$$P = \frac{\Omega}{V} = \frac{kT}{V}\ln[\mathcal{Z}(T, V, \mu)] = z\left(\frac{2\pi m}{h^2}\right)^{3/2}(kT)^{5/2}$$

and from this we obtain the fugacity for an ideal gas in terms of temperature and pressure as

$$z_{\mathrm{g}}(T, P) = P(kT)^{-5/2}\left(\frac{h^2}{2\pi m}\right)^{3/2}.$$

(b) From the grand canonical point of view, we regard the gas and the absorbing walls as being separately in equilibrium with a particle reservoir. The chemical potential μ, the temperature T and hence also the fugacity z are properties of this reservoir and therefore the same for both the gas and the absorbed molecules. In fact, each absorbing site is separately in equilibrium with the reservoir and has a grand canonical partition function given by

$$\mathcal{Z}_{\text{site}} = \sum_{N=0}^{1} z^N \, \mathrm{e}^{N\beta\varepsilon} = 1 + z\,\mathrm{e}^{\beta\varepsilon}.$$

The average number of absorbed molecules per site is

$$\langle N\rangle_{\text{site}} = z\left(\frac{\partial(\ln \mathcal{Z}_{\text{site}})}{\partial z}\right)_T = \frac{1}{1 + z^{-1}\,\mathrm{e}^{-\beta\varepsilon}}.$$

So, for the N_0 sites, we have

$$\langle N\rangle = \frac{N_0}{1 + z^{-1}\,\mathrm{e}^{-\beta\varepsilon}}.$$

This average number of absorbed molecules can now be expressed in terms of the temperature and pressure of the gas by substituting for z the function $z_{\mathrm{g}}(T, P)$ found above:

$$\langle N\rangle = \frac{N_0}{1 + (2\pi m/h^2)^{3/2} P^{-1} (kT)^{5/2}\,\mathrm{e}^{-\beta\varepsilon}}.$$

At a fixed temperature, this varies with pressure in a manner that might intuitively be expected. At high pressures, the density is high, molecules frequently approach the absorbing sites and $\langle N\rangle \simeq N_0$. At low pressures, the opposite is true and $\langle N\rangle \simeq 0$. The effect of temperature is also apparent. At high temperatures, absorbed molecules are easily dislodged, and a high pressure is required to keep the sites filled. At low temperatures, a more modest pressure is sufficient.

• **Problem 3.26** In principle, the grand canonical ensemble describes the gas of free particles and each absorbing site as an open system, in equilibrium with a particle reservoir characterized by its temperature T and chemical potential μ. The partition functions (see problem 3.25 for details) are

$$\mathcal{Z}_{\text{free}}(T, V, \mu) = \exp\left[\mathrm{e}^{\beta\mu} V \left(\frac{2\pi mkT}{h^2}\right)^{3/2}\right]$$

$$\mathcal{Z}_{\text{site}}(T, \mu) = \sum_{N=0}^{2} (\mathrm{e}^{\beta\mu})^N \, \mathrm{e}^{N\beta\varepsilon} = 1 + \mathrm{e}^{\beta(\mu+\varepsilon)} + \mathrm{e}^{2\beta(\mu+\varepsilon)}$$

and from these we obtain expressions for the pressure of the gas and for the average numbers of free and absorbed particles:

$$P = \frac{kT}{V}\ln \mathcal{Z}_{\text{free}} = e^{\beta\mu}\left(\frac{2\pi m}{h^2}\right)^{3/2}(kT)^{5/2}$$

$$\langle N_{\text{free}}\rangle = \left(\frac{\partial(\ln \mathcal{Z}_{\text{free}})}{\partial(\beta\mu)}\right)_{T,V} = e^{\beta\mu}\,V\left(\frac{2\pi mkT}{h^2}\right)^{3/2} = \frac{PV}{kT}$$

$$\langle N_{\text{abs}}\rangle = N_0\left(\frac{\partial \mathcal{Z}_{\text{site}}}{\partial(\beta\mu)}\right)_T = N_0\frac{e^{\beta(\mu+\varepsilon)} + 2\,e^{2\beta(\mu+\varepsilon)}}{1 + e^{\beta(\mu+\varepsilon)} + e^{2\beta(\mu+\varepsilon)}}.$$

To apply these results to a system with a fixed total number N of particles, we adjust the chemical potential μ so that the grand canonical average of the total number of particles is N. That is, we determine μ as the solution of the constraint equation

$$\langle N_{\text{free}}\rangle + \langle N_{\text{abs}}\rangle = N.$$

This will be justified if the relative fluctuations in average particle numbers are small, which will be true if $\langle N_{\text{free}}\rangle$ and $\langle N_{\text{abs}}\rangle$ are both large. We must therefore assume that N, N_0 and $N - 2N_0$ are all very large, although $N/2N_0$ need not be much greater than 1. The above result for P may be solved for $e^{\beta\mu}$ as a function of P and T. When this is done, the constraint equation and the average number of absorbed particles can be expressed as

$$\frac{PV}{NkT} + \frac{N_0}{N}\frac{Pf + 2(Pf)^2}{1 + Pf + (Pf)^2} = 1$$

$$\langle N_{\text{abs}}\rangle = N_0\frac{Pf + 2(Pf)^2}{1 + Pf + (Pf)^2}$$

where $f = (h^2/2\pi m)^{3/2}(kT)^{-5/2}\,e^{\beta\varepsilon}$ is a function of T only. In this form, the constraint equation is clearly the equation of state, which reproduces the usual ideal gas equation $PV = NkT$, with a correction due to the absorbing walls.

To discover how $\langle N_{\text{abs}}\rangle$ varies with temperature, we need to know the temperature dependence of P (to be found from the equation of state) as well as that of f. The second term in the equation of state is clearly just $\langle N_{\text{abs}}\rangle/N$, which is positive and less than unity. Therefore, in the limit of either high or low temperatures, P/kT must approach a finite non-zero constant. So, in either limit, we have $Pf \sim kTf \sim (kT)^{-3/2}\,e^{\varepsilon/kT}$. At high temperatures, we find that $Pf \to 0$ and $\langle N_{\text{abs}}\rangle \to 0$ while, at low temperatures, $Pf \to \infty$ and $\langle N_{\text{abs}}\rangle \to 2N_0$. These results are easily understood by thinking of the equilibrium state as that which minimizes the Helmholtz free energy $F = U - TS$. At high temperatures, the gain in entropy of a particle which becomes dislodged from the wall outweighs the increase in its potential energy, so few particles remain absorbed. At very low temperatures, the entropic contribution to F is negligible, and $\langle N_{\text{abs}}\rangle$ rises to its maximum value of $2N_0$, so as to minimize the energy.

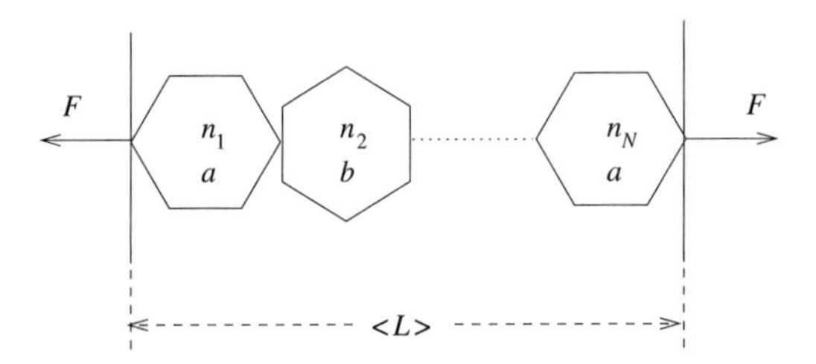

Figure 3.5. A possible state of an elastic molecule under the action of a tension F.

• **Problem 3.27** The chain held at a fixed tension is analogous to a gas maintained at a fixed pressure, rather than confined to a fixed volume, the tension F playing the role of $-P$ (cf problem 2.8). For this problem, the most convenient ensemble is the isobaric ensemble, with the partition function

$$\Xi(T, F, N) = \sum_{\text{microstates}} \mathrm{e}^{\beta F L}\, \mathrm{e}^{-\beta E}$$

where $L = \sum_i l_i$ is the total length of the chain, the index $i = 1, \ldots, N$ labelling the links. The advantage of this ensemble is that, since the total length of the chain is unconstrained, the states of individual links are independent, and the partition function is a product of N identical factors associated with individual links. A possible state of the whole chain is illustrated in figure 3.5.

Thus, $\Xi(T, F, N) = [\Xi_{\text{link}}(T, F)]^N$, where

$$\begin{aligned}
\Xi_{\text{link}}(T, F) &= \sum_{n=0}^{\infty} \sum_{l=a,b} \mathrm{e}^{\beta F l}\, \mathrm{e}^{-\beta(n+1/2)\hbar\omega_l} \\
&= \sum_{n=0}^{\infty} (\mathrm{e}^{\beta F a}\, \mathrm{e}^{-\beta(n+1/2)\hbar\omega_a} + \mathrm{e}^{\beta F b}\, \mathrm{e}^{-\beta(n+1/2)\hbar\omega_b}) \\
&= \frac{\mathrm{e}^{\beta F a}}{2\sinh(\beta\hbar\omega_a/2)} + \frac{\mathrm{e}^{\beta F b}}{2\sinh(\beta\hbar\omega_b/2)}.
\end{aligned}$$

The mean energy and length of the chain are now easily obtained:

$$\begin{aligned}
\langle E \rangle &= -\left(\frac{\partial(\ln \Xi)}{\partial\beta}\right)_{\beta F} = -N\left(\frac{\partial(\ln \Xi_{\text{link}})}{\partial\beta}\right)_{\beta F} \\
&= \frac{N\hbar}{2} \frac{\omega_a\, \mathrm{e}^{\beta F a}\, c(x_a) s(x_b) + \omega_b\, \mathrm{e}^{\beta F b}\, c(x_b) s(x_a)}{\mathrm{e}^{\beta F a}\, s(x_b) + \mathrm{e}^{\beta F b}\, s(x_a)} \\
\langle L \rangle &= \left(\frac{\partial(\ln \Xi)}{\partial(\beta F)}\right)_{\beta} = N\left(\frac{\partial(\ln \Xi_{\text{link}})}{\partial(\beta F)}\right)_{\beta} \\
&= N \frac{a\, \mathrm{e}^{\beta F a}\, s(x_b) + b\, \mathrm{e}^{\beta F b}\, s(x_a)}{\mathrm{e}^{\beta F a}\, s(x_b) + \mathrm{e}^{\beta F b}\, s(x_a)}
\end{aligned}$$

where we have used the abbreviations $x_l = \beta\hbar\omega_l/2$, $s(x_l) = \sinh(x_l)$ and $c(x_l) = \coth(x_l)$.

At low temperatures, we have $x_l \to \infty$, $s(x_l) \to \mathrm{e}^{x_l}$ and $c(x_l) \to 1$. In this limit, we find for the mean energy

$$\langle E\rangle \simeq \frac{N\hbar}{2}\frac{\omega_a\,\mathrm{e}^{\beta(Fa+\hbar\omega_b/2)} + \omega_b\,\mathrm{e}^{\beta(Fb+\hbar\omega_a/2)}}{\mathrm{e}^{\beta(Fa+\hbar\omega_b/2)} + \mathrm{e}^{\beta(Fb+\hbar\omega_a/2)}}$$
$$\simeq \begin{cases} \dfrac{N\hbar\omega_a}{2} & \text{if } a + \dfrac{\hbar\omega_b}{2F} > b + \dfrac{\hbar\omega_a}{2F} \\ \dfrac{N\hbar\omega_b}{2} & \text{if } a + \dfrac{\hbar\omega_b}{2F} < b + \dfrac{\hbar\omega_a}{2F} \end{cases}$$

while the mean length of the chain becomes

$$\langle L\rangle \simeq N\frac{a\,\mathrm{e}^{\beta(Fa+\hbar\omega_b/2)} + b\,\mathrm{e}^{\beta(Fb+\hbar\omega_a/2)}}{\mathrm{e}^{\beta(Fa+\hbar\omega_b/2)} + \mathrm{e}^{\beta(Fb+\hbar\omega_a/2)}}$$
$$\simeq \begin{cases} Na & \text{if } a + \dfrac{\hbar\omega_b}{2F} > b + \dfrac{\hbar\omega_a}{2F} \\ Nb & \text{if } a + \dfrac{\hbar\omega_b}{2F} < b + \dfrac{\hbar\omega_a}{2F}. \end{cases}$$

Clearly, every link has the same length, and its energy is the zero-point energy corresponding to that length. Suppose, for definiteness, that $a > b$. If the frequencies ω_a and ω_b are equal, we see that each link has its maximum length, a, as might be expected. If the frequencies are unequal, each link also has its maximum length unless $(a-b)F < \hbar(\omega_a - \omega_b)/2$, in which case it has its minimum length b. In the latter case, the work $(a-b)F$ that would be done by the external force F in extending the link is insufficient to supply the extra zero point energy $\hbar(\omega_a - \omega_b)/2$.

In the limit of high temperatures, we have $\mathrm{e}^{\beta Fl} \to 1$, $s(x_l) \simeq x_l$ and $c(x_l) \simeq 1/x_l$. Then the mean energy and length are

$$\langle E\rangle \simeq NkT \qquad \langle L\rangle \simeq N\frac{a\omega_b + b\omega_a}{\omega_b + \omega_a}.$$

The mean energy clearly agrees with the classical equipartition theorem, as expected. The result for $\langle L\rangle$ is that which maximizes the entropy. When the oscillator frequencies are equal, the entropy is maximized when half the links have length a and half have length b, giving an overall chain length of $N(a+b)/2$. However, when the frequencies are different, the vibrational entropies of long and short links are different, and the entropy can be increased by having more links of the length with greater vibrational entropy. At high temperatures, indeed, the tension is unimportant. The entropy of the length distribution is easily found to be $S_l = k(N\ln N - N_a\ln N_a - N_b\ln N_b)$, where N_a is the number of links of length a and $N_b = N - N_a$ is the number with length b, provided that all these numbers are large. The entropy of a harmonic oscillator is also easy to obtain from the canonical ensemble, and at high temperatures is given by $S_{\text{osc}} \simeq k[1 + \ln(kT/\hbar\omega)]$. Consequently, the entropy of the chain at high temperatures is

$$\begin{aligned} S &= k\,[N \ln N - N_a \ln N_a - (N - N_a)\ln(N - N_a)] \\ &\quad + k\left[N + N_a \ln\left(\frac{kT}{\hbar\omega_a}\right) + (N - N_a)\ln\left(\frac{kT}{\hbar\omega_b}\right)\right] \end{aligned}$$

and the above chain length corresponds to the value $N_a = N\omega_b/(\omega_b+\omega_a)$ which maximizes this entropy.

• **Problem 3.28** In solving this problem, it will be sufficient to consider systems of fixed volume V, and for simplicity we shall ignore the dependence of all thermodynamic functions on V. It will also prove convenient to use the parameter $\alpha = \mu/T$ rather than the chemical potential μ itself. For the canonical ensemble, we have

$$\begin{aligned} (\Delta_{\text{C}}E)^2 &= \langle E^2\rangle_{\text{C}} - \langle E\rangle_{\text{C}}^2 \\ &= \frac{1}{Z}\left(\frac{\partial^2 Z}{\partial\beta^2}\right)_N - \left[\frac{1}{Z}\left(\frac{\partial Z}{\partial\beta}\right)_N\right]^2 = \left(\frac{\partial^2(\ln Z)}{\partial\beta^2}\right)_N \\ &= kT^2\left(\frac{\partial U}{\partial T}\right)_N . \end{aligned}$$

For the grand canonical ensemble, we find in the same way that

$$\begin{aligned} (\Delta_{\text{G}}E)^2 &= \left(\frac{\partial^2(\ln \mathcal{Z})}{\partial\beta^2}\right)_\alpha = kT^2\left(\frac{\partial U}{\partial T}\right)_\alpha \\ (\Delta_{\text{G}}N)^2 &= \left(\frac{\partial^2(\ln \mathcal{Z})}{\partial(\beta\mu)^2}\right)_T = k\left(\frac{\partial N}{\partial\alpha}\right)_T . \end{aligned}$$

Regarding U as a function of T and N, we have

$$\left(\frac{\partial U}{\partial T}\right)_\alpha = \left(\frac{\partial U}{\partial T}\right)_N + \left(\frac{\partial U}{\partial N}\right)_T\left(\frac{\partial N}{\partial T}\right)_\alpha$$

and therefore

$$(\Delta_{\text{G}}E)^2 - (\Delta_{\text{C}}E)^2 = kT^2\left(\frac{\partial U}{\partial N}\right)_T\left(\frac{\partial N}{\partial T}\right)_\alpha .$$

To relate this to $(\Delta_{\text{G}}N)^2$, consider the thermodynamic potential Ω asssociated with the grand canonical ensemble, which obeys the differential relation $\mathrm{d}\Omega = P\,\mathrm{d}V + S\,\mathrm{d}T + N\,\mathrm{d}\mu$. Written in terms of α with $\mathrm{d}V = 0$ and $\mathrm{d}\mu = \alpha\,\mathrm{d}T + T\,\mathrm{d}\alpha$, this is

$$\mathrm{d}\Omega = (S + \alpha N)\,\mathrm{d}T + NT\,\mathrm{d}\alpha$$

from which we obtain a relation of the Maxwell type:

$$\left(\frac{\partial(S+\alpha N)}{\partial\alpha}\right)_T = \left(\frac{\partial(NT)}{\partial T}\right)_\alpha = N + T\left(\frac{\partial N}{\partial T}\right)_\alpha .$$

The expression on the left can be evaluated by rewriting the first law $\mathrm{d}U = T\,\mathrm{d}S - P\,\mathrm{d}V + \mu\,\mathrm{d}N$ as

$$\mathrm{d}(S+\alpha N) = \frac{1}{T}\,\mathrm{d}U + N\,\mathrm{d}\alpha$$

from which we deduce that

$$\left(\frac{\partial(S+\alpha N)}{\partial\alpha}\right)_T = N + \frac{1}{T}\left(\frac{\partial U}{\partial\alpha}\right)_T$$

and using our previous result

$$T^2\left(\frac{\partial N}{\partial T}\right)_\alpha = \left(\frac{\partial U}{\partial\alpha}\right)_T = \left(\frac{\partial U}{\partial N}\right)_T\left(\frac{\partial N}{\partial\alpha}\right)_T = \frac{1}{k}\left(\frac{\partial U}{\partial N}\right)_T(\Delta_{\mathrm{G}}N)^2.$$

Thus, we obtain the desired relation

$$(\Delta_{\mathrm{G}}E)^2 - (\Delta_{\mathrm{C}}E)^2 = (\Delta_{\mathrm{G}}N)^2\left[\left(\frac{\partial U}{\partial N}\right)_T\right]^2 .$$

The expression on the right is clearly positive and can loosely be interpreted as

$$[(\text{average number of exchanged particles}) \times (\text{energy per particle})]^2.$$

• **Problem 3.29** The canonical partition function for N non-interacting particles is

$$Z(T,V,N) = [Z_1(T,V)]^N.$$

From it, we immediately obtain

$$\begin{aligned}
U_{\mathrm{C}}(T,V,N) &= -\left(\frac{\partial(\ln Z)}{\partial\beta}\right)_{V,N} = NkT^2\frac{Z_1'}{Z_1}\\
F_{\mathrm{C}}(T,V,N) &= -kT\ln Z = -NkT\ln Z_1\\
S_{\mathrm{C}}(T,V,N) &= \frac{U_{\mathrm{C}}-F_{\mathrm{C}}}{T} = Nk\left(T\frac{Z_1'}{Z_1} + \ln Z_1\right)
\end{aligned}$$

where $Z_1' = (\partial Z_1/\partial T)_V$.

The grand canonical partition function is

$$\begin{aligned}
\mathcal{Z}(T,V,\mu) &= \sum_{N=0}^{\infty} \mathrm{e}^{\beta\mu N} Z(T,V,N) = \sum_{N=0}^{\infty}[\mathrm{e}^{\beta\mu} Z_1(T,V)]^N\\
&= \frac{1}{1-\mathrm{e}^{\beta\mu} Z_1(T,V)}.
\end{aligned}$$

To obtain thermodynamic quantities as functions of N, we must adjust the chemical potential μ so that

$$N = \langle N \rangle_{\mathrm{G}} = \left(\frac{\partial (\ln \mathcal{Z})}{\partial (\beta\mu)} \right)_{T,V} = \frac{\mathrm{e}^{\beta\mu} Z_1}{1 - \mathrm{e}^{\beta\mu} Z_1}$$

which implies that

$$\mathrm{e}^{\beta\mu} Z_1 = \frac{N}{N+1} \qquad \mu = -kT \ln \left(\frac{(N+1)Z_1}{N} \right).$$

We then find

$$\begin{aligned}
\Omega_{\mathrm{G}}(T, V, N) &= kT \ln \mathcal{Z} = kT \ln(N+1) \\
F_{\mathrm{G}}(T, V, N) &= \mu N - \Omega_{\mathrm{G}} = kT[N \ln N - (N+1)\ln(N+1) - N \ln Z_1] \\
U_{\mathrm{G}}(T, V, N) &= -\left(\frac{\partial (\ln \mathcal{Z})}{\partial \beta} \right)_{\beta\mu, V} = NkT^2 \frac{Z_1'}{Z_1} \\
S_{\mathrm{G}}(T, V, N) &= \frac{U_{\mathrm{G}} - F_{\mathrm{G}}}{T} \\
&= NkT \frac{Z_1'}{Z_1} - k[N \ln N - (N+1)\ln(N+1) - N \ln Z_1].
\end{aligned}$$

Clearly, we have $U_{\mathrm{G}}(T, V, N) = U_{\mathrm{C}}(T, V, N)$ and

$$\begin{aligned}
(S_{\mathrm{G}} - S_{\mathrm{C}})/k &= -\frac{(F_{\mathrm{G}} - F_{\mathrm{C}})}{kT} \\
&= (N+1)\ln(N+1) - N \ln N \\
&= \ln N + 1 + \frac{1}{2N} + \cdots.
\end{aligned}$$

On dividing this last result by N and taking N to be large, we get

$$\frac{s_{\mathrm{G}} - s_{\mathrm{C}}}{k} = -\frac{f_{\mathrm{G}} - f_{\mathrm{C}}}{kT} \simeq \frac{\ln N}{N}.$$

• Problem 3.30

(a) *Microcanonical ensemble*. This ensemble describes an isolated system with a fixed number N of particles and a fixed total energy E. The numbers of particles n_+ and n_- with energies ε and $-\varepsilon$ respectively are clearly constrained by the two relations $E = n_+\varepsilon - n_-\varepsilon$ and $N = n_+ + n_-$. Defining the quantity $x = E/N\varepsilon$ (which is proportional to the energy per particle), we easily find that $n_+ = N(1+x)/2$ and $n_- = N(1-x)/2$. The partition function is just the number of microstates consistent with these constraints and is equal to the number of ways of choosing, say, n_+ particles from N, namely

$$\Omega(E, N) = \binom{N}{n_+} = \frac{N!}{n_+! n_-!}.$$

The thermodynamic interpretation of this ensemble is through the definition of the entropy $S_{\mathrm{M}}(E, N) = k \ln[\Omega(E, N)]$. In the limit when N is very large, we may use Stirling's approximation $\ln N! \simeq N \ln N - N$ to obtain the entropy per particle as

$$s = \lim_{N\to\infty}\left(\frac{S_{\mathrm{M}}}{N}\right) = k[\ln 2 - \tfrac{1}{2}(1+x)\ln(1+x) - \tfrac{1}{2}(1-x)\ln(1-x)].$$

(b) *Canonical ensemble.* This ensemble describes a system with a fixed number of particles in contact with a heat bath at a fixed temperature. The partition function is

$$Z(T, N) = \sum_{n_+=0}^{N}\binom{N}{n_+}\mathrm{e}^{-\beta\varepsilon n_+}\,\mathrm{e}^{\beta\varepsilon(N-n_+)} = (\mathrm{e}^{-\beta\varepsilon} + \mathrm{e}^{\beta\varepsilon})^N$$

and the thermodynamic interpretation is through the Helmholtz free energy $F(T, N) = -kT\ln[Z(T, N)]$. The entropy in this ensemble is obtained from the thermodynamic relation $F = U - TS$, where the internal energy is identified as the mean value

$$U_{\mathrm{C}} = \langle E\rangle_{\mathrm{C}} = -\left(\frac{\partial(\ln Z)}{\partial\beta}\right)_N = -N\varepsilon\tanh(\beta\varepsilon).$$

The mean energy per particle is now identified as U_{C}/N and the parameter x introduced above is $x = U_{\mathrm{C}}/N\varepsilon = -\tanh(\beta\varepsilon)$. In order to obtain the entropy as a function of x, we must first solve this relation to obtain the temperature in terms of x:

$$\beta\varepsilon = \frac{1}{2}\ln\left(\frac{1-x}{1+x}\right) = \tfrac{1}{2}\left[\ln(1-x) - \ln(1+x)\right].$$

With this result in hand, we find that

$$\begin{aligned}
U_{\mathrm{C}} &= N\varepsilon x\\
F_{\mathrm{C}} &= -kT\ln Z = -NkT[\ln 2 - \tfrac{1}{2}\ln(1+x) - \tfrac{1}{2}\ln(1-x)]\\
S_{\mathrm{C}} &= \frac{U_{\mathrm{C}} - F_{\mathrm{C}}}{T} = Nk[\ln 2 - \tfrac{1}{2}(1+x)\ln(1+x) - \tfrac{1}{2}(1-x)\ln(1-x)]
\end{aligned}$$

and the entropy per particle $s = S_{\mathrm{C}}/N$ obviously reproduces the microcanonical result.

(c) *Grand canonical ensemble.* This ensemble describes a system in equilibrium with a reservoir, with which it can exchange both energy and particles, so both the energy and the number of particles fluctuate. The reservoir is characterized by a definite temperature $T = 1/k\beta$ and chemical potential μ or fugacity $z = \mathrm{e}^{\beta\mu}$. The partition function is

$$\mathcal{Z}(T, \mu) = \sum_{N=0}^{\infty} z^N Z(T, N) = \frac{1}{1 - z(\mathrm{e}^{\beta\varepsilon} + \mathrm{e}^{-\beta\varepsilon})}$$

and the thermodynamic interpretation is through the grand potential $\Omega_{\rm G}(T,\mu) = kT\ln[\mathcal{Z}(T,\mu)]$. The entropy will be obtained through the thermodynamic relation $\Omega = TS - U + \mu N$, where the particle number and internal energy are identified as the mean values

$$N_{\rm G} = \langle N\rangle_{\rm G} = z\left(\frac{\partial(\ln\mathcal{Z})}{\partial z}\right)_\beta = \frac{z(e^{\beta\varepsilon}+e^{-\beta\varepsilon})}{1-z(e^{\beta\varepsilon}+e^{-\beta\varepsilon})}$$

$$U_{\rm G} = \langle E\rangle_{\rm G} = -\left(\frac{\partial(\ln\mathcal{Z})}{\partial\beta}\right)_z = -\varepsilon\frac{z(e^{\beta\varepsilon}-e^{-\beta\varepsilon})}{1-z(e^{\beta\varepsilon}+e^{-\beta\varepsilon})} = -N_{\rm G}\varepsilon\tanh(\beta\varepsilon).$$

As above, we would like to express thermodynamic functions in terms of the parameter $x = U_{\rm G}/N_{\rm G}\varepsilon = -\tanh(\beta\varepsilon)$ and $N_{\rm G}$, rather than T and μ, so we first solve the above equations to obtain

$$\beta\varepsilon = \tfrac{1}{2}[\ln(1-x)-\ln(1+x)]$$

$$\beta\mu = \ln\left(\frac{N_{\rm G}}{N_{\rm G}+1}\right) - \ln 2 + \tfrac{1}{2}[\ln(1+x)+\ln(1-x)].$$

With these results in hand, we can calculate

$$\begin{aligned}
U_{\rm G} &= N_{\rm G}\varepsilon x\\
\Omega_{\rm G} &= kT\ln\mathcal{Z} = kT\ln(N_{\rm G}+1)\\
S_{\rm G} &= \frac{1}{T}(\Omega_{\rm G}+U_{\rm G}-\mu N_{\rm G})\\
&= kN_{\rm G}\left[\frac{\ln(N_{\rm G}+1)}{N_{\rm G}} + \ln\left(1+\frac{1}{N_{\rm G}}\right)\right.\\
&\qquad\left.+\ln 2 - \tfrac{1}{2}(1+x)\ln(1+x) - \tfrac{1}{2}(1-x)\ln(1-x)\right].
\end{aligned}$$

On taking the limit that $N_{\rm G}$ is very large, we find for the entropy per particle that

$$s = \lim_{N_{\rm G}\to\infty}\left(\frac{S_{\rm G}}{N_{\rm G}}\right) = k[\ln 2 - \tfrac{1}{2}(1+x)\ln(1+x) - \tfrac{1}{2}(1-x)\ln(1-x)]$$

which of course agrees with the canonical and microcanonical versions.

We see that in the thermodynamic limit, the thermodynamic functions given by all three ensembles are identical, provided that one deals with intensive quantities such as the entropy and energy per particle, which have finite and non-zero limits. This is true for a wide variety of systems, the principal requirement being the absence of long-range interactions, so that each particle interacts significantly with only a few near neighbours. (In our example, of course, there are no interactions at all.) We can therefore understand intuitively that a large system in effect provides a heat bath and particle reservoir for any local region contained within it, and it becomes irrelevant whether the whole system is contained within a still larger reservoir or not. As seen in the example of problem 3.29, the relative fluctuations in the grand canonical energy and particle number become very small for a large system. Consequently, it becomes possible to think of E, N, T and μ as all having definite (but not independently specifiable) values. Thus, provided that T and μ are appropriately adjusted, the average energy and particle number in the grand canonical ensemble can loosely be identified with the fixed energy and particle number of a corresponding microcanonical ensemble, and so on. In practice, one will generally make use of the ensemble in which a given problem can most easily be solved.

4

QUANTUM STATISTICS

In this chapter, the ensembles of equilibrium statistical mechanics are applied to quantum-mechanical systems whose constituent particles are bosons or fermions. Problems 4.1–4.3 compare simple properties of systems of indistinguishable particles that obey Bose–Einstein and Fermi–Dirac statistics, and also the fictitious Maxwell–Boltzmann statistics, which afford a convenient quasi-classical description at high temperatures and low densities. Problems 4.4 and 4.5 deal with black-body radiation, while problems 4.6–4.11 provide examples of electrons and phonons in solids with different dispersion relations. Bose–Einstein condensation is covered in problems 4.12–4.16, and Fermi gases are dealt with in problems 4.17–4.26. Finally the last three problems, 4.27–4.29, compare some physical quantities derived for bosonic and fermionic systems.

4.1 Questions

• **Problem 4.1** Two non-interacting particles inhabit a potential well such that the orbital motion of each particle gives rise to an energy spectrum $E(n) = n\varepsilon$, the nth energy level having a degeneracy $g(n) = 2n + 1$. Consider spin-0 particles obeying Bose–Einstein statistics, spin-$\frac{1}{2}$ particles obeying Fermi–Dirac statistics and spin-s particles obeying Maxwell–Boltzmann statistics. Assuming that the energy is spin independent,

(a) find the microcanonical partition function of the system when the total energy has the fixed value $E = N\varepsilon$,

(b) find the canonical partition function of the system when it is in contact with a heat reservoir at fixed temperature T and

(c) investigate the relationship between the partition functions obtained for the three types of statistics.

• **Problem 4.2** Consider a system of indistinguishable particles whose states can be specified in the following way.

(i) There are single particle states, labelled by an index i, with energy ε_i, which will be degenerate (ε_i may have the same value for several values of i) for particles with non-zero spin.

(ii) Each multiparticle state corresponds to a set of occupation numbers $\{n_i\}$, where n_i counts the number of particles occupying the ith single-particle state and has values from 0 to M. Each distinct set of occupation numbers corresponds to a distinct state.

(a) Use the grand canonical ensemble to find expressions for the pressure P, the average number $\langle N \rangle$ of particles and the internal energy $U = \langle E \rangle$ in the form $T^\alpha f(z)$, where $z = e^{\beta\mu}$ is the fugacity and the function f is defined by a suitable integral. Take the single-particle energies to be $\varepsilon_i = |\boldsymbol{p}_i|^2/2m$ and space to be d dimensional.

(b) Obtain the probability $\mathcal{P}_i(n_i, T, \mu)$ of finding n_i particles in the ith state at temperature T. Compute the mean-square fluctuation $(\Delta n_i)^2 \equiv \langle (n_i - \langle n_i \rangle)^2 \rangle$ in terms of temperature.

(c) Obtain the average occupation number $\langle n(\varepsilon) \rangle$ of an energy level of energy ε and evaluate it at zero temperature.

(d) How, if at all, can we recover from this model the thermodynamics of particles obeying Bose–Einstein, Fermi–Dirac and Maxwell–Boltzmann statistics?

• **Problem 4.3** For a system of non-interacting indistinguishable particles obeying Bose–Einstein, Fermi–Dirac or Maxwell–Boltzmann statistics, obtain the probability distribution $\mathcal{P}_i(n_i, T, \mu)$ for finding n_i particles in a given single-particle state, labelled by i, when the system is in equilibrium with a particle reservoir with temperature T and chemical potential μ. Making use of this distribution, find the average occupation number $\langle n_i \rangle$ and express $\mathcal{P}_i$ as a function of n_i and $\langle n_i \rangle$. Compute the relative fluctuation in the occupation number $\Delta n_i / \langle n_i \rangle$, where $\Delta n_i = \sqrt{\langle (n_i - \langle n_i \rangle)^2 \rangle}$.

• **Problem 4.4** Find expressions for the pressure P, energy density u, entropy density s and specific heat C_V per unit volume of black-body radiation in a d-dimensional cavity at temperature T. Evaluate these quantities explicitly for $d = 3$.

• **Problem 4.5** Consider an atom in a cavity containing black-body radiation. When the atom is in a state a with energy E_a, there is a probability per unit time that it will make a transition to a state b with energy $E_b < E_a$, emitting a photon of frequency $\omega = (E_a - E_b)/\hbar$ in a state k. Denote this probability per unit time by $\mathcal{P}[(a, n_k) \to (b, n_k + 1_k)]$ when there are already n_k photons in state k, and the probability per unit time for the reverse process of absorption by $\mathcal{P}[(b, n_k) \to (a, n_k - 1_k)]$. From quantum electrodynamics (QED), one can deduce (roughly speaking) that

$$\mathcal{P}[(a, n_k) \to (b, n_k + 1_k)] = (n_k + 1)\mathcal{P}[(a, 0_k) \to (b, 1_k)]$$
$$\mathcal{P}[(b, n_k) \to (a, n_k - 1_k)] = n_k \mathcal{P}[(b, 1_k) \to (a, 0_k)].$$

(a) Use the condition of detailed balance for thermal equilibrium and the above results of QED to derive Planck's law of black-body radiation.

(b) Assume that Planck's law and the condition of detailed balance are valid. From these assumptions, is it possible to show that the above relations ought to be true in QED? If not, what further information is needed?

• **Problem 4.6** Electronic states near the top of an energy band in a simple cubic metal can be treated as free-particle states with energies given by a dispersion relation of the form $\varepsilon(\boldsymbol{k}) = \varepsilon_0 - A|\boldsymbol{k}|^2$, where A is a positive constant. Find the density of states $g(\varepsilon)$ for energies just below ε_0 in a cubic crystal of side L. The wavefunctions might be considered to vanish at the boundaries of the crystal or to satisfy periodic boundary conditions. How does $g(\varepsilon)$ depend on the choice of boundary conditions?

• **Problem 4.7** Consider a linear chain of $N+1$ atoms, each of mass m, with harmonic forces (which might be modelled by N springs of spring constant K) acting between nearest neighbours. Taking the atoms at each end of the chain to be fixed, find the dispersion relation for longitudinal waves on the chain. For the case when N is very large, find the density of states and an expression for the internal energy at temperature T. Evaluate the internal energy explicitly in the high- and low-temperature limits.

• **Problem 4.8** For a d-dimensional medium in which the lowest-lying vibrational modes of wavelength λ have frequencies given by a dispersion relation of the form $\omega \sim \lambda^{-s}$, find expressions for the internal energy and the specific heat at constant volume using the Debye approximation. What is the behaviour of $C_V(T)$ near absolute zero?

• **Problem 4.9** For the Debye model of phonons in a d-dimensional isotropic solid, obtain expressions for the limiting behaviour of C_V at very high and very low temperatures and for the leading corrections to these limits. Evaluate these expressions explicitly for $d = 3$.

• **Problem 4.10** Consider a three-dimensional isotropic solid formed by N atoms as an ensemble of harmonic oscillators, with a density of states given by the Debye approximation.

(a) Show that the zero point energy of the solid is $E_0 = 9NkT_\mathrm{D}/8$, where T_D is the Debye temperature.

(b) Show that $\int_0^\infty [C_V(\infty) - C_V(T)]\,\mathrm{d}T = E_0$.

• **Problem 4.11** A linear chain consists of $N+1$ atoms of type A, whose mass is m and $N+1$ atoms of type B, whose mass is M. The two types of atom alternate along the chain. The type-A atom at one end and the type-B atom at the other end are fixed a distance $L = (2N + 1)a$ apart, while the remaining atoms oscillate in the direction along the chain. Harmonic forces characterized by a 'spring constant' K act between neighbouring atoms.

(a) Find the dispersion relation for the normal modes of vibration of this chain. You should find that it has a low-frequency branch (called the acoustic branch) and a high-frequency branch (called the optical branch).

(b) Investigate the behaviour of the dispersion relation when m becomes equal to M.

(c) Identify circumstances under which the frequency of the optical branch is almost independent of wavelength.

Under the conditions of (c), we construct a simplified version of the excitation spectrum by assuming that all the optical modes have the same frequency ω_0, while the acoustic modes can be treated in the Debye approximation. Assuming that $\omega_0 > \omega_D$, the density of states is

$$g(\omega) = \begin{cases} \dfrac{L}{\pi c} & \omega < \omega_D \\ N\delta(\omega - \omega_0) & \omega > \omega_D \end{cases}$$

where c is the speed of sound.

(d) Find the Debye frequency ω_D for this model.

(e) Find the dependence of the specific heat on temperature for $kT \ll \hbar\omega_D$ and $kT \gg \hbar\omega_0$.

• **Problem 4.12** Consider a gas of N spin-zero bosons in a d-dimensional container of volume V, with a dispersion relation

$$\varepsilon_p = \alpha |\boldsymbol{p}|^s$$

where the constant α and the index s are both positive.

(a) Find expressions for the mean number of particles per unit volume in the ground state and the mean total number of particles in the excited states, in terms of the temperature T and the fugacity $z = e^{\beta\mu}$.

(b) Find the conditions on s and d for which Bose–Einstein condensation takes place.

(c) Find the equation of state for this gas.

(d) Find the relative population of the ground state N_0/N as a function of temperature, assuming that N/V is fixed.

(e) Find the entropy per unit volume of the gas in terms of T and z.

(f) Assuming that N/V is fixed, evaluate the discontinuity in the specific heat at the critical temperature. Show that for $d = 3$, $s = 2$, there is no discontinuity.

(g) Evaluate the discontinuity in the derivative of the specific heat at the critical temperature for the case $d = 3$, $s = 2$.

(h) When there is condensation, it is possible to regard the particles in the ground state and those in excited states as two distinct phases coexisting in the same container. Find the latent heat per particle for the transition between these two phases and verify that the Clausius–Clapeyron equation is obeyed.

• **Problem 4.13** Find the Helmholtz free energy and the Gibbs free energy of a gas of free spin-0 bosons in three dimensions. Show that, as the critical temperature is approached from above, the isothermal compressibility κ_T diverges as $\kappa_T \sim (T - T_c)^{-1}$.

• **Problem 4.14** The first observation of Bose–Einstein condensation in a dilute atomic vapour was achieved by M Anderson *et al* [*Science* **269** 198 (1995)]. The condensed state was produced in a vapour of ^{87}Rb that was confined by magnetic fields and evaporatively cooled. The first evidence of condensation appeared at a temperature of about 170 nK with 2.5×10^{12} atoms cm^{-3}. Compare the condensation temperature obtained in this seminal experiment with the critical temperature for an ideal Bose gas at the same density.

• **Problem 4.15** Consider a three-dimensional gas of bosons for which the single-particle energy is given by

$$\varepsilon_{p,n} = \frac{|\boldsymbol{p}|^2}{2m} + \alpha n$$

where α is a positive constant and $n = -j, \ldots, j$ is an integer.

(a) Find expressions, valid in the thermodynamic limit, for the pressure P and the mean number of particles per unit volume N/V in terms of the temperature T and the fugacity $z = \mathrm{e}^{\beta\mu}$.

(b) Write down the condition for Bose–Einstein condensation.

• **Problem 4.16** A model due to Landau treats ^{4}He at low temperatures as consisting of a superfluid (associated with, but not identical to, a Bose condensate) with energy E_s, say, together with 'quasiparticle' excitations whose energy $\hbar\omega(q)$ is given as a function of wavevector $q = |\boldsymbol{q}|$ by the dispersion relation sketched in figure 4.1. Neutron scattering data suggest that the slope of this curve at small q corresponds to a speed of sound $c = \omega/q \simeq 238$ m s^{-1}. The minimum at $q_R \simeq 1.9$ Å^{-1} is approximately described by $\Delta \equiv \hbar\omega(q_R)/k \simeq 8.6$ K and $\hbar\omega(q) \simeq \hbar\omega(q_R) + \hbar^2(q - q_R)^2/2m_R$, where m_R is 0.16 times the mass of a ^{4}He atom.

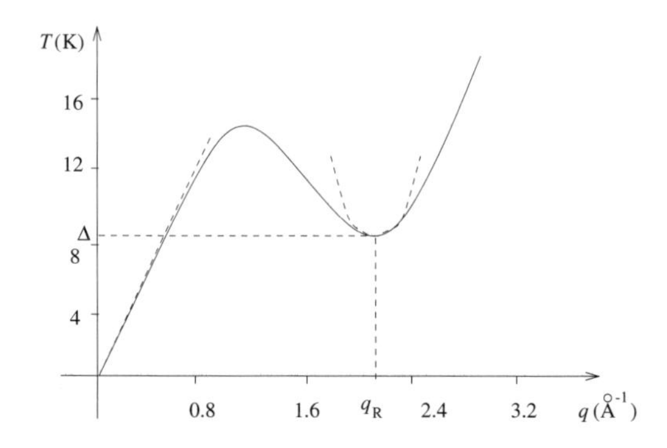

Figure 4.1. Dispersion relation for quasiparticle excitations in superfluid ^{4}He.

(a) Find an expression for the Helmholtz free energy of this system.

(b) For small q, the dispersion curve can be treated as linear and is therefore similar to the dispersion curve for phonons in a solid. Estimate the contribution of these phonon-like excitations to the specific heat $C_V^{\mathrm{ph}}(T)$ at constant volume.

(c) For values of q near q_{R}, the dispersion curve can be treated as parabolic, and excitations in this region are called *rotons*. Estimate the contribution of rotons to the specific heat at constant volume, $C_V^{\mathrm{rot}}(T)$.

(d) Compare the numerical values of C_V^{ph} and C_V^{rot} as functions of temperature.

• Problem 4.17
(a) For a gas of free electrons in d dimensions, compute the isothermal compressibility $\kappa_T(0)$ at zero temperature, in terms of the mean number of particles per unit volume, $n = \langle N \rangle / V$, and the Fermi energy ε_{F}.

(b) Estimate the Fermi temperature for metallic copper by treating the electrons as a gas of free particles. The density of copper is 8920 kg m^{-3}, its atomic weight is 63.5 and it may be assumed that there is one conduction electron per atom. Why is it correct, for all practical purposes, to treat the electrons in copper as a degenerate Fermi gas?

• Problem 4.18 Suppose that one has a gas of N free spin-$\frac{1}{2}$ fermions of mass m on a planar surface of area A. Obtain an explicit expression for the chemical potential of this gas as a function of temperature.

• Problem 4.19 An electron in a magnetic field H has energy $\varepsilon \mp \mu_{\mathrm{B}} H$, where ε is its kinetic energy and μ_{B} is the Bohr magneton, depending on whether its magnetic moment is parallel or antiparallel to the field. Find the zero-field paramagnetic susceptibility of a gas of electrons which is *nearly* degenerate, including the leading correction to the zero-temperature value.

• Problem 4.20

(a) A three-dimensional container of fixed total volume V is divided into two compartments by a wall which is freely movable and heat-conducting, but impermeable to gas particles. One compartment contains N particles of a non-relativistic Fermi gas A, which have spin $s_A = \frac{1}{2}$ and the other contains N particles of a gas B, which have spin $s_B = \frac{3}{2}$. To simplify matters, we make the somewhat unrealistic assumption that the two species have identical masses $m_A = m_B = m$. Write down the conditions for this system to be in thermal and mechanical equilibrium and use these conditions to find the volume ratio V_A/V_B at very high and very low temperatures.

(b) A Fermi gas of N spin-$\frac{1}{2}$ particles is initially confined, at zero temperature, to a compartment of volume V_0 in a thermally isolated container of total volume $V_0 + \Delta V$, the remaining volume ΔV being empty. When the partition separating the two compartments is removed, the gas expands to fill the whole volume, and eventually reaches a new equilibrium state. Estimate the temperature of this final state, assuming that $\Delta V/V_0 \ll 1$.

• Problem 4.21 Consider an ideal gas in a d-dimensional box of volume V. The energy of each particle is $\varepsilon_{\boldsymbol{p}} = \alpha|\boldsymbol{p}|^s$, with $\alpha > 0$ and $s > 0$.

(a) Show that, at all temperatures, the internal energy of the gas is related to its pressure by $PV = \sigma U$, where σ is a constant which has the same value for bosons, fermions and Maxwell–Boltzmann particles. How is radiation pressure related to energy density in d dimensions?

(b) For the case of fermions, find the average energy per particle at low temperatures.

• Problem 4.22 A d-dimensional container is divided into two regions A and B by a fixed wall. The two regions contain identical Fermi gases of spin-$\frac{1}{2}$ particles which have a magnetic moment τ. In region A there is a magnetic field of strength H, but there is no field in region B. Initially, the entire system is at zero temperature, and the numbers of particles per unit volume are the same in both regions. If the wall is now removed, particles may flow from one region to the other. Determine the direction in which particles begin to flow, and how the answer depends on the dimensionality d.

• Problem 4.23 Solution of the Schrödinger equation for an electron in a uniform magnetic field $\boldsymbol{B} = B\hat{\boldsymbol{z}}$ yields a set of energy levels (known as Landau levels) given by

$$\varepsilon(p_z, j) = \frac{p_z^2}{2m} + \frac{\hbar e B}{mc}\left(j + \tfrac{1}{2}\right)$$

if the electron's magnetic moment is neglected. Here e is the electronic charge, p_z $(-\infty < p_z < \infty)$ is the momentum of the electron in the $\hat{z}$ direction and j is a quantum number that takes the values $0, 1, \ldots, \infty$. These energy levels are degenerate, the number of states for a given level being $g = 2V^{2/3}eB/2\pi\hbar c$.

(a) Write down the grand canonical partition function for a gas of electrons in a magnetic field.

(b) Find the leading behaviour of the fugacity $z = e^{\beta\mu}$ in the high-temperature limit.

(c) Also in that limit, show that the zero-field diamagnetic susceptibility varies as $-1/T$.

• **Problem 4.24** A cavity containing a gas of electrons has a small hole, of area δA, through which electrons can escape. External electrodes are so arranged that, if the potential energy of an electron inside the cavity is taken as zero, then its potential energy outside the cavity is $\mathcal{V}$. Estimate the current carried by the escaping electrons, assuming (i) that a constant number density of electrons is maintained inside the cavity (for example by thermionic emission from a heated filament), (ii) that these electrons are in thermal equilibrium at a temperature T and chemical potential μ, (iii) that electrons moving towards the hole escape if they have an energy greater than $\mathcal{V}$ and (iv) that $\mathcal{V} - \mu \gg kT$.

• **Problem 4.25** Consider a gas of N electrons contained in a box of volume V, whose walls have N_0 absorbent sites, each of which can absorb one electron. Let $-\varepsilon_0$ be the energy of an electron absorbed at one of these sites and $|\boldsymbol{p}|^2/2m$ the energy of a free electron.

(a) For $N > N_0$, find the limits as $T \to 0$ and $T \to \infty$ of the number N_a of absorbed electrons and the number N_f of free electrons.

(b) For $N = N_0$, find the chemical potential $\mu(T)$ and the particle numbers $N_a(T)$ and $N_f(T)$ at low temperatures.

• **Problem 4.26** The Dirac wave equation for relativistic spin-$\frac{1}{2}$ particles has solutions with energies extending both to $+\infty$ and to $-\infty$. An interpretation suggested by Dirac is that the vacuum state is one in which all the negative-energy states are filled, while all the positive-energy states are empty. We shall call such a state a 'Dirac sea'. A negative-energy particle which absorbs enough energy can jump to a positive-energy state, leaving a 'hole' in the sea. While this picture is ultimately unsatisfactory as a model of the relativistic vacuum, its statistical-mechanical properties are reasonably well defined. Indeed, it can serve as a simple model for electrons in the almost full valence band and the almost empty conduction band in a semiconductor, in circumstances where only those states near the band edges are important.

Suppose, then, that we have a continuous spectrum of energies ε whose density of states is

$$g(\varepsilon) = \begin{cases} A(\varepsilon - \varepsilon_0)^{1/2} & \varepsilon_0 < \varepsilon < \infty \\ B(-\varepsilon)^{1/2} & -\infty < \varepsilon < 0 \end{cases}$$

where A and B are constants, there being no states with $0 < \varepsilon < \varepsilon_0$.

(a) When this system is at a temperature T and has chemical potential μ, show that the probability of finding a state of energy $\mu + \alpha$ occupied by a particle is equal to that of finding a state of energy $\mu - \alpha$ unoccupied.

(b) If the system is a Dirac sea at $T = 0$, find an equation for the chemical potential μ at all temperatures. For the case $A = B$, obtain explicit expressions for the numbers of particles and holes which are present at low temperatures.

(c) Suppose now that at $T = 0$, in addition to the Dirac sea, there are N_0 particles occupying positive-energy states ($\varepsilon > \varepsilon_0$), which we shall call a 'Fermi sea'. Find the chemical potential μ at $T = 0$. Obtain the chemical potential for low temperatures and identify the correction due to the Dirac sea.

• **Problem 4.27**
(a) Consider a highly relativistic gas composed of several species of bosons, fermions and their antiparticles. Assume that particles which are their own antiparticles can be created and annihilated in arbitrary numbers by collisions, while particles which are not their own antiparticles can be created and annihilated in particle–antiparticle pairs. For the case in which equal numbers of particles and antiparticles are present, obtain general expressions for the energy density, the entropy density and the total number density of particles in thermal equilibrium.

(b) Consider a neutral plasma at temperatures such that electrons are highly relativistic, but the nuclei to which they would be bound at low temperatures are non-relativistic. The plasma also contains positrons and photons maintained in thermal equilibrium by the reaction $e^+ + e^- \rightleftharpoons 2\gamma$. Find the leading correction to the energy density of relativistic particles due to the excess of electrons which balance the charge of the nuclei.

• **Problem 4.28** Let $P^{\mathrm{BE}}(T, V, \mu)$ and $P^{\mathrm{FD}}(T, V, \mu)$ be the pressures of ideal non-relativistic Bose–Einstein and Fermi–Dirac gases respectively. Show that, at equal temperature and volume, there exists a relation of the form

$$P^{\mathrm{BE}}(T, V, \mu) = \sum_{k=0}^{\infty} a_k P^{\mathrm{FD}}(T, V, \mu_k)$$

and find the values of a_k and μ_k.

• **Problem 4.29** Find the first quantum correction to the classical equation of state of an ideal gas expressed as a virial expansion:

$$\frac{PV}{NkT} = 1 + B_2 v^{-1} + \cdots$$

where v is the specific volume. Consider both a bosonic and a fermionic gas and interpret the difference between the signs of B_2^{BE} and B_2^{FD}.

4.2 Answers

• Problem 4.1

(a) Let us denote the orbital wavefunctions of the nth energy level by $\psi_{n,m}$ for $m = 1, \ldots, 2n+1$. If the two particles have energies $n_1\varepsilon$ and $n_2\varepsilon$, there are symmetric states of the form $\psi_{n_1,m_1}(1)\psi_{n_2,m_2}(2) + \psi_{n_2,m_2}(1)\psi_{n_1,m_1}(2)$ and antisymmetric states of the form $\psi_{n_1,m_1}(1)\psi_{n_2,m_2}(2) - \psi_{n_2,m_2}(1)\psi_{n_1,m_1}(2)$. If $n_1 \neq n_2$, then there are $(2n_1+1)(2n_2+1)$ states of each kind while, if $n_1 = n_2 = n$, then there are $(n+1)(2n+1)$ symmetric states and $n(2n+1)$ antisymmetric states. The microcanonical partition function $\Omega(E)$ is the number of distinct states of total energy $E = N\varepsilon$, so we need to consider pairs of quantum numbers of the form $n_1 = n$ and $n_2 = N - n$, with $n = 0, \ldots, N/2$ if N is even, or $n = 0, \ldots, (N-1)/2$ if N is odd. Larger values of n correspond to interchanging n_1 and n_2 and and hence to a relabelling of the same set of states.

Consider first the case when N is odd. Then we never encounter equal values of n_1 and n_2. For spin-0 bosons, we use the summation formulae given in appendix A to obtain

$$\Omega_{\text{odd}}^{\text{BE}}(E) = \sum_{n=0}^{(N-1)/2} (2n+1)[2(N-n)+1] = \tfrac{1}{3}N^3 + N^2 + \tfrac{7}{6}N + \tfrac{1}{2}.$$

For spin-$\frac{1}{2}$ fermions, we must take into account that the four possible spin states can be grouped into a triplet of symmetric states and a single antisymmetric state. The total wavefunction must be the product of a symmetric orbital state with an antisymmetric spin state or vice versa. Since for odd N the numbers of symmetric and antisymmetric orbital states are the same, we find that

$$\Omega_{\text{odd}}^{\text{FD}}(E) = 4\sum_{n=0}^{(N-1)/2} (2n+1)[2(N-n)+1] = 4\Omega_{\text{odd}}^{\text{BE}}(E).$$

When N is even, we have to take special account of the case that $n_1 = n_2 = N/2$. From the above discussion, we find that

$$\begin{aligned}
\Omega_{\text{even}}^{\text{BE}} &= \sum_{n=0}^{N/2-1} (2n+1)[2(N-n)+1] + \frac{(N+1)(N+2)}{2} \\
&= \tfrac{1}{3}N^3 + N^2 + \tfrac{5}{3}N + 1 \\
\Omega_{\text{even}}^{\text{FD}} &= 4\sum_{n=0}^{N/2-1} (2n+1)[2(N-n)+1] + 3\frac{N(N+1)}{2} + \frac{(N+1)(N+2)}{2} \\
&= 4(\tfrac{1}{3}N^3 + N^2 + \tfrac{11}{12}N + \tfrac{1}{4}).
\end{aligned}$$

The artificial device of Maxwell–Boltzmann statistics treats identical particles as having a quantum-mechanical spectrum, but being indistinguishable in the classical sense (see problem 3.4 for a related discussion). Thus, we can first find the partition function for distinguishable particles and then divide by 2! to obtain the 'correct Boltzmann counting' of states. The states of two distinguishable particles can be described by the $(2n_1+1)(2n_2+1)$ wavefunctions $\psi_{n_1,m_1}(1)\psi_{n_2,m_2}(2)$, where now all values of n_1 and n_2 correspond to distinct states, with a further spin-degeneracy factor of $(2s+1)^2$. Thus, for both even and odd values of N, we obtain

$$\begin{aligned}\Omega^{\mathrm{MB}}(E) &= \tfrac{1}{2}(2s+1)^2 \sum_{n=0}^{N}(2n+1)[2(N-n)+1] \\ &= (2s+1)^2(\tfrac{1}{3}N^3 + N^2 + \tfrac{7}{6}N + \tfrac{1}{2}).\end{aligned}$$

(b) At this point, it is convenient to divide each microcanonical partition function by the appropriate spin-degeneracy factor, defining $\tilde{\Omega} = (2s+1)^{-2}\Omega$, and to observe that

$$\tilde{\Omega}^{\mathrm{BE}}_{\mathrm{odd}}(E) = \tilde{\Omega}^{\mathrm{FD}}_{\mathrm{odd}}(E) = \tilde{\Omega}^{\mathrm{MB}}(E)$$

while

$$\tilde{\Omega}^{\mathrm{BE}}_{\mathrm{even}}(E) = \tilde{\Omega}^{\mathrm{MB}}(E) + \frac{N+1}{2} \qquad \text{and} \qquad \tilde{\Omega}^{\mathrm{FD}}_{\mathrm{even}}(E) = \tilde{\Omega}^{\mathrm{MB}}(E) - \frac{N+1}{4}.$$

The canonical partition functions are given by

$$Z(T) = \sum_E \Omega(E)\,\mathrm{e}^{-\beta E} = \sum_{N=0}^{\infty} \Omega(E)\,\mathrm{e}^{-N\beta\varepsilon}$$

since the degeneracy of each energy level for the whole system is, of course, just the microcanonical partition function. The required sums can be evaluated by means of the formula

$$\sum_{N=0}^{\infty} N^n\,\mathrm{e}^{-\lambda N} = \left(-\frac{\partial}{\partial\lambda}\right)^n \sum_{N=0}^{\infty} \mathrm{e}^{-\lambda N} = \left(-\frac{\partial}{\partial\lambda}\right)^n \frac{1}{1-\mathrm{e}^{-\lambda}}.$$

In this way, we find that

$$\begin{aligned}\tilde{Z}^{\mathrm{MB}}(T) &= \frac{1}{2}\frac{(1+\mathrm{e}^{-\beta\varepsilon})^2}{(1-\mathrm{e}^{-\beta\varepsilon})^4} \\ \tilde{Z}^{\mathrm{BE}}(T) &= \tilde{Z}^{\mathrm{MB}}(T) + \frac{\delta Z(T)}{2} \\ \tilde{Z}^{\mathrm{FD}}(T) &= \tilde{Z}^{\mathrm{MB}}(T) - \frac{\delta Z(T)}{4}\end{aligned}$$

where

$$\delta Z(T) = \sum_{N\ \mathrm{even}} (N+1)\,\mathrm{e}^{-N\beta\varepsilon} = \sum_{M=0}^{\infty}(2M+1)\,\mathrm{e}^{-2M\beta\varepsilon} = \frac{1+\mathrm{e}^{-2\beta\varepsilon}}{(1-\mathrm{e}^{-2\beta\varepsilon})^2}$$

and, in each case, $Z(T) = (2s+1)^2\tilde{Z}(T)$.

(c) For the microcanonical ensemble, we clearly have

$$\tilde{\Omega}^{\text{BE}}(E) \geq \tilde{\Omega}^{\text{MB}}(E) \geq \tilde{\Omega}^{\text{FD}}(E).$$

This result is rather general and arises simply from the fact that, for a given set of single-particle states, the number of symmetric multiparticle combinations is greater than or equal to the number of antisymmetric combinations. Clearly, the same inequalities hold for the canonical partition functions. For the microcanonical ensemble, the terms of order $N^3 = (E/\varepsilon)^3$ and $N^2 = (E/\varepsilon)^2$ are identical for each of the reduced partition functions $\tilde{\Omega}$, so, when the energy is large, all three partition functions become approximately identical, apart from the spin-degeneracy factors. This is because there are many degenerate states, only a small fraction of which are affected by the symmetry restrictions. Similarly, at high temperatures, the canonical partition functions have leading contributions of order T^4, while they differ only by terms of order T^2. It is under these circumstances that the fictitious Maxwell–Boltzmann statistics provide a useful approximation.

• Problem 4.2

(a) Using the occupation number representation, the energy of a multiparticle state is $E = \sum_i n_i \varepsilon_i$ and the number of particles is $N = \sum_i n_i$, so the grand canonical partition function is

$$\mathcal{Z} = \sum_{\{n_i\}} \exp\left(-\beta \sum_i n_i(\varepsilon_i - \mu)\right) = \prod_i \sum_{n_i=0}^{M} \mathrm{e}^{-\beta(\varepsilon_i-\mu)n_i} = \prod_i \frac{1-q_i^{M+1}}{1-q_i}$$

where $q_i = \mathrm{e}^{-\beta(\varepsilon_i-\mu)}$.

Thermodynamic functions are obtained from $\ln \mathcal{Z}$. Assuming that the energy levels are very closely spaced, the sum over these energy levels is accurately approximated by an integral. We obtain

$$\ln \mathcal{Z} = \sum_i \ln\left(\frac{1-q_i^{M+1}}{1-q_i}\right) \simeq g_s \int_0^\infty \mathrm{d}\varepsilon\, g(\varepsilon) \ln\left(\frac{1-q(\varepsilon)^{M+1}}{1-q(\varepsilon)}\right)$$

where, for particles of spin s, $g_s = 2s+1$ is the spin degeneracy of each energy level and $g(\varepsilon)$ is the density of states—the number of energy levels per unit interval of energy. To evaluate this quantity, we suppose that our system is enclosed in a d-dimensional hypercubic box of side L and impose as a boundary condition that wavefunctions vanish at the walls. Under these conditions, the allowed values of the momentum are $\boldsymbol{p} = (\pi\hbar/L)\boldsymbol{n}$, where $\boldsymbol{n}$ is a vector of d positive integers, in terms of which the single-particle energy is $\varepsilon = |\boldsymbol{n}|^2\pi^2\hbar^2/2mL^2$. The number of allowed states with energy less than or equal to ε is 2^{-d} times the volume of a d-dimensional sphere of radius

$R = \sqrt{2mL^2\varepsilon/\pi^2\hbar^2}$, which is (see appendix A) $2\pi^{d/2}R^d/d\Gamma(d/2)$. Thus we have[1]

$$g(\varepsilon) = \frac{1}{2^d}\frac{\mathrm{d}}{\mathrm{d}\varepsilon}\left(\frac{2\pi^{d/2}R^d}{d\Gamma(d/2)}\right) = c_d V \varepsilon^{d/2-1}$$

where $c_d = (m/2\pi\hbar^2)^{d/2}/\Gamma(d/2)$ and $V = L^d$ is the volume. With the change of variable $\varepsilon = kTx$, we obtain

$$\ln \mathcal{Z} = c_d g_s V(kT)^{d/2} f_M(z)$$

$$f_M(z) = \int_0^\infty \mathrm{d}x\, x^{d/2-1} \ln\left(\frac{1-(z\,\mathrm{e}^{-x})^{M+1}}{1-z\,\mathrm{e}^{-x}}\right)$$

and the thermodynamic functions are given by

$$P = \frac{kT}{V}\ln \mathcal{Z} = c_d g_s (kT)^{d/2+1} f_M(z)$$

$$U = -\left(\frac{\partial(\ln \mathcal{Z})}{\partial\beta}\right)_{z,V} = \frac{d}{2} c_d g_s V(kT)^{d/2+1} f_M(z)$$

$$\langle N\rangle = z\left(\frac{\partial(\ln \mathcal{Z})}{\partial z}\right)_{\beta,V} = c_d g_s V(kT)^{d/2} z \frac{\mathrm{d}[f_M(z)]}{\mathrm{d}z}.$$

We note the relation $U = (d/2)PV$, valid for both classical and quantum gases when the particles have only translational energy. Also, in the limit of high temperatures, with a fixed average number of particles per unit volume, we must have $z\,\mathrm{d}f/\mathrm{d}z \to 0$. Since $\mathrm{d}f/\mathrm{d}z > 0$, this implies that $z \to 0$ and then we discover that $f(z) \to 0$ also. We then have $f_M(z) \simeq \Gamma(d/2)z$ and we recover the equation of state $PV = \langle N\rangle kT$ of a classical ideal gas.

(b) The expression given in (a) for $\mathcal{Z}$ is a product of independent unnormalized probabilities for all the occupation numbers, so it is easy to write down the normalized probability distribution for one of them:

$$\mathcal{P}_i(n_i, T, \mu) = \frac{\mathrm{e}^{-\beta(\varepsilon_i-\mu)n_i}}{\sum_{n=0}^M \mathrm{e}^{-\beta(\varepsilon_i-\mu)n}} = \frac{\mathrm{e}^{-\beta(\varepsilon_i-\mu)n_i}}{L(q_i)}$$

where $L(q_i) = (1-q_i^{M+1})/(1-q_i)$. Evidently, $L(q_i)$ plays the role of a partition function for the distribution of occupation numbers of a single-particle state. The variance of this distribution is

$$(\Delta n_i)^2 = \sum_{n_i=0}^M n_i^2 \mathcal{P}_i(n_i, T, \mu) - \left(\sum_{n_i=0}^M n_i \mathcal{P}_i(n_i, T, \mu)\right)^2.$$

[1] Note that, if we use periodic boundary conditions, the allowed momentum values are $\boldsymbol{p} = (2\pi\hbar/L)\boldsymbol{n}$, the integers $\boldsymbol{n}$ taking both positive and negative values. Then we use the whole volume of a sphere of radius $R = \sqrt{2mL^2\varepsilon/4\pi^2\hbar^2}$, obtaining the same result for $g(\varepsilon)$.

This can be expressed as

$$(\Delta n_i)^2 = \left(\frac{\partial^2\{\ln[L(q_i)]\}}{\partial(\beta\varepsilon_i)^2}\right)_z = \frac{q_i^2 L(q_i)L''(q_i) + q_i L(q_i)L'(q_i) - q_i^2 L'(q_i)^2}{L(q_i)^2}$$

where the prime denotes $\mathrm{d}/\mathrm{d}q_i$. As will be seen in (d), the physically interesting cases are $M = 1$, corresponding to Fermi–Dirac statistics, and $M \to \infty$ with $q < 1$, corresponding to Bose–Einstein statistics. For these cases, we find that $(\Delta n_i)^2_{\mathrm{FD}} = q_i/(1 + q_i)^2$ and $(\Delta n_i)^2_{\mathrm{BE}} = q_i/(1 - q_i)^2$.

(c) The occupation number $\langle n(\varepsilon)\rangle$ is the average number of particles with energy ε. Of course, this will be zero unless ε is one of the eigenvalues ε_i, in which case it is equal to $g_s\langle n_i\rangle$, which we evaluate as

$$\langle n(\varepsilon)\rangle = g_s \sum_{n=0}^{M} n\mathcal{P}(n, T, \mu) = g_s q \frac{L'(q)}{L(q)} = \frac{g_s}{\mathrm{e}^{\beta(\varepsilon-\mu)} - 1} - \frac{g_s(M+1)}{\mathrm{e}^{(M+1)\beta(\varepsilon-\mu)} - 1}.$$

In finding the zero-temperature limit of this occupation number, we must first consider that, in order to maintain a fixed average number of particles per unit volume $\langle N\rangle/V$, the chemical potential μ must be allowed to vary appropriately with temperature. From the expression found in (a) for $\langle N\rangle$, we see that this means that $\mathcal{N}(z) \propto (kT)^{-d/2}$, where $\mathcal{N}(z) = z\,\mathrm{d}[f_M(z)]/\mathrm{d}z$, and thus that z must become large as $T \to 0$. To see how large z must be, we first rewrite the integration variable x as $x = (\ln z)y$, which gives

$$\mathcal{N}(z) = (\ln z)^{d/2} \int_0^\infty \mathrm{d}y\, y^{d/2-1} \left(\frac{1}{z^{y-1} - 1} - \frac{M+1}{z^{(M+1)(y-1)} - 1}\right).$$

Inside the remaining integral, we can take the limit $z \to \infty$ by noting that both z^{y-1} and $z^{(M+1)(y-1)}$ tend to 0 if $y < 1$ or to ∞ if $y > 1$. In this limit, therefore, we have

$$\mathcal{N}(z) \simeq (\ln z)^{d/2} \int_0^1 \mathrm{d}y\, y^{d/2-1}\, M = \frac{2M}{d}(\ln z)^{d/2}.$$

We therefore find that $\ln z = \mu/kT \propto 1/kT$, so μ approaches a positive constant μ_0. The integral for $\mathcal{N}(z)$ involves, of course, the occupation number $\langle n(\varepsilon)\rangle$ in a different notation, so, by the same argument, we find that $\langle n(\varepsilon)\rangle \to g_s M\theta(\varepsilon - \mu_0)$. As might be expected, enough low-energy states are fully occupied to accommodate the number of particles in the gas, while higher-energy states are all unoccupied. This is a generalized 'Fermi sea'.

(d) This model reproduces a Fermi–Dirac gas if we set the spin s to an half-odd-integer value and take $M = 1$. In particular, the average occupation number becomes $\langle n(\varepsilon)\rangle_{\mathrm{FD}} = g_s/(\mathrm{e}^{\beta(\varepsilon-\mu)} + 1)$. To recover a Bose–Einstein gas, we must set s to an integer value and take $M \to \infty$. For bosons, μ is always less than or equal to the minimum single-particle energy, $\mu \le \epsilon_{\mathrm{min}}$, which is zero in our case. Hence μ is always negative, so $q = \mathrm{e}^{\beta(\varepsilon-\mu)} > 1$ and $\langle n(\varepsilon)\rangle_{\mathrm{BE}} = g_s/(\mathrm{e}^{\beta(\varepsilon-\mu)} - 1)$.

In this case, $\mu \to 0$ as $T \to 0$ and $\mathcal{N}(z)$ becomes large on account of the singularity in the integrand at $y = 1$ when $z = 1$. The ground states then become macroscopically occupied, which is the phenomenon of Bose–Einstein condensation. In either case, the specification of the states described in the question is in accord with the rules of quantum mechanics. Given a set of occupation numbers, we can form a product of single-particle wavefunctions $\psi_1(1)\psi_1(2)\cdots\psi_1(n_1)\psi_2(n_1+1)\cdots$, in which ψ_1 appears n_1 times, ψ_2 appears n_2 times and so on. For any values of the n_i, there is exactly one totally symmetric wavefunction that can be formed by permuting the particle labels in this product and this gives a bosonic state. On the other hand, there is exactly one totally antisymmetric wavefunction that can be formed if all the n_i are either zero or one, but none if any n_i is greater than one, so this yields a fermionic state, consistent with the Pauli exclusion principle.

We cannot recover a Maxwell–Boltzmann gas, because the counting of states works differently in this case. For distinguishable particles, the simple product wavefunction gives a valid state and, for a total of $N = \sum_i n_i$ particles, there are $N!/(n_1!n_2!\cdots)$ permutations of particle labels which give distinct states. Consider for simplicity, the canonical partition function for distinguishable spin-0 particles

$$Z_{\text{dist}}(N,T) = \sum_{\{n_i\}} \frac{N!}{n_1!n_2!\cdots} \exp\left(-\beta\sum_i \varepsilon_i n_i\right) = \left(\sum_i \mathrm{e}^{-\beta\varepsilon_i}\right)^N .$$

For Maxwell–Boltzmann statistics, we define $Z_{\text{MB}}(N,T) = Z_{\text{dist}}(N,T)/N!$. From a quantum-mechanical point of view, this seems to assign a peculiar fractional degeneracy $g = 1/(n_1!n_2!\cdots) < 1$ to each state. However, the final expression for $Z_{\text{dist}}(N,T)$ is in fact a product of N single-particle partition functions, such as we encounter for a classical gas. Maxwell–Boltzmann statistics therefore treat these N particles as indistinguishable in a classical sense, since dividing by $N!$ does yield the 'correct Boltzmann counting' of states of a classical gas. Clearly, however, the model studied here cannot reproduce this method of counting.

• **Problem 4.3** We begin by writing the grand canonical partition function as a sum over multi-particle states specified by occupation numbers of single-particle states. For bosons and fermions, it is

$$\mathcal{Z} = \sum_{\{n_i\}} \exp\left(-\beta\sum_i n_i(\varepsilon_i - \mu)\right) = \prod_i \sum_{n_i} \mathrm{e}^{-\beta(\varepsilon_i-\mu)n_i}$$

while for Maxwell–Boltzmann particles (see problem 4.2 for a discussion of the counting of states) it is

$$\mathcal{Z} = \prod_i \sum_{n_i} \frac{1}{n_i!} \mathrm{e}^{-\beta(\varepsilon_i-\mu)n_i} .$$

In each case, we see that $\mathcal{Z}$ has the form $\prod_i L_i$, where L_i can be regarded as the partition function for the occupancy of the ith single-particle state. Consequently, we can pick out the probabilities

$$\mathcal{P}_i(n_i, T, \mu) = \begin{cases} \dfrac{1}{L_i}\,\mathrm{e}^{-\beta(\varepsilon_i-\mu)n_i} & \text{for bosons and fermions} \\ \dfrac{1}{L_i}\dfrac{\mathrm{e}^{-\beta(\varepsilon_i-\mu)n_i}}{n_i!} & \text{for Maxwell–Boltzmann particles} \end{cases}$$

with ($a = 1$ for fermions and $a = -1$ for bosons)

$$L_i = \sum_{n_i} \mathrm{e}^{-\beta(\varepsilon_i-\mu)n_i} = (1 + a\,\mathrm{e}^{-\beta(\varepsilon_i-\mu)})^a$$

for bosons and fermions and

$$L_i = \sum_{n_i} \frac{1}{n_i!}\,\mathrm{e}^{-\beta(\varepsilon_i-\mu)n_i} = \exp(\mathrm{e}^{-\beta(\varepsilon_i-\mu)})$$

for Maxwell–Boltzmann particles. For fermions, of course, the sum is over $n_i = 0, 1$ while, for bosons and Maxwell–Boltzmann particles $n_i = 0, 1, \ldots, \infty$.

The average occupation numbers are

$$\langle n_i\rangle = \sum_{n_i} n_i \mathcal{P}_i = \frac{\partial L_i}{\partial(\beta\mu)} = \begin{cases} \dfrac{1}{\mathrm{e}^{\beta(\varepsilon_i-\mu)} + a} & \text{for bosons and fermions} \\ \mathrm{e}^{-\beta(\varepsilon_i-\mu)} & \text{for Maxwell–Boltzmann particles.} \end{cases}$$

In each case, it is a simple matter to express $\mathrm{e}^{\beta(\varepsilon_i-\mu)}$ and hence $\mathcal{P}_i$ in terms of $\langle n_i\rangle$, with the results

$$\mathcal{P}_i^{\mathrm{BE}} = \frac{1}{1+\langle n_i\rangle}\left(\frac{\langle n_i\rangle}{1+\langle n_i\rangle}\right)^{n_i} \qquad \mathcal{P}_i^{\mathrm{FD}} = (1-\langle n_i\rangle)\left(\frac{\langle n_i\rangle}{1-\langle n_i\rangle}\right)^{n_i}$$

$$\mathcal{P}_i^{\mathrm{MB}} = \frac{\mathrm{e}^{-\langle n_i\rangle}\,\langle n_i\rangle^{n_i}}{n_i!}.$$

We see that the probability distribution for Maxwell–Boltzmann particles has the Poisson form associated with purely random processes, while those for bosons and fermions are rather different. We can understand this in a rough way by imagining a process of populating the ith state in an ensemble of identical systems with particles from a reservoir. A Maxwell–Boltzmann particle can enter any system in the ensemble with equal probability. A fermion, of course, can enter only a system in which the state is not already occupied, while a boson has an enhanced probability of entering a system in which the state is already occupied.

The variance of the distribution is

$$(\Delta n_i)^2 = \sum_{n_i} n_i^2 \mathcal{P}_i - \left(\sum_{n_i} n_i \mathcal{P}_i\right)^2 = \frac{\partial^2 L_i}{\partial(\beta\mu)^2}.$$

We easily find that

$$\left(\frac{\Delta n_i}{\langle n_i\rangle}\right)_{\mathrm{BE}} = \sqrt{\frac{1}{\langle n_i\rangle}+1} \qquad \left(\frac{\Delta n_i}{\langle n_i\rangle}\right)_{\mathrm{FD}} = \sqrt{\frac{1}{\langle n_i\rangle}-1}$$

$$\left(\frac{\Delta n_i}{\langle n_i\rangle}\right)_{\mathrm{MB}} = \sqrt{\frac{1}{\langle n_i\rangle}}.$$

For Maxwell–Boltzmann particles, we have the familiar result for random processes that the relative fluctuations in particle numbers are proportional to $1/\sqrt{\langle n\rangle}$. For bosons and fermions (which are genuine quantum-mechanical particles), this is approximately true only when $\langle n\rangle$ is very small.

• **Problem 4.4** All the desired results can be obtained as derivatives of the grand canonical partition function which, for photons, is

$$\mathcal{Z}(T,V) = \sum_{\{n_i\}} \mathrm{e}^{-\beta\varepsilon_i n_i} = \prod_i \sum_{n_i=0}^{\infty} \mathrm{e}^{-\beta\varepsilon_i n_i} = \prod_i (1-\mathrm{e}^{-\beta\varepsilon_i})^{-1}$$

where i labels single-particle states of energy ε_i. Photons have zero chemical potential because the number of photons in the cavity is not conserved and cannot appear in the equilibrium probability density. Since the energy levels of photons in a cavity of macroscopic size are very closely spaced, we can write

$$\ln[\mathcal{Z}(T,V)] = -\sum_i \ln(1-\mathrm{e}^{-\beta\varepsilon_i}) = -\int_0^\infty \mathrm{d}\varepsilon\, g(\varepsilon)\ln(1-\mathrm{e}^{-\beta\varepsilon})$$

where $g(\varepsilon)$ is the number of states per unit energy interval. Assuming that photon wavefunctions vanish at the walls of a hypercubical cavity of side L, the allowed momenta are $\boldsymbol{p} = (\pi\hbar/L)\boldsymbol{n}$, where $\boldsymbol{n}$ is a vector of positive integers. Since the energy of a photon is $|\boldsymbol{p}|c$, the number of states with energy less than or equal to ε is $(d-1)(1/2^d)$ multiplied by the volume of a d-dimensional sphere of radius $L\varepsilon/\pi\hbar c$, where the factor $d-1$ allows for the polarization states. Thus, we have

$$g(\varepsilon) = \frac{\mathrm{d}}{\mathrm{d}\varepsilon}\left[\frac{d-1}{2^{d-1}}\frac{\pi^{d/2}}{d\Gamma(d/2)}\left(\frac{L\varepsilon}{\pi\hbar c}\right)^d\right] = \frac{(d-1)\,V\varepsilon^{d-1}}{2^{d-1}\,\Gamma(d/2)(\hbar c\sqrt{\pi})^d} \equiv a_d V\varepsilon^{d-1}$$

and

$$\ln[\mathcal{Z}(T,V)] = -a_d V\int_0^\infty \mathrm{d}\varepsilon\,\varepsilon^{d-1}\ln(1-\mathrm{e}^{-\beta\varepsilon}) = a_d V I_d (kT)^d$$

where

$$I_d = -\int_0^\infty \mathrm{d}x\, x^{d-1}\ln(1-\mathrm{e}^{-x}).$$

The thermodynamic quantities are

$$P = \left(\frac{kT}{V}\right) \ln \mathcal{Z} = a_d I_d (kT)^{d+1}$$

$$u = -\frac{1}{V}\frac{\partial(\ln \mathcal{Z})}{\partial \beta} = d a_d I_d (kT)^{d+1}$$

$$s = \frac{1}{V}\left(\frac{\partial(PV)}{\partial T}\right)_V = (d+1) a_d I_d k (kT)^d$$

$$C_V = \left(\frac{\partial u}{\partial T}\right)_V = d(d+1) a_d I_d k (kT)^d = T\left(\frac{\partial s}{\partial T}\right)_V .$$

In three dimensions, we have $I_3 = \pi^4/45$ and $a_3 = 1/\pi^2(\hbar c)^3$, so

$$P = \frac{\pi^2}{45(\hbar c)^3}(kT)^4 = \tfrac{1}{3}u \qquad s = \frac{4\pi^2}{45}\left(\frac{kT}{\hbar c}\right)^3 k = \tfrac{1}{3}C_V .$$

These results (which are sometimes expressed in terms of the Stefan–Boltzmann constant $\sigma = \pi^2 k^4/60\hbar^3 c^2 \simeq 5.67 \times 10^{-8}$ W m^{-2} K^{-4}) are useful, for example, in the study of cosmological models where much of the matter content of the Universe is in the form of photons and other highly relativistic particles.

• Problem 4.5

(a) In general, there will be many photon states k into which the photon can be emitted, and from which a photon can be absorbed during a particular atomic transition. Indeed, it is only under these circumstances that a transition rate per unit time can be approximately defined. Equally, there may well be several atomic states of energy E_a or E_b. In order for thermal equilibrium to exist, the net rate of emissions $a \to b +$ photon must equal the net rate of absorptions $b +$ photon $\to a$. The net transition rate is the product of the probability of finding the atom in the specified initial state and the transition probability per unit time. Since the probabilities of finding the atom in states a and b are in the ratio of the Boltzmann factors $\mathrm{e}^{-\beta E_a}$ and $\mathrm{e}^{-\beta E_b}$, we have

$$\mathrm{e}^{-\beta E_a} \sum_{b,k} \mathcal{P}[(a, n_k) \to (b, n_k + 1_k)] = \mathrm{e}^{-\beta E_b} \sum_{b,k} \mathcal{P}[(b, n_k) \to (a, n_k - 1_k)]$$

assuming that the initial state for each transition contains the equilibrium number of photons n_k. The principle of detailed balance is the assumption that, for complete equilibrium to be achieved, transitions between each possible pair of states must balance individually. In that case, using also the given relations for transition probabilities involving multi-photon states, we obtain

$$\mathrm{e}^{-\beta E_a}(n_k + 1)\mathcal{P}[(a, 0_k) \to (b, 1_k)] = \mathrm{e}^{-\beta E_b} n_k \mathcal{P}[(b, 1_k) \to (a, 0_k)].$$

Finally, we must assume (as can be verified by explicit, although approximate, calculations) that $\mathcal{P}[(a, 0_k) \rightarrow (b, 1_k)] = \mathcal{P}[(b, 1_k) \rightarrow (a, 0_k)]$. We then have

$$\mathrm{e}^{-\beta E_a}(n_k + 1) = \mathrm{e}^{-\beta E_b} n_k$$

which is easily solved for n_k, giving $n_k = 1/(\mathrm{e}^{\hbar\omega/kT} - 1)$. From these equilibrium occupation numbers and the density $g(\hbar\omega)$ of photon states given in problem 4.4, we obtain the Planck distribution for the energy spectrum of the radiation:

$$u(\omega)\,\mathrm{d}\omega = \frac{1}{V} g(\hbar\omega)\hbar\omega\,\mathrm{d}(\hbar\omega)\,n(\omega) = \frac{\hbar\omega^3\,\mathrm{d}\omega}{\pi^2 c^3(\mathrm{e}^{\hbar\omega/kT} - 1)}.$$

(b) We can attempt to construct the reverse argument as follows. From the Planck spectrum, we deduce the above occupation numbers, which are related to the atomic Boltzmann factors by

$$\mathrm{e}^{-\beta E_a}(n_k + 1) = \mathrm{e}^{-\beta E_b} n_k.$$

The condition of detailed balance therefore is

$$n_k \mathcal{P}[(a, n_k) \rightarrow (b, n_k + 1_k)] = (n_k + 1)\mathcal{P}[(b, n_k) \rightarrow (a, n_k - 1_k)]$$

and from this it follows that

$$\mathcal{P}[(a, n_k) \rightarrow (b, n_k + 1_k)] = (n_k + 1)\alpha(n_k)$$
$$\mathcal{P}[(b, n_k) \rightarrow (a, n_k - 1_k)] = n_k \alpha(n_k)$$

where $\alpha(n_k)$ is an undetermined function. In the semiclassical theory of radiation by atoms, where the electromagnetic field is treated classically, n_k can be interpreted as proportional to the intensity of radiation in the cavity. The second of the above probabilities is that for absorption of radiation. It might be supposed to be simply proportional to the intensity, and this is confirmed by explicit calculation. Thus, α should be independent of n_k. If so, then we can set $n_k = 0$ in the first result to obtain $\alpha = \mathcal{P}[(a, 0_k) \rightarrow (b, 1_k)]$ and in this way we recover the original QED relations. The quantity $\mathcal{P}[(a, 0_k) \rightarrow (b, 1_k)]$ is, of course, the probability for spontaneous emission of a photon by an excited atom. The semiclassical theory provides no direct way of calculating it. As first observed by Einstein, the requirement of obtaining the Planck spectrum in thermal equilibrium provides a means of determining the probability of spontaneous emission in terms of the calculable probabilities of absorption and stimulated emission.

Knowledgeable readers will recognize that we have considerably simplified the quantum mechanics of radiation processes, but the essential argument seems to be correct.

• **Problem 4.6** The density of states $g(\varepsilon)$ is defined so that $g(\varepsilon)\,\mathrm{d}\varepsilon$ is the number of states with energy between ε and $\varepsilon + \mathrm{d}\varepsilon$. If we take wavefunctions to vanish at the boundaries of a cube of side L, then the wavevectors are restricted to be of the form $\boldsymbol{k} = (\pi/L)\boldsymbol{n}$, where $\boldsymbol{n}$ is a vector of positive integers. The states with energy between ε and ε_0 are those for which $|\boldsymbol{n}|^2 = L^2|\boldsymbol{k}|^2/\pi^2 \leq L^2(\varepsilon_0 - \varepsilon)/\pi^2 A \equiv R^2(\varepsilon)$. The number of vectors $\boldsymbol{n}$ satisfying this condition is equal to the volume of a sphere of radius R if positive and negative integers are included, and is a fraction $1/2^3$ of this number if only positive integers are included. Including also a factor of 2 for the number of spin polarizations of an electron, we find that the number of states with energy between ε and ε_0 is

$$\mathcal{N}(\varepsilon) = 2 \times \frac{1}{8} \times \frac{4\pi}{3} R^3(\varepsilon) = \frac{L^3}{3\pi^2 A^{3/2}} (\varepsilon_0 - \varepsilon)^{3/2}.$$

Using periodic boundary conditions, the allowed wavevectors are $\boldsymbol{k} = (2\pi/L)\boldsymbol{n}$, but both positive and negative integers are allowed. In this case, we take the whole volume of a sphere of radius $R/2$, which gives the same result. The number $\mathcal{N}(\varepsilon + \mathrm{d}\varepsilon)$ of states with energy between $\varepsilon + \mathrm{d}\varepsilon$ and ε_0 is, of course, for $\mathrm{d}\varepsilon > 0$, *smaller* than $\mathcal{N}(\varepsilon)$ by $g(\varepsilon)\,\mathrm{d}\varepsilon$; so we find that

$$g(\varepsilon) = -\frac{\partial \mathcal{N}(\varepsilon)}{\partial \varepsilon} = \frac{L^3}{2\pi^2 A^{3/2}} (\varepsilon_0 - \varepsilon)^{1/2}.$$

Near the bottom of a band, where the dispersion relation is, say, of the form $\varepsilon = \varepsilon_0 + B|\boldsymbol{k}|^2$, a similar calculation gives $g(\varepsilon) = (L^3/2\pi^2 B^{3/2})(\varepsilon - \varepsilon_0)^{1/2}$.

• **Problem 4.7** We begin by treating the chain as a classical system. If η_ν represents the displacement of the νth atom from its equilibrium position, then the Lagrangian is

$$\mathcal{L} = \frac{1}{2} m \sum_{\nu=1}^{N-1} \dot{\eta}_\nu^2 - \frac{1}{2} K \sum_{\nu=0}^{N-1} (\eta_{\nu+1} - \eta_\nu)^2$$

where, since the atoms at the ends are fixed, $\eta_0 = \eta_N = 0$. For the remaining atoms, $\nu = 1, \ldots, N-1$, the equations of motion are

$$m\ddot{\eta}_\nu + 2K\eta_\nu - K(\eta_{\nu+1} + \eta_{\nu-1}) = 0.$$

These equations can be solved by expressing the η_ν in terms of $N-1$ normal coordinates x_j as

$$\eta_\nu = \sqrt{\frac{2}{N}} \sum_{j=1}^{N-1} x_j \sin\left(\frac{\nu j \pi}{N}\right)$$

which automatically yields $\eta_0 = \eta_N = 0$. With this representation, the equations of motion are

$$\sum_{j=1}^{N-1} \left[m\ddot{x}_j + 4K \sin^2\left(\frac{j\pi}{2N}\right) x_j \right] \sin\left(\frac{\nu j \pi}{N}\right) = 0.$$

These are equivalent to

$$\ddot{x}_j = -\omega_j^2 x_j \qquad \text{with } \omega_j = \sqrt{\frac{4K}{m}} \sin\left(\frac{j\pi}{2N}\right)$$

showing that the system is equivalent to a set of $N-1$ independent harmonic oscillators, with frequencies ω_j. Indeed, we can substitute the above expression for η_ν into $\mathcal{L}$. If $\sin(\nu j\pi/N)$ is expressed in terms of exponentials, the sum over ν leads to several geometric series and, after straightforward if tedious algebra, we obtain

$$\mathcal{L} = \tfrac{1}{2}\sum_{j=1}^{N-1}(m\dot{x}_j^2 - m\omega_j^2 x_j^2).$$

It is now simple to quantize the system and find that its energy eigenstates are specified by sets of integers $\{n_j\}$ as

$$E(\{n_j\}) = \sum_{j=1}^{N-1}(n_j + \tfrac{1}{2})\hbar\omega_j.$$

The *canonical* partition function for this system of oscillators is

$$Z(T) = \sum_{\{n_j\}} \mathrm{e}^{-\beta E(\{n_j\})} = \prod_{j=1}^{N-1}\sum_{n_j=0}^{\infty} \mathrm{e}^{-\beta\hbar\omega_j/2}\,\mathrm{e}^{-\beta\hbar\omega_j n_j} = \prod_{j=1}^{N-1}\frac{\mathrm{e}^{-\beta\hbar\omega_j/2}}{1-\mathrm{e}^{-\beta\hbar\omega_j}}.$$

Alternatively, let $L = Na$ be the total length of the chain, and $k_j = j\pi/L$ the wavenumbers of the standing waves whose amplitudes are the normal coordinates x_j. The n_j are the occupation numbers of single-particle states of momentum $\hbar k_j$. Ignoring the zero-point energy of the oscillators, we can interpret $Z(T)$ as the *grand canonical* partition function (with chemical potential $\mu = 0$) for a system of bosonic particles (phonons) whose single-particle energies are given by the dispersion relation

$$\varepsilon(k) = \hbar\omega(k) = \left(\frac{4\hbar^2 K}{m}\right)^{1/2} \sin\left(\frac{ka}{2}\right).$$

For large N, these states are closely spaced in energy. The density of states $g(\omega)$, defined as the number of states per unit interval in frequency (it could equally be defined in terms of energy by including appropriate factors of $\hbar$), is given simply by

$$g(\omega) = \left(\frac{\partial\omega_j}{\partial j}\right)^{-1} = \left[\frac{\pi}{2N}\left(\frac{4K}{m}\right)^{1/2}\cos\left(\frac{j\pi}{2N}\right)\right]^{-1} = \frac{2N}{\pi\sqrt{\omega_{\max}^2 - \omega^2}}$$

where $\omega_{\max} = \sqrt{4K/m}$ is the greatest possible frequency.

The internal energy can be expressed as

$$U(T) = -\frac{\partial\{\ln[Z(T)]\}}{\partial\beta} = \int_0^{\omega_{\max}} \mathrm{d}\omega\, g(\omega)\left(\frac{1}{2}\hbar\omega + \frac{\hbar\omega}{\mathrm{e}^{\beta\hbar\omega}-1}\right)$$
$$= \frac{N}{\pi}\hbar\omega_{\max} + \frac{2N}{\pi}\int_0^{\omega_{\max}} \mathrm{d}\omega\, \frac{\hbar\omega}{(\mathrm{e}^{\beta\hbar\omega}-1)\sqrt{\omega_{\max}^2-\omega^2}}$$

where the sum over discrete frequencies has been approximated by an integral. At high temperatures such that $\beta\hbar\omega_{\max} \ll 1$, we can use the approximation $(\mathrm{e}^{\beta\hbar\omega}-1)^{-1} = (\beta\hbar\omega)^{-1} - \frac{1}{2} + \mathrm{O}(1/T)$ to obtain $U = NkT + \mathrm{O}(1/T)$.

To estimate $U(T)$ at low temperatures, we make the change in integration variable $\omega = (kT/\hbar)y$ to obtain

$$U(T) \simeq \frac{N}{\pi}\hbar\omega_{\max} + \frac{2N}{\pi}\frac{(kT)^2}{\hbar\omega_{\max}}\int_0^{\infty}\frac{y\,\mathrm{d}y}{\mathrm{e}^y-1}$$
$$= \frac{N}{\pi}\hbar\omega_{\max}\left[1 + \frac{\pi^2}{3}\left(\frac{kT}{\hbar\omega_{\max}}\right)^2\right].$$

In the coefficient of $(kT)^2$, we have taken the limit $T \to 0$ and the remaining integral over y is equal to $\zeta(2) = \pi^2/6$.

• **Problem 4.8** To make the problem tractable, we assume that the normal modes of vibration can be treated in a harmonic approximation, such that the Hamiltonian is equivalent to that of a set of harmonic oscillators. This will generally be a reasonable approximation if the amplitudes of vibration are not too large. According to Debye's strategy, we take the dispersion relation to be an exact power law, up to a cut-off frequency ω_{D}. The internal energy is then given by (see problem 4.7 for further details)

$$U(T) = \int_0^{\omega_{\mathrm{D}}} g(\omega)\,\mathrm{d}\omega\,\frac{\hbar\omega}{\mathrm{e}^{\beta\hbar\omega}-1}$$

where $g(\omega)$ is the density of normal modes. For a d-dimensional medium with the dispersion relation $\omega \sim \lambda^{-s} \sim |\boldsymbol{k}|^s$, the method illustrated in problems 4.2 and 4.4 yields $g(\omega) = \kappa\omega^{d/s-1}$, where κ is a constant which cannot be determined without knowing the exact dispersion relation. Assuming our system to consist of N atoms, each of which can oscillate in d directions, the total number of normal modes must be Nd, so, if κ were known, we would determine the Debye frequency ω_{D} from the condition

$$Nd = \int_0^{\omega_{\mathrm{D}}} g(\omega)\,\mathrm{d}\omega = \frac{\kappa s}{d}\omega_{\mathrm{D}}^{d/s}.$$

Here, we can write $\kappa = (Nd^2/s)\omega_{\mathrm{D}}^{-d/s}$ and use ω_{D} as our undetermined constant. The internal energy is then

$$U(T) = \frac{Nd^2\hbar}{s}\int_0^{\omega_{\mathrm{D}}} \mathrm{d}\omega\,\frac{(\omega/\omega_{\mathrm{D}})^{d/s}}{\mathrm{e}^{\beta\hbar\omega}-1}$$

and we immediately obtain the specific heat as

$$C_V(T) = -\frac{1}{kT^2}\frac{\mathrm{d}U}{\mathrm{d}\beta} = \frac{Nd^2}{s}k\left(\frac{kT}{\hbar\omega_\mathrm{D}}\right)^{d/s}\int_0^{\beta\hbar\omega_\mathrm{D}} \frac{x^{d/s+1}\,\mathrm{e}^x}{(\mathrm{e}^x-1)^2}\,\mathrm{d}x.$$

Note that a change in volume would affect ω_D both through a change in the normal frequencies of vibration and through an explicit factor of V in $g(\omega)$, so, by keeping ω_D fixed, we do indeed obtain C_V. At low temperatures, we can take the upper limit of integration to infinity, to obtain $C_V(T) \sim T^{d/s}$. For phonons ($s = 1$) in a three-dimensional material, we naturally recover Debye's celebrated result $C_V \sim T^3$, but the more general calculation shows how both the spatial dimension and the dispersion relation contribute to this result.

• **Problem 4.9** The internal energy in the Debye model is given by

$$U(T) = \sum_i \frac{\hbar\omega_i}{\mathrm{e}^{\beta\hbar\omega_i}-1} \simeq \int_0^{\omega_\mathrm{D}} g(\omega)\,\mathrm{d}\omega\,\frac{\hbar\omega}{\mathrm{e}^{\beta\hbar\omega}-1}$$

where $g(\omega)\,\mathrm{d}\omega$ is the number of modes having frequencies between ω and $\omega + \mathrm{d}\omega$ while ω_D is the Debye frequency. If the solid contains N atoms, each able to oscillate in d directions, then ω_D is determined by the constraint $\int_0^{\omega_\mathrm{D}} g(\omega)\,\mathrm{d}\omega = Nd$. The dispersion relation for phonons is $\omega = c|\boldsymbol{k}|$, so, using the method illustrated in problems 4.2 and 4.4, we find that $g(\omega) \propto \omega^{d-1}$ or, using the constraint,

$$g(\omega) = \frac{Nd^2\omega^{d-1}}{\omega_\mathrm{D}^d}.$$

Then the internal energy can be expressed as

$$U(T) = Nd^2\hbar\omega_\mathrm{D}\alpha^{-(d+1)}f(\alpha)$$

where $\alpha = \hbar\omega_\mathrm{D}/kT$ and

$$f(\alpha) = \int_0^\alpha \frac{x^d}{\mathrm{e}^x-1}\,\mathrm{d}x.$$

We need to evaluate $f(\alpha)$ approximately when α is very small (the high-temperature limit $kT \gg \hbar\omega_\mathrm{D}$) or very large (the low-temperature limit $kT \ll \hbar\omega_\mathrm{D}$). When α is small, the integration variable x is small in the whole range of integration, so we can use the power series expansion

$$\begin{aligned} f(\alpha) &= \int_0^\alpha \mathrm{d}x\,x^{d-1}\left(1 - \frac{x}{2} + \frac{x^2}{12} + \cdots\right) \\ &= \frac{\alpha^d}{d} - \frac{\alpha^{d+1}}{2(d+1)} + \frac{\alpha^{d+2}}{12(d+2)} + \cdots \end{aligned}$$

to find

$$U(T) = Nd\hbar\omega_{\rm D}\left(\frac{kT}{\hbar\omega_{\rm D}} - \frac{d}{2(d+1)} + \frac{d}{12(d+2)}\frac{\hbar\omega_{\rm D}}{kT} + {\rm O}(T^{-2})\right)$$

$$C_V(T) = \frac{{\rm d}U(T)}{{\rm d}T} = Ndk\left[1 - \frac{d}{12(d+2)}\left(\frac{\hbar\omega_{\rm D}}{kT}\right)^2 + {\rm O}(T^{-3})\right].$$

When α is large, we write $f(\alpha) = f(\infty) - \Delta f(\alpha)$, with

$$f(\infty) = \int_0^\infty \frac{x^d}{{\rm e}^x - 1}\,{\rm d}x = \Gamma(d+1)\zeta(d+1)$$

$$\Delta f(\alpha) = \int_\alpha^\infty \frac{x^d}{{\rm e}^x - 1}\,{\rm d}x = {\rm e}^{-\alpha}\,\alpha^d \int_0^\infty \frac{{\rm e}^{-y}(1+\alpha^{-1}y)^d}{1 - {\rm e}^{-\alpha}\,{\rm e}^{-y}}\,{\rm d}y$$

where $y = x - \alpha$ and $\zeta(\cdot)$ is the Riemann zeta function. We see that $\Delta f(\alpha)$ involves powers of both ${\rm e}^{-\alpha}$ and α itself. Since, when α is large, $\alpha^n\,{\rm e}^{-\alpha} \to 0$ for any power n, the appropriate expansion parameter is ${\rm e}^{-\alpha}$. The coefficient of ${\rm e}^{-n\alpha}$ can be expressed as α^d times a power series in α^{-1}. Thus we have

$$\Delta f(\alpha) = {\rm e}^{-\alpha}\,\alpha^d[1 + {\rm O}(\alpha^{-1})] + {\rm O}({\rm e}^{-2\alpha})$$

and the leading behaviour of the internal energy and specific heat is

$$U(T) = Nd^2\hbar\omega_{\rm D}[\Gamma(d+1)\zeta(d+1)\alpha^{-(d+1)} - {\rm e}^{-\alpha}\,\alpha^{-1} + \cdots]$$

$$C_V(T) = -\frac{\alpha}{T}\frac{\partial U}{\partial\alpha} = Nd^2k[(d+1)\Gamma(d+1)\zeta(d+1)\alpha^{-d} - \alpha\,{\rm e}^{-\alpha} + \cdots].$$

In three dimensions, with $\zeta(4) = \pi^4/90$, we finally obtain

$$C_V(T) = 3Nk\left[1 - \frac{1}{20}\left(\frac{\hbar\omega_{\rm D}}{kT}\right)^2 + \cdots\right]$$

if $kT/\hbar\omega_{\rm D} \gg 1$ and

$$C_V(T) = Nk\left[\frac{12\pi^4}{5}\left(\frac{kT}{\hbar\omega_{\rm D}}\right)^3 - 9\frac{\hbar\omega_{\rm D}}{kT}\,{\rm e}^{-\hbar\omega_{\rm D}/kT} + \cdots\right]$$

if $kT/\hbar\omega_{\rm D} \ll 1$. The variation in C_V with temperature is sketched in figure 4.2.

• Problem 4.10

(a) The density of states for low-frequency phonons in a three-dimensional solid of volume V is

$$g(\omega) = \frac{3V\omega^2}{2\pi^2 c^3}$$

where c is the speed of sound in the solid. (The calculation is the same as that given for photons in problem 4.4, except that we allow for three polarizations:

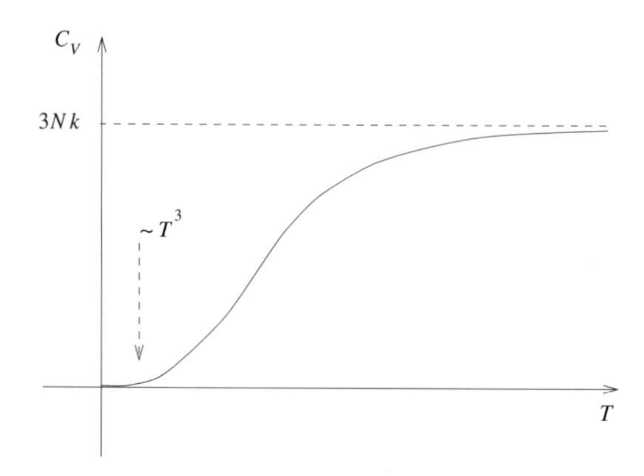

Figure 4.2. Specific heat for a three-dimensional isotropic solid described by the Debye model.

two transverse and one longitudinal.) As discussed in previous problems, the Debye approximation consists in adopting this density of states for all frequencies up to a cut-off ω_D, determined by the condition that the total number of phonon modes be equal to the number of normal modes of oscillation of N atoms in three dimensions, namely $3N$:

$$\int_0^{\omega_\mathrm{D}} g(\omega)\,\mathrm{d}\omega = 3N \qquad \text{which implies that} \quad \omega_\mathrm{D}^3 = \frac{6N\pi^2 c^3}{V}.$$

The internal energy *of the system of phonons* is

$$U(T) = \int_0^{\omega_\mathrm{D}} g(\omega)\,n(\omega,T)\,\hbar\omega\,\mathrm{d}\omega = \frac{6V}{\hbar c}(kT)^4 \int_0^{T_\mathrm{D}/T} \frac{x^3\,\mathrm{d}x}{\mathrm{e}^x - 1}$$

where $n(\omega,T) = (\mathrm{e}^{\hbar\omega/kT} - 1)^{-1}$ is the occupation number for each phonon mode and $T_\mathrm{D} = \hbar\omega_\mathrm{D}/k$ is the Debye temperature, and this vanishes at $T = 0$. However, as seen explicitly in problem 4.7, each of the $3N$ atomic oscillators, when treated quantum-mechanically, has a zero-point energy $\hbar\omega/2$ in addition to the excitations which we interpret as phonons. In the Debye approximation, the total zero-point energy for the solid is

$$E_0 = \int_0^{\omega_\mathrm{D}} \frac{\hbar\omega}{2}\,g(\omega)\,\mathrm{d}\omega = \frac{3V\hbar}{16\pi^2 c^3}\omega_\mathrm{D}^4 = \tfrac{9}{8}N\hbar\omega_\mathrm{D} = \tfrac{9}{8}NkT_\mathrm{D}$$

and the total internal energy *of the solid* is $U_\mathrm{tot}(T) = E_0 + U(T)$. It is of some interest to note that quantum field theory in effect treats the electromagnetic field as an assembly of harmonic oscillators which might also be regarded as having a zero-point energy in addition to the excitations which we interpret as photons. In this case, there is no natural cut-off frequency and the zero-point energy is infinite. For most purposes, this 'vacuum energy' can be ignored, on the grounds that total energy is defined only up to an arbitrary additive constant, and the energy of the vacuum may naturally be assigned as zero. According to QED, however, there is a measurable force (the Casimir force) between two conducting plates in empty space, which arises from the vacuum energy. In general-relativistic cosmology too, a constant vacuum energy (the 'cosmological constant') would have observable gravitational effects. Although actual observations indicate that any such effect is extremely small, the correct theoretical treatment of the zero-point energy is a matter of some controversy, but that is well beyond the scope of this book.

(b) Using the definition of the specific heat $C_V(T) = \mathrm{d}U(T)/\mathrm{d}T$ and the limiting behaviour of $C_V(T)$ and $U(T)$ established in problem 4.9, it is straightforward to show that

$$\int_0^\infty [C_V(\infty) - C_V(T)]\,\mathrm{d}T = \lim_{T\to\infty}[3NkT - U(T)] + U(0) = \tfrac{9}{8}NkT_\mathrm{D} = E_0.$$

The origin of this amusing fact can be seen by writing

$$\begin{aligned}\int_0^\infty [C_V(\infty) - C_V(T)]\,\mathrm{d}T &= \lim_{T\to\infty}[TC_V(T) - U(T)] \\ &= \int_0^{\omega_\mathrm{D}} g(\omega)\hbar\omega\rho(\omega)\,\mathrm{d}\omega \\ \rho(\omega) &= \lim_{T\to\infty}\left(T\frac{\partial n(\omega,T)}{\partial T} - n(\omega,T)\right).\end{aligned}$$

A simple calculation shows that $\rho(\omega) = \frac{1}{2}$, so the above integral is identical with that which defines E_0.

• Problem 4.11

(a) To be specific, suppose that the fixed leftmost atom is of type B and the fixed rightmost atom is of type A. The equilibrium positions of the N oscillating type-A atoms are then $x_\nu^{(\mathrm{A})} = (2\nu - 1)a$ for $\nu = 1, \ldots, N$ and those of the type-B atoms are $x_\nu^{(\mathrm{B})} = 2\nu a$. We denote the displacements from these equilibrium positions by ξ_ν for the type-A atoms and by η_ν for the type-B atoms. For the two fixed atoms, $\eta_0 = 0$ and $\xi_{N+1} = 0$. Then, for $\nu = 1, \ldots, N$, the classical equations of motion are

$$\begin{aligned}m\ddot{\xi}_\nu &= K(\eta_\nu - \xi_\nu) - K(\xi_\nu - \eta_{\nu-1}) \\ M\ddot{\eta}_\nu &= K(\xi_{\nu+1} - \eta_\nu) - K(\eta_\nu - \xi_\nu).\end{aligned}$$

We look for normal-mode solutions, labelled by an integer n, of the form

$$\xi_\nu^{(n)}(t) = A_n \cos(\omega_n t) \sin\left(\frac{(2\nu - 1)n\pi}{2N+1}\right) = A_n \cos(\omega_n t) \sin(k_n x_\nu^{(\mathrm{A})})$$
$$\eta_\nu^{(n)}(t) = B_n \cos(\omega_n t) \sin\left(\frac{2\nu n\pi}{2N+1}\right) = A_n \cos(\omega_n t) \sin(k_n x_\nu^{(\mathrm{B})})$$

where $k_n = n\pi/L$ and $L = (2N+1)a$ is the total length of the chain. Since there are $2N$ oscillating atoms, there must be $2N$ linearly independent normal modes, so one might suppose that these correspond to $n = 1, \ldots, 2N$. However, if we define $\bar{n} = 2N + 1 - n$, it is easy to see that $\xi_\nu^{(\bar{n})}/A_{\bar{n}} = -\xi_\nu^{(n)}/A_n$ and $\eta_\nu^{(\bar{n})}/B_{\bar{n}} = -\eta_\nu^{(n)}/B_n$. This means that the modes labelled by n and $2N+1-n$ are really the same: only the set of modes with $n = 1, \ldots, N$ are independent. We shall shortly discover, however, that there are in fact two modes corresponding to each n, giving the expected total of $2N$ modes, with normal coordinates which are linear combinations of the $2N$ amplitudes A_n and B_n. The maximum wavenumber is k_N. When N is large, we have $k_N \simeq \pi/2a$, corresponding to the fact that a unit cell in the chain contains two atoms and has a length of $2a$. (In the band theory of electrons in a regular lattice of ions, these considerations generalize to the fact that an electronic state with wavevector $\boldsymbol{k}$ can equally well be assigned a wavevector $\boldsymbol{k} + \boldsymbol{g}$, where $\boldsymbol{g}$ is any reciprocal-lattice vector.)

On substituting our trial solutions into the equations of motion, we find that these are satisfied provided that, for each n, the amplitudes satisfy

$$(2K - m\omega_n^2)A_n - 2K\cos(k_n a)\, B_n = 0$$
$$2K\cos(k_n a)\, A_n - (2K - M\omega_n^2)B_n = 0.$$

The condition for these simultaneous equations to have a non-trivial solution is that the determinant of coefficients vanishes:

$$(2K - m\omega_n^2)(2K - M\omega_n^2) - 4K^2\cos^2(k_n a) = 0.$$

This is a quadratic equation for ω_n^2, whose solution gives the dispersion relation

$$\omega_n^{(\pm)} = \sqrt{\frac{K}{\mu}}\{1 \pm [1 - \sigma\sin^2(k_n a)]^{1/2}\}^{1/2}.$$

In writing this solution, we have defined the reduced mass $\mu = mM/(m+M)$ and the parameter $\sigma = 4mM/(m+M)^2$, which is equal to unity if $m = M$ and very small if $m \ll M$ or $m \gg M$. For definiteness, we shall suppose that $m \leq M$. Taking N to be large, we can regard this result as defining two functions $\omega^{(\pm)}(k)$. Clearly, the dispersion relation has a high-frequency (optical) branch $\omega^{(+)}(k)$ and a low-frequency (acoustic) branch $\omega^{(-)}(k)$, which are sketched in figure 4.3.

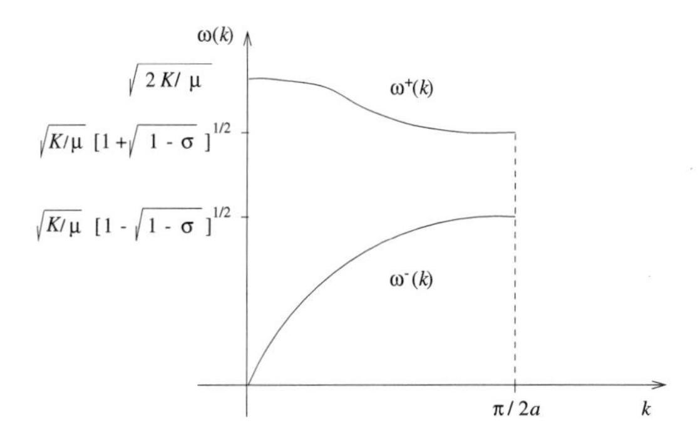

Figure 4.3. Dispersion relation for a one-dimensional chain composed of two kinds of atom with fixed boundary conditions at both ends. The upper and lower curves are the optical and acoustic branches respectively.

(b) When $m = M$, all the atoms in the chain are identical. We might expect that standing waves on the chain have a minimum wavelength of $2a$ (corresponding to neighbouring atoms moving exactly in antiphase) and hence a maximum wavenumber $k_{\max} = 2\pi/\lambda_{\min} = \pi/a$. Indeed, this situation was considered in problem 4.7, where we obtained the dispersion relation $\omega(k) = \sqrt{4K/m}\sin(ka/2)$, for $k = 0, \ldots, \pi/a$. According to the present description, however, we can set $\sigma = 1$ to obtain, for $k = 0, \ldots, \pi/2a$,

$$\omega^{(-)}(k) = \sqrt{\frac{4K}{m}}\sin\left(\frac{ka}{2}\right) \qquad \omega^{(+)}(k) = \sqrt{\frac{4K}{m}}\cos\left(\frac{ka}{2}\right).$$

We can resolve this apparent inconsistency and see what happens when m and M are only slightly different by recalling that modes labelled by n and $2N+1-n$ are equivalent, which means that wavenumbers k and $\pi/a-k$ are also equivalent. We can therefore define a single dispersion relation

$$\omega(k) = \begin{cases} \omega^{(-)}(k) & 0 < k < \dfrac{\pi}{2a} \\ \omega^{(+)}\left(\dfrac{\pi}{a-k}\right) & \dfrac{\pi}{2a} < k < \dfrac{\pi}{a} \end{cases}$$

When $m = M$, we recover the expected result since $\omega^{(+)}(\pi/a - k) \sim \cos(\pi/2 - ka/2) = \sin(ka/2)$. When $M > m$, there is a discontinuity (or band gap) at $k = \pi/2a$, given by

$$\Delta\omega = \omega^{(+)}\left(\frac{\pi}{2a}\right) - \omega^{(-)}\left(\frac{\pi}{2a}\right) = \sqrt{\frac{2K}{m}} - \sqrt{\frac{2K}{M}}.$$

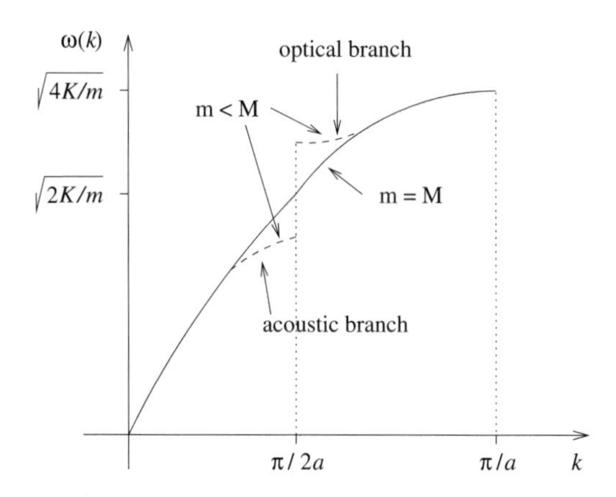

Figure 4.4. Dispersion relation for phonon excitations of a one-dimensional chain composed of a single atomic species (solid line) or of two atomic species with different masses $m < M$ which alternate along the chain (dashed line).

Clearly, this discontinuity decreases to zero as m becomes equal to M. This description of the dispersion relation is sketched in figure 4.4.

One might ask, if wavenumbers k and $\pi/a - k$ are equivalent, what the actual wavelength of a given mode is. This is necessarily somewhat ambiguous, since the wave profile is sampled only at discrete points, but a rough answer is the following. A snapshot of all the atoms would reveal wave profiles with wavelengths down to $\lambda_{\min} = 2a$, but these will be sinusoidal only for a chain of identical atoms. For such a chain, this is the only natural description. For atoms with different masses, we could instead consider unit cells of the chain, each containing one atom of each kind. The wave profile formed by the centres of mass of the unit cells has a minimum wavelength $\lambda_{\min} = 4a$ ($k_{\max} = \pi/2a$) but, to describe the complete state of the chain, we need also to consider the internal state of each unit cell. For states in the acoustic band, the two atoms in a unit cell tend to move in the same direction, so the wave profile formed by individual atoms is much the same as that formed by the unit cells. For states in the optical band, the two atoms tend to move in opposite directions, so the wave profile formed by the atoms has a short-wavelength component with $4a > \lambda > 2a$.

(c) The ratio of the maximum and minimum frequencies in the optical band is

$$\frac{\omega_{\max}}{\omega_{\min}} = \frac{\omega^{(+)}(0)}{\omega^{(+)}(\pi/2a)} = \frac{\sqrt{2K/\mu}}{\sqrt{2K/m}} = \sqrt{1 + \frac{m}{M}}.$$

Consequently, the frequency can be considered approximately independent of wavelength when $M \gg m$. This can be understood qualitatively by considering the heavy type-B atoms to be almost stationary at their equilibrium positions. Then the light type-A atoms between them vibrate almost independently, with approximately equal frequencies. The N modes in the optical band are distinguished by the relative phases of these vibrations.

(d) For the acoustic band, the dispersion relation for the lowest frequencies is $\omega \simeq ck$, where

$$c = \lim_{k \to 0} \left(\frac{\omega(k)}{k} \right) \simeq \sqrt{\frac{2Ka^2}{M}}$$

is (for $M \gg m$) the speed of the waves. Using $k = n\pi/L$, we easily find the density of states for these low-frequency modes:

$$g(\omega) = \frac{\mathrm{d}n}{\mathrm{d}\omega} = \frac{\mathrm{d}n}{\mathrm{d}k}\frac{\mathrm{d}k}{\mathrm{d}\omega} = \frac{L}{\pi c}$$

as given in the question. Using the Debye approximation, we assume that the dispersion relation is exactly linear, with a cut-off frequency given by

$$\int_0^{\omega_\mathrm{D}} g(\omega)\,\mathrm{d}\omega = \frac{L}{\pi c}\omega_\mathrm{D} = N$$

since there are N states in the acoustic band. When N is large, and $L \simeq 2Na$, this yields

$$\omega_\mathrm{D} = \frac{N\pi c}{L} \simeq \frac{\pi c}{2a} \simeq \frac{\pi}{2}\sqrt{\frac{2K}{M}}$$

which may be compared with the true maximum frequency in the acoustic band $\omega_{\mathrm{max}} = \omega^{(-)}(\pi/2a) = \sqrt{2K/M} = (2/\pi)\omega_\mathrm{D}$.

(e) The internal energy is

$$U(T) = \int_0^\infty g(\omega)\,\mathrm{d}\omega \frac{\hbar\omega}{\mathrm{e}^{\beta\hbar\omega} - 1} = \frac{L}{\pi\hbar c}(kT)^2 \int_0^{\beta\hbar\omega_\mathrm{D}} \frac{x}{\mathrm{e}^x - 1}\,\mathrm{d}x + \frac{N\hbar\omega_0}{\mathrm{e}^{\beta\hbar\omega_0} - 1}.$$

At low temperatures, $kT \ll \hbar\omega_\mathrm{D} < \hbar\omega_0$, we can take the upper limit of integration to infinity, to find

$$U(T) \simeq \text{constant} \times T^2 \qquad C_V \simeq \text{constant} \times T.$$

As might be expected, the second term, due to the optical modes, is suppressed by a factor $\mathrm{e}^{-\beta\hbar\omega_0}$ and makes no significant contribution. At high temperatures, $kT \gg \hbar\omega_0 > \hbar\omega_\mathrm{D}$, we can expand the integrand of the first term in powers of x (it is $1 + \mathrm{O}(x)$) and expand the second term in powers of $\beta\hbar\omega_0$ to get

$$U(T) \simeq \frac{L}{\pi\hbar c}(kT)^2\beta\hbar\omega_\mathrm{D} + NkT = \frac{L\omega_\mathrm{D}}{\pi c}kT + NkT = 2NkT.$$

Again as expected, the N acoustic modes and the N optical modes make equal contributions at high temperatures and, of course, we find that $C_V \simeq 2Nk$, in agreement with the classical equipartition theorem.

• Problem 4.12

(a) As usual for quantum gases, this problem is most conveniently treated using the grand canonical ensemble. Taking the system to be confined to a box of volume V, the single-particle states correspond to a discrete set of momenta $\boldsymbol{p}$, so the partition function for the system of free spinless bosons is

$$\mathcal{Z} = \prod_{\boldsymbol{p}} \sum_{n_p=0}^{\infty} \mathrm{e}^{-\beta n_p(\varepsilon_p - \mu)} = \prod_{\boldsymbol{p}} \frac{1}{1 - \mathrm{e}^{-\beta(\varepsilon_p - \mu)}}$$

and the mean number of particles is given by

$$\langle N \rangle = \left(\frac{\partial(\ln \mathcal{Z})}{\partial(\beta\mu)} \right)_\beta = \sum_{\boldsymbol{p}} \frac{1}{z^{-1}\,\mathrm{e}^{\beta\varepsilon_p} - 1}$$

where $z = \mathrm{e}^{\beta\mu}$ is the fugacity.

If we quantize the system of spinless bosons in a box of linear size L (so $V = L^d$) and impose *periodic* conditions on the single-particle wavefunctions $\psi \sim \mathrm{e}^{\mathrm{i}\boldsymbol{k}\cdot\boldsymbol{r}}$ ($\boldsymbol{p} = \hbar\boldsymbol{k}$), then each component of the wavevector takes the values $k_i = (2\pi/L)n_i$, where n_i is an integer. For a macroscopic system, the values k_i are closely spaced, and it is usually possible to approximate the sum by an integral over single-particle energies:

$$\sum_{\boldsymbol{p}} \to \int_0^\infty \mathrm{d}\varepsilon\, g(\varepsilon)$$

where the density of states $g(\varepsilon)$ is the number of states with energy between ε and $\varepsilon + \mathrm{d}\varepsilon$. For a given state, the set of integers $\boldsymbol{n}$ is related to the energy by $|\boldsymbol{n}|^2 = (L/h)^2(\varepsilon/\alpha)^{2/s}$, so the number of states with energy less than or equal to ε is the volume $\mathcal{V}(\varepsilon) = 2\pi^{d/2}R^d(\varepsilon)/d\,\Gamma(d/2)$ of a sphere of radius $R(\varepsilon) = (L/h)(\varepsilon/\alpha)^{1/s}$, and the density of states is

$$g(\varepsilon) = \frac{\mathrm{d}\mathcal{V}(\varepsilon)}{\mathrm{d}\varepsilon} = \frac{\Omega_d V}{s h^d \alpha} \left(\frac{\varepsilon}{\alpha} \right)^{d/s-1}.$$

Here we have written $\Omega_d = 2\pi^{d/2}/\Gamma(d/2)$ for the surface area of the unit sphere in d dimensions. This approximation of the discrete sum by an integral is valid only when each state contains an infinitesimal fraction of the total number of particles. However, Bose condensation occurs when a finite fraction of the particles occupy the ground state $\varepsilon = 0$. To allow for this possibility, we add to the integral an extra term representing the particles in the ground state, so the mean number of particles per unit volume becomes

$$\frac{\langle N \rangle}{V} = \frac{\Omega_d}{s h^d \alpha^{d/s}} \int_0^\infty \frac{\varepsilon^{d/s-1}}{z^{-1}\,\mathrm{e}^{\beta\varepsilon} - 1}\,\mathrm{d}\varepsilon + \frac{1}{V}\frac{z}{1-z}.$$

In the thermodynamic limit $V \to \infty$ and $N \to \infty$ such that N/V is fixed, the second term is negligible except when z is very close to unity. On the other hand, when this term is significant, the infinitesimal fraction of particles assigned to the ground state by the integral leads only to a negligible overcounting of these particles.

With the change of variables $x = \beta\varepsilon$, the total number of particles per unit volume can be written in terms of the number N_0 in the ground state and the number N_{exc} in excited states:

$$\frac{\langle N\rangle}{V} = \frac{\langle N_0\rangle}{V} + \frac{\langle N_{\text{exc}}\rangle}{V}$$

$$\frac{\langle N_0\rangle}{V} = \frac{1}{V}\frac{z}{1-z}$$

$$\frac{\langle N_{\text{exc}}\rangle}{V} = C_{d,s}(T)g_{d/s}(z)$$

where $C_{d,s}(T) = (\Omega_d/sh^d)(kT/\alpha)^{d/s}\Gamma(d/s)$ and

$$g_n(z) = \frac{1}{\Gamma(n)}\int_0^\infty \frac{x^{n-1}}{z^{-1}\,\mathrm{e}^x - 1}\,\mathrm{d}x = \sum_{l=1}^{\infty}\frac{z^l}{l^n}.$$

(b) Bose–Einstein condensation is a condensation in momentum space and takes place when the ground state ($\varepsilon = 0$ in this problem) becomes macroscopically occupied at temperatures below some non-zero critical temperature T_c. Macroscopic occupation means that $\langle N_0\rangle/V$ has a finite non-zero limit, say, n_0 as $V \to \infty$. As usual, it is convenient to consider a system with a fixed total number of particles N, and to regard the fugacity as a function $z(T, V, N)$ determined implicitly by the constraint that the expression for $\langle N\rangle/V$ derived above be equal to N/V. When Bose–Einstein condensation occurs, we see that

$$z = 1 - \frac{1}{n_0 V} + \mathrm{O}(V^{-2}).$$

At $T = 0$, we expect all the particles to be in the ground state, so $n_0 = N/V$. As T is increased towards T_c, n_0 decreases but, so long as $n_0 \neq 0$, z differs from 1 only by an amount of order $1/V$. Therefore, in the thermodynamic limit below T_c, we can write the constraint as

$$\frac{N}{V} = n_0 + C_{d,s}(T)g_{d/s}(1).$$

If this is to hold at a non-zero temperature (for which $C_{d,s}(T) > 0$), then $g_{d/s}(1)$ must have a finite value. For $z = 1$, however, the denominator in the integrand vanishes at $x = 0$ and the integral is potentially infinite. In fact, near $x = 0$, the integrand is approximately $x^{d/s-2}$, so the integral is finite only if $d/s > 1$. We conclude that true Bose condensation can occur only if $d/s > 1$ while, if $d/s \leq 1$, the ground state becomes macroscopically occupied only at $T = 0$.

When Bose condensation does occur, we can identify the critical temperature by taking the limit $n_0 \to 0$. Then we have $N/V = C_{d,s}(T_c)g_{d/s}(1)$ or

$$kT_c = \alpha \left(\frac{N}{V} \frac{sh^d}{\Omega_d \Gamma(d/s)\zeta(d/s)} \right)^{s/d}$$

where $\zeta(d/s) = g_{d/s}(1)$ is the Riemann zeta function.

The role of the thermodynamic limit in this discussion is worth emphasizing. In this limit (where $V \to \infty$ and $N \to \infty$ with N/V fixed), there is a sharp distinction between a normal phase in which $z < 1$ and $N_0/V \to 0$ and a condensed phase in which $z \to 1$ and $N_0/V \to n_0 > 0$. It is only in this latter limit that the transition temperature T_c is well defined. To determine T_c, indeed, it is necessary first to take $V \to \infty$ with $n_0 > 0$, so that $z = 1 - 1/n_0 V + \cdots \to 1$ and then set $n_0 = 0$ in the remaining expression for N/V: only this sequence of limits gives the correct result. For a large but finite system, a numerical solution of the constraint equation would show that N_0/V changes rapidly with temperature, from being very small but still non-zero at temperatures a little above T_c, to being an appreciable fraction of N/V at temperatures a little below T_c, but with no sharp transition between these two states. Similar remarks apply very generally to phase transitions of all kinds.

(c) In the thermodynamic limit, we expect to obtain an equation of state in the form $P = f(n, T)$, where $n = N/V$. Clearly, our first move is to obtain an expression for the pressure. Following the same steps as in (a), we obtain

$$P = \frac{kT}{V} \ln \mathcal{Z} = kTC_{d,s}(T)g_{d/s+1}(z) - \frac{kT}{V} \ln(1 - z)$$

where the second term is the pressure due to the condensed particles. In the thermodynamic limit, this contribution is negligible both above and below T_c since, in the condensed phase, $\ln(1 - z) \simeq -\ln(n_0 V)$ and $\lim_{V\to\infty}[V^{-1}\ln(n_0 V)] = 0$. At all temperatures, therefore, we have

$$P = kTC_{d,s}(T)g_{d/s+1}(z).$$

Above T_c, this must be supplemented by the constraint equation

$$n = C_{d,s}(T)g_{d/s}(z)$$

which determines z as a function of n and T. Alternatively, this pair of equations can be regarded as a parametric representation of the quantity $P/kTC_{d,s}(T)$ as a function of $n/C_{d,s}(T)$. Below T_c, we have $z = 1$, so the equation of state is simply

$$P = kTC_{d,s}(T)g_{d/s+1}(1)$$

and P is a function of T only. In interpreting these equations, it must be borne in mind that T_c depends on n. In fact, we have $T_c(n) = \text{constant} \times n^{s/d}$,

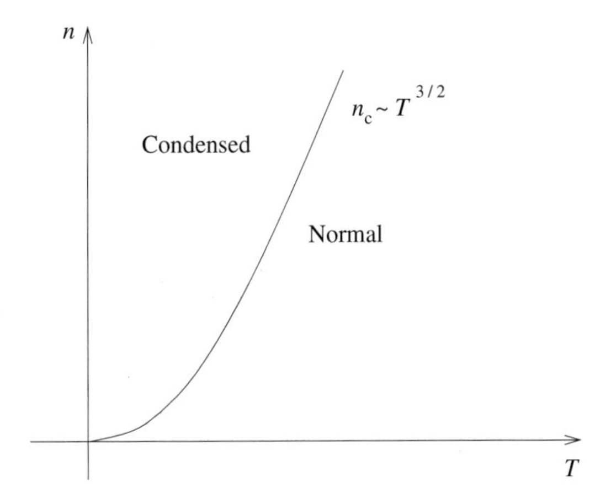

Figure 4.5. Condensed and normal phases in the n–T plane for Bose condensation.

which might also be written in terms of a critical number density $n_c(T) = \text{constant} \times T^{d/s}$. Thus, the condensed and normal phases are separated by a locus in the n–T plane as illustrated in figure 4.5 for the physically relevant case $d = 3$, $s = 2$. The corresponding locus in the P–v plane (specific volume $v = 1/n$), namely $P \sim T_c^{d/s+1} \sim v^{-(s/d+1)}$ is sketched in figure 4.6, which also shows isotherms corresponding to the equation of state derived above. The nature of the transition between the condensed and normal phases can be understood in two different ways, which are explored in the following parts of the problem.

(d) For $T > T_c$, we have already seen that the fraction of particles in the ground state is zero in the thermodynamic limit. For $T < T_c$, we have $n = n_0 + C_{d,s}(T)g_{d/s}(1)$. If $n = N/V$ is fixed, then $n = C_{d,s}(T_c)g_{d/s}(1)$ and we easily find that

$$\frac{N_0}{N} = \frac{n_0}{n} = 1 - \frac{C_{d,s}(T)}{C_{d,s}(T_c)} = 1 - \left(\frac{T}{T_c}\right)^{d/s}.$$

As T approaches T_c from below, the fraction of condensed particles behaves as

$$\frac{N_0}{N} \simeq \frac{d}{s}\left(1 - \frac{T}{T_c}\right).$$

This behaviour, illustrated in figure 4.7, is typical of a second-order phase transition, in which the order parameter N_0/N approaches zero continuously as a power of $T_c - T$ which is independent of parameters such as α which describe the detailed constitution of the system.

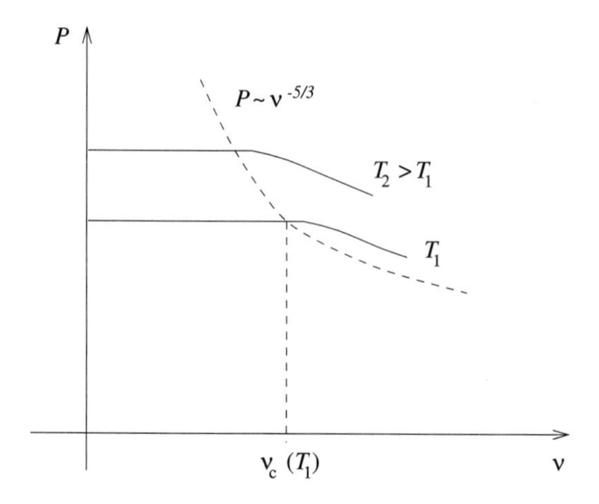

Figure 4.6. Locus in the P–v plane separating condensed and normal phases for Bose condensation.

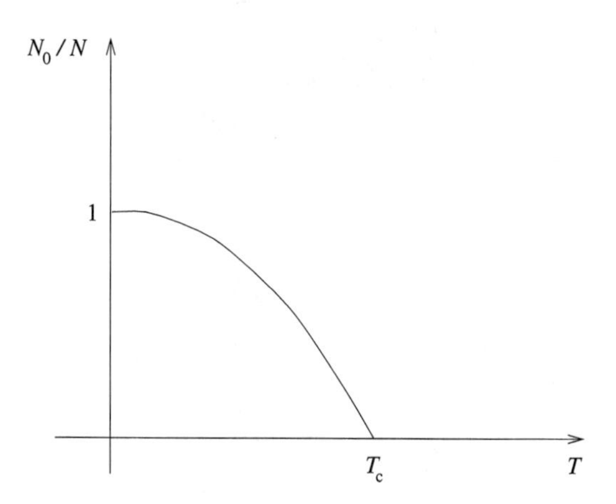

Figure 4.7. Fraction of condensed particles as a function of temperature for a fixed specific volume corresponding to Bose condensation.

(e) The entropy per unit volume is given by

$$\frac{S}{V} = \frac{1}{V}\left(\frac{\partial \Omega}{\partial T}\right)_{\mu,V} = \left(\frac{\partial P}{\partial T}\right)_{\mu,V} = \left(\frac{\partial P}{\partial T}\right)_{z,V} + \left(\frac{\partial P}{\partial z}\right)_{T,V}\left(\frac{\partial z}{\partial T}\right)_{\mu}.$$

Using the results

$$\frac{\partial C_{d,s}(T)}{\partial T} = \frac{d}{s}\frac{C_{d,s}(T)}{T} \qquad \left(\frac{\partial z}{\partial T}\right)_\mu = -\frac{z\ln z}{T} \qquad z\frac{\mathrm{d}g_n(z)}{\mathrm{d}z} = g_{n-1}(z)$$

we obtain

$$\begin{aligned}\frac{S}{V} &= kC_{d,s}(T)\left[\left(\frac{d}{s}+1\right)g_{d/s+1}(z) - \ln z\, g_{d/s}(z)\right] \\ &\quad - \frac{k}{V}\left(\ln(1-z) + \frac{z\ln z}{1-z}\right) \\ &= k\left(\frac{d}{s}+1\right)C_{d,s}(T)g_{d/s+1}(z) - k\frac{N}{V}\ln z - \frac{k}{V}\ln(1-z).\end{aligned}$$

The last term vanishes in the thermodynamic limit but (beware!) its derivatives do not. We see that, for fixed N/V, the entropy is continuous at the transition temperature T_c, so, as expected for a second-order transition, there is no latent heat.

(f) When N/V is fixed, we can calculate the specific heat at constant volume, per unit volume, as

$$\begin{aligned}C_V &= \frac{T}{V}\left(\frac{\partial S}{\partial T}\right)_{N,V} \\ &= \frac{T}{V}\left(\frac{\partial S}{\partial T}\right)_{N,V,z} + \frac{T}{V}\left(\frac{\partial S}{\partial z}\right)_{N,V,T}\left(\frac{\partial z}{\partial T}\right)_{N,V} \\ &= k\frac{d}{s}C_{d,s}(T)\left[\left(\frac{d}{s}+1\right)g_{d/s+1}(z) + g_{d/s}(z)\frac{T}{z}\left(\frac{\partial z}{\partial T}\right)_{N,V}\right].\end{aligned}$$

To obtain the last expression, we have used the constraint equation in the form

$$\frac{N}{V} = C_{d,s}(T)g_{d/s}(z) + \frac{1}{V}\frac{z}{1-z}$$

and by differentiating this equation we find that

$$\frac{T}{z}\left(\frac{\partial z}{\partial T}\right)_{N,V} = -\frac{d}{s}C_{d,s}(T)g_{d/s}(z)\left(C_{d,s}(T)g_{d/s-1}(z) + \frac{1}{V}\frac{z}{(1-z)^2}\right)^{-1}.$$

The thermodynamic limit of this quantity has different values above and below T_c. Above T_c, with $z < 1$, we have $V^{-1}[kz/(1-z)^2] \to 0$. Below T_c, with $1 - z \simeq 1/n_0 V$, we find instead that $V^{-1}[kz/(1-z)^2] \to \infty$. In the latter case, this yields $\partial z/\partial T \to 0$, which is expected since $z = 1$ in the thermodynamic limit below T_c. By first taking the thermodynamic limit and then approaching T_c from above and below, we obtain the discontinuity

$$\Delta C_V = \lim_{T\to T_c^-}[C_V(T)] - \lim_{T\to T_c^+}[C_V(T)] = \lim_{z\to 1}\left[k\frac{N}{V}\left(\frac{d}{s}\right)^2\frac{g_{d/s}(z)}{g_{d/s-1}(z)}\right].$$

Now, the limit of $g_n(z)$ as $z \to 1$ is finite (and equal to $\zeta(n)$) for $n > 1$. For $n = 1$, the integral given in (a) can easily be evaluated to give $g_1(z) = -\ln(1-z)$. For $n < 1$, we substitute $x = (1-z)y$ to obtain

$$g_n(z) \simeq \frac{(1-z)^{n-1}}{\Gamma(n)} \int_0^\infty \mathrm{d}y \, \frac{y^{n-1}}{1+y} = \Gamma(1-n)(1-z)^{n-1}$$

since the remaining integral is equal to the beta function $B(n, 1-n) = \Gamma(n)\Gamma(1-n)$. We see, therefore, that ΔC_V vanishes when $(d/s) \leq 2$, which includes the physically relevant case of a non-relativistic gas with $s = 2$ and $d = 3$.

(g) For $d = 3$ and $s = 2$, and above T_c, the specific heat in the thermodynamic limit is

$$C_V = \frac{3}{2}k\frac{N}{V}\left(\frac{5}{2}\frac{g_{5/2}(z)}{g_{3/2}(z)} - \frac{3}{2}\frac{g_{3/2}(z)}{g_{1/2}(z)}\right)$$

and we also have

$$\frac{T}{z}\left(\frac{\partial z}{\partial T}\right)_{N,V} = -\frac{3}{2}\frac{g_{3/2}(z)}{g_{1/2}(z)}.$$

Using our earlier results, we find that

$$\begin{aligned} T\left(\frac{\partial C_V}{\partial T}\right)_{N,V} &= \frac{T}{z}\left(\frac{\partial z}{\partial T}\right)_{N,V} z\left(\frac{\partial C_V}{\partial z}\right)_{N,V} \\ &= \frac{9}{4}k\frac{N}{V}\left(\frac{5}{2}\frac{g_{5/2}(z)}{g_{3/2}(z)} - \frac{g_{3/2}(z)}{g_{1/2}(z)} - \frac{3}{2}\frac{g_{3/2}^2(z)g_{-1/2}(z)}{g_{1/2}^3(z)}\right). \end{aligned}$$

Just below T_c, our previous results give

$$\begin{aligned} C_V &= \tfrac{15}{4}kg_{5/2}(1)C_{3,2}(T) \\ \frac{N}{V} &= g_{3/2}(1)C_{3,2}(T) + n_0 \\ n_0 &\simeq \frac{3}{2}\frac{N}{V}\left(1 - \frac{T}{T_c}\right). \end{aligned}$$

So we can write

$$C_V \simeq \frac{15}{4}k\frac{N}{V}\frac{g_{5/2}(1)}{g_{3/2}(1)}\left(\frac{3}{2}\frac{T}{T_c} - \frac{1}{2}\right) \qquad T_c\left(\frac{\partial C_V}{\partial T}\right)_{N,V} \simeq \frac{45}{8}k\frac{N}{V}\frac{g_{5/2}(1)}{g_{3/2}(1)}.$$

Finally, with the limiting forms of $g_z(z)$ derived above and $\Gamma(\frac{1}{2}) = \sqrt{\pi}$, we obtain

$$\lim_{T\to T_c^-}\left(\frac{\partial C_V}{\partial T}\right)_{N,V} - \lim_{T\to T_c^+}\left(\frac{\partial C_V}{\partial T}\right)_{N,V} = \frac{27}{16\pi}g_{3/2}^2(1)\frac{k}{T_c}\frac{N}{V}.$$

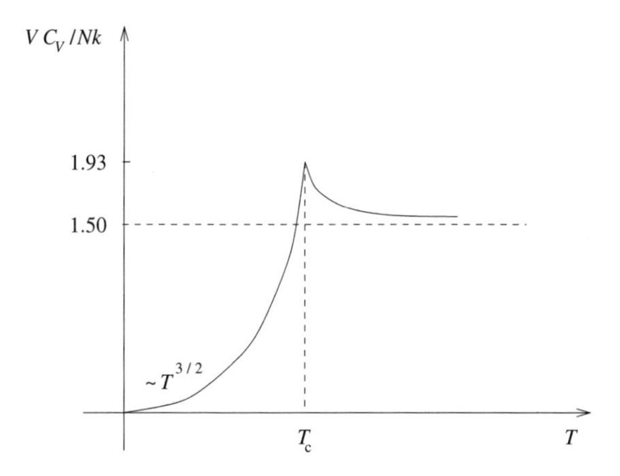

Figure 4.8. Specific heat of an ideal Bose gas as a function of temperature.

At very low temperatures, we get $C_V \sim T^{3/2}$ while, at high temperatures, with $z \to 0$ and $g_n(z) \simeq z$, we recover the classical result $C_V \simeq \frac{3}{2}Nk/V$. At $T = T_c$, we get $C_V = [15g_{5/2}(1)/4g_{3/2}(1)](Nk/V) \simeq 1.93Nk/V$ (since $g_{3/2}(1) \simeq 2.61$ and $g_{5/2}(1) \simeq 1.34$), which is a little higher than the classical value. Collecting all these results, we can sketch the behaviour of C_V over the whole temperature range as in figure 4.8.

(h) Up to this point, we have identified two phases of the Bose–Einstein gas: a condensed phase in which the ground state is occupied by some finite fraction of the particles and a normal phase in which there is no macroscopic occupation of the ground state. Keeping the total number of particles per unit volume constant, these two phases become identical at the critical temperature T_c in a continuous manner, and we observe a second-order phase transition somewhat analogous to that in a ferromagnet, where the magnetization (analogous, strictly speaking, to $\sqrt{n_0}$) approaches zero continuously at the Curie temperature. We now take a different point of view, identifying the particles in the ground state and those in the excited states as constituting two coexisting phases. Since these phases coexist, the transition between them must be of first order, somewhat analogous to that between the liquid and vapour phases of a simple fluid, which can also coexist in the same container.

An important difference between these two situations is that the condensed and normal phases of the Bose–Einstein gas occupy the same region of space, in contrast with a liquid coexisting with its vapour in different regions of their container. Nevertheless, we can keep the temperature fixed and cause particles to condense or 'evaporate' by changing the volume. In order to work in the thermodynamic limit, we deal with the number of particles per unit volume, $n = N/V$, or with the specific volume $v = V/N$. From our earlier discussion, we have a constraint equation relating the total number density n to the number density n_0 of condensed particles and an expression for the total entropy per unit volume. When there is condensation, these equations are

$$n = g_{d/s}(1)C_{d,s}(T) + n_0$$
$$\frac{S}{V} = k\left(\frac{d}{s}+1\right) g_{d/s+1}(1)C_{d,s}(T)$$

and from them we obtain the entropy per particle

$$s = \frac{S/V}{n} = k\frac{(d/s+1)g_{d/s+1}(1)C_{d,s}(T)}{g_{d/s}(1)C_{d,s}(T) + n_0}.$$

We now look for the boundaries of the region of coexistence, where all the particles have condensed, or they have all evaporated into the normal phase. When all the particles are condensed, we have $n_0 = n$ and the constraint equation shows that $n_0 = \infty$ or $v_0 = 0$. This must clearly be regarded as a fictitious limiting case, in contrast with an ordinary fluid which condenses into a non-zero volume. Evidently, the entropy per particle in this limit is $s_0 = 0$. All the particles have evaporated when $n_0 = 0$, or when n reaches its critical value $n_{\mathrm{c}}(T) = g_{d/s}(1)C_{d,s}(T)$. The specific volume of this normal phase is $v_{\mathrm{n}}(T) = 1/n_{\mathrm{c}}(T)$ and the entropy per particle is

$$s_{\mathrm{n}} = k\left(\frac{d}{s}+1\right)\frac{g_{d/s+1}(1)}{g_{d/s}(1)}.$$

The latent heat per particle absorbed during the evaporation is

$$l = T\,(s_{\mathrm{n}} - s_0) = kT\left(\frac{d}{s}+1\right)\frac{g_{d/s+1}(1)}{g_{d/s}(1)}.$$

The Clausius–Clapeyron equation which applies to an ordinary first-order transition is

$$\frac{\mathrm{d}P_{\mathrm{coex}}(T)}{\mathrm{d}T} = \frac{s_{\mathrm{n}} - s_0}{v_{\mathrm{n}} - v_0} = s_{\mathrm{n}}n_{\mathrm{c}}(T) = k\left(\frac{d}{s}+1\right) g_{d/s+1}(1)C_{d,s}(T)$$

where $P_{\mathrm{coex}}(T)$ is the pressure at which the phases coexist. In our case, this is just the pressure $P(T) = kTC_{d,s}(T)g_{d/s+1}(1)$ obtained in (c) and it is simple to verify that the Clausius–Clapeyron equation holds. For an ordinary fluid, $P_{\mathrm{coex}}(T)$ is the vapour pressure. The fluid exists as a gas when $P < P_{\mathrm{coex}}(T)$ and as a liquid when $P > P_{\mathrm{coex}}(T)$. Here, the situation is somewhat different. To condense all the particles, we must compress the gas to $v = 0$, and $P(T)$ is the greatest pressure that the gas can sustain. In this connection, it is worth emphasizing that, when an ordinary liquid coexists with its vapour, there is a specific volume v_l for the liquid, which is the volume occupied by liquid divided by the number of particles in the liquid, and a corresponding specific volume v_v for the vapour. In our case, there is only one specific volume v, the total volume divided by the total number of particles, which has limiting values $v_0 = 0$ and v_n, so the Clausius–Clapeyron equation has a slightly different meaning. On the other hand, the entropy per particle s_n is just the total entropy divided by the number of particles in the excited states. We can therefore say that the condensed particles have no entropy, and the latent heat l per particle is the heat absorbed in evaporating each particle and not merely the total latent heat divided by the total number of particles.

• **Problem 4.13** The Helmholtz free energy $F(T, V, N)$ can be expressed as $F = U - TS = -PV + \mu N$. It is most readily obtained from the grand canonical partition function $\mathcal{Z}$, using $PV = kT \ln \mathcal{Z}$. From the general results obtained in problem 4.12, we have

$$P(T, V, \mu) = kT\lambda_T^{-3} g_{5/2}(z)$$

where, for particles of mass m, the thermal de Broglie wavelength is $\lambda_T = h/\sqrt{2\pi mkT}$, the fugacity is $z = \mathrm{e}^{\mu/kT}$ and the functions $g_n(z)$ are defined by

$$g_n(z) = \frac{1}{\Gamma(n)} \int_0^\infty \frac{x^{n-1}}{z^{-1}\,\mathrm{e}^x - 1}\,\mathrm{d}x = \sum_{l=1}^{\infty} \frac{z^l}{l^n}.$$

For a gas of N particles in a volume V, the chemical potential μ (or the fugacity z) is determined by the requirement that the grand canonical average $\langle N \rangle$ be equal to N, which leads (in the thermodynamic limit) to the constraint equation

$$\frac{N}{V} = \begin{cases} \lambda_T^{-3} g_{3/2}(z) & T > T_\mathrm{c} \\ n_0 + \lambda_T^{-3} g_{3/2}(1) & T < T_\mathrm{c} \end{cases}$$

where n_0 is the number of particles per unit volume occupying the ground state. Below T_c, we have $z = 1$ (or $\mu = 0$) and the transition temperature corresponding to the borderline state $z = 1, n_0 = 0$ is given by

$$\lambda_{T_\mathrm{c}}^3 = g_{3/2}(1)V/N.$$

Clearly, the Helmholtz free energy per particle is given by

$$N^{-1}F(T, V, N) = \begin{cases} -\lambda_T^{-3} g_{5/2}(z)v + kT \ln z & T > T_\mathrm{c} \\ -\lambda_T^{-3} g_{5/2}(1) & T < T_\mathrm{c} \end{cases}$$

where $v = V/N$ is the specific volume and, for $T > T_\mathrm{c}$, z is determined as a function of (T, V, N) by the above constraint equation. The Gibbs free energy per particle is given by $G = F + PV$, or

$$N^{-1}G(T, P, N) = \begin{cases} kT \ln z & T > T_c \\ 0 & T < T_c \end{cases}$$

and is, of course, equal to the chemical potential. A minor subtlety is that, above T_c, z should now be regarded as a function of (T, P, N), determined by the above expression for the pressure P.

By regarding G as a function of the natural variables (T, P, N), we can use the differential relation $\mathrm{d}G = -S\,\mathrm{d}T + V\,\mathrm{d}P + \mu\,\mathrm{d}N$ to identify the isothermal compressibility as

$$\kappa_T^{-1} = -V\left(\frac{\partial P}{\partial V}\right)_{T,N} = -\left(\frac{\partial G}{\partial V}\right)_{T,N} = -\frac{kT}{z}\left(\frac{\partial z}{\partial v}\right)_{T,N}.$$

To evaluate $(\partial z/\partial v)_{T,N}$, however, we clearly regard z as a function of (T, v, N) and, by differentiating the constraint equation, obtain

$$\frac{1}{z}\left(\frac{\partial z}{\partial v}\right)_{T,N} = -\frac{\lambda_T^3}{v^2 z\,\mathrm{d}g_{3/2}(z)/\mathrm{d}z} = -\frac{\lambda_T^3}{v^2 g_{1/2}(z)}.$$

As $T \to T_c$ and $z \to 1$, we found in problem 4.12 that $g_{3/2}(z)$ has a finite limit, $g_{3/2}(1) \simeq 2.61$, while $g_{1/2}(z) \simeq \Gamma(\frac{1}{2})(1-z)^{-1/2} = \sqrt{\pi}(1-z)^{-1/2}$. We therefore find that

$$\kappa_T^{-1} = \frac{kT\lambda_T^3}{v^2 g_{1/2}(z)} \simeq \frac{kT_c\lambda_{T_c}^3}{\sqrt{\pi}v^2}(1-z)^{1/2}.$$

To express the divergence of κ_T in terms of temperature, we use the constraint equation once again, together with $z\,\mathrm{d}g_{3/2}(z)/\mathrm{d}z = g_{1/2}(z)$, to write

$$\begin{aligned} \frac{\lambda_T^3}{v} = g_{3/2}(z) &\simeq g_{3/2}(1) + \int_1^z \frac{g_{1/2}(z')}{z'}\,\mathrm{d}z' \\ &\simeq g_{3/2}(1) - 2\sqrt{\pi}(1-z)^{1/2} \\ &\simeq \frac{\lambda_{T_c}^3}{v} - 2\sqrt{\pi}(1-z)^{1/2} \end{aligned}$$

or

$$\begin{aligned} (1-z)^{1/2} \simeq \frac{\lambda_{T_c}^3 - \lambda_T^3}{2\sqrt{\pi}v} &\simeq \frac{\lambda_T^3}{2\sqrt{\pi}v}\left[\left(\frac{T}{T_c}\right)^{3/2} - 1\right] \\ &\simeq \frac{3\lambda_{T_c}^3}{4\sqrt{\pi}v}\frac{T - T_c}{T_c}. \end{aligned}$$

For the compressibility, this finally yields

$$\kappa_T \simeq \frac{4\pi v}{3kg_{3/2}^2(1)}(T - T_c)^{-1}.$$

A linear divergence of the compressibility is also found for a classical gas near its critical point, if one uses a mean-field approximation such as the van der Waals model (see problem 5.8). In the present case, κ_T remains infinite below T_c, since $z = 1$ and $\partial z/\partial v = 0$, corresponding to the fact that P is independent of the volume. This does not happen for a classical gas.

• **Problem 4.14** The critical temperature obtained in problem 4.12 for an ideal Bose gas in three dimensions can be written as

$$T_c = \frac{h^2}{2\pi mk}\left(\frac{n}{g_{3/2}(1)}\right)^{2/3}$$

where m is the mass of each atom, n is the number of atoms per unit volume and $g_{3/2}(1) = \zeta(3/2) \simeq 2.61$. For ^{87}Rb, we have $m \simeq 87 \times 1.66 \times 10^{-27}$ kg, and using $n = 2.5 \times 10^{18}$ m^{-3} we find that $T_c \simeq 3.4 \times 10^{-8}$ K, which is about a fifth of the measured value. The difference arises from two sources. First, the gases studied experimentally are not homogeneous, as in the idealized case considered here and in previous problems, but are trapped in an external potential, usually harmonic. In this case the critical temperature for the trapped ideal Bose gas turns out to be $T_c = (\hbar\omega_0/k)(N/1.202)^{1/3}$, where ω_0 is the frequency associated with the trapping potential. Below the critical temperature, the condensate fraction varies as $N_0/N = 1 - (T/T_c)^3$; note that the exponent is twice that of the untrapped gas. Second, further modifications arise due to the weak interactions between the bosons, which raise the transition temperature.

• **Problem 4.15**

(a) As usual, we quantize the system in a three-dimensional box of volume V. Each single-particle state is characterized by its momentum $\boldsymbol{p}$ and the integer n, which can be interpreted as the spin component of a spin-j particle in, say, the direction $\hat{z}$, the energy αn being that due to the interaction of an associated magnetic moment with an external field in the $\hat{z}$ direction. The grand canonical partition function is therefore a product of partition sums for single-particle states labelled by both $\boldsymbol{p}$ and n, so the pressure and number density are

$$P = \frac{kT}{V}\ln\mathcal{Z} = -\frac{kT}{V}\sum_{\boldsymbol{p}}\sum_{n=-j}^{j}\ln(1 - z\,\mathrm{e}^{-\beta\alpha n}\,\mathrm{e}^{-\beta\varepsilon(\boldsymbol{p})})$$

$$\frac{N}{V} = \frac{z}{V}\frac{\partial(\ln\mathcal{Z})}{\partial z} = V^{-1}\sum_{\boldsymbol{p}}\sum_{n=-j}^{j}(z^{-1}\,\mathrm{e}^{\beta\alpha n}\,\mathrm{e}^{\beta\varepsilon(\boldsymbol{p})} - 1)^{-1}$$

where $\varepsilon(\boldsymbol{p}) = |\boldsymbol{p}|^2/2m$ is the kinetic energy. In the thermodynamic limit, the sum over momenta can be replaced by an integral. However, to allow for a possible macroscopic occupation of the ground state, the corresponding term in the sum for N/V (although not in the sum for P) must be explicitly retained (see problem 4.12 for a detailed discussion). Since the state of lowest energy is that with $\boldsymbol{p} = \boldsymbol{0}$ and $n = -j$, we eventually obtain

$$P = kT\left(\frac{2\pi mkT}{h^2}\right)^{3/2}\sum_{n=-j}^{j} g_{5/2}(z\,\mathrm{e}^{-\beta\alpha n})$$

$$\frac{N}{V} = \left(\frac{2\pi mkT}{h^2}\right)^{3/2}\sum_{n=-j}^{j} g_{3/2}(z\,\mathrm{e}^{-\beta\alpha n}) + \frac{1}{V}\frac{z\,\mathrm{e}^{\beta\alpha j}}{1 - z\,\mathrm{e}^{\beta\alpha j}}.$$

(b) The ground state becomes macroscopically occupied, say, by n_0V particles, when z approaches its maximum value $z_{\mathrm{max}} = \mathrm{e}^{-\beta\alpha j}$. More specifically, in the thermodynamic limit, we set $z \simeq z_{\mathrm{max}} - 1/n_0V$ and take the limit $V \to \infty$ with the number density N/V fixed. The critical temperature T_{c} which marks the onset of condensation can then be identified by setting $n_0 = 0$ in the remaining equation for N/V. With the change in summation variable $\nu = n + j$, we obtain

$$\frac{N}{V} = \left(\frac{2\pi mkT_{\mathrm{c}}}{h^2}\right)^{3/2}\sum_{\nu=0}^{2j} g_{3/2}(\mathrm{e}^{-\alpha\nu/kT_{\mathrm{c}}})$$

which is an implicit equation for the condensation temperature. For $\alpha = 0$, this equation can be solved explicitly, giving the critical temperature for free bosons of spin j as

$$T_{\mathrm{c}} = (2j+1)^{-2/3}T_{\mathrm{c}}^0$$

where T_{c}^0 is the critical temperature for spin-0 bosons. Note, however, that in this case the $2j+1$ degenerate ground states all become macroscopically occupied below T_{c}, and the (identical) terms for all these states should be retained in the equation for N/V.

• Problem 4.16

(a) The partition function for this model is

$$Z(T,V) = \mathrm{e}^{-\beta E_{\mathrm{s}}}\prod_{q}\sum_{n_q=0}^{\infty}\mathrm{e}^{-\beta n_q\hbar\omega(q)} = \mathrm{e}^{-\beta E_{\mathrm{s}}}\prod_{q}(1 - \mathrm{e}^{-\beta\hbar\omega(q)})^{-1}.$$

It can be interpreted either as the canonical partition function for a collection of normal mode oscillators, the oscillator labelled by $\boldsymbol{q}$ containing n_q quanta of energy $\hbar\omega(\boldsymbol{q})$, or as the grand canonical partition function with chemical potential $\mu = 0$ for a gas of bosonic quasiparticles (cf problem 4.7). Assuming, as usual, that our system is contained in a cubic box of side L with periodic boundary conditions, so that the allowed values of the wavevector component q_i are $q_i = 2\pi\nu_i/L$, where ν_i is an integer, we find for the Helmholtz free energy

$$F(T,V) = -kT\ln[Z(T,V)] = E_{\mathrm{s}} + \frac{kTV}{2\pi^2}\int_0^{\infty}\mathrm{d}q\,q^2\ln(1 - \mathrm{e}^{-\beta\hbar\omega(q)})$$

where $V = L^3$ and the sum over $\boldsymbol{q}$ has been replaced by an integral. Note that modes for which $\beta\hbar\omega(q) \gg 1$ make a very small contribution to $F(T, V)$ and need not be treated accurately.

(b) The phonon modes have a dispersion relation $\omega(q) \simeq cq$, with $c \simeq 238 \text{ m s}^{-1}$. Their contribution to the Helmholtz free energy is

$$F^{\mathrm{ph}}(T, V) = \frac{kTV}{2\pi^2} \int_0^{q_0} \mathrm{d}q\, q^2 \ln(1 - \mathrm{e}^{-\beta\hbar c q})$$

where q_0 is the largest wavevector for which the linear dispersion relation is a reasonable approximation. This might be taken as $q_0 = 0.7 \text{ Å}^{-1}$. However, at the temperature $T_\lambda = 2.18$ K where superfluidity sets in, we find that $\mathrm{e}^{-\beta_\lambda \hbar c q_0} \simeq 3 \times 10^{-3}$, so that phonons with $q \geq q_0$ contribute much less to the free energy than those with $q \simeq 0$ and it will be a fair approximation to replace q_0 by infinity. At lower temperatures, this approximation will be even better justified. In this way, with the change of variable $x = \beta\hbar c q$, we obtain

$$F^{\mathrm{ph}}(T, V) = \frac{(kT)^4 V}{2\pi^2 (\hbar c)^3} \int_0^\infty \mathrm{d}x\, x^2 \ln(1 - \mathrm{e}^{-x}) = -\frac{\pi^2}{90} \frac{(kT)^4 V}{(\hbar c)^3}$$

and the phonon contribution to the specific heat per unit volume is

$$C_V^{\mathrm{ph}} = -\frac{T}{V} \left(\frac{\partial^2 F^{\mathrm{ph}}}{\partial T^2} \right)_V = k \frac{2\pi^2}{15} \left(\frac{kT}{\hbar c} \right)^3 \simeq (3 \times 10^3 \text{ J K}^{-4} \text{ m}^{-3}) T^3.$$

(c) The dispersion relation for the roton modes is $\hbar\omega(\boldsymbol{q}) \simeq k\Delta + \hbar^2 (q - q_{\mathrm{R}})^2 / 2m_{\mathrm{R}}$. The contribution of these modes to the free energy is

$$F^{\mathrm{rot}}(T, V) = \frac{kTV}{2\pi^2} \int_{q_1}^{q_2} \mathrm{d}q\, q^2 \ln(1 - \mathrm{e}^{-\Delta/T}\, \mathrm{e}^{-\beta\hbar^2 (q - q_{\mathrm{R}})^2 / 2m_{\mathrm{R}}})$$

where q_1 and q_2 are the lowest and highest values respectively of q for which the quadratic approximation to the dispersion curve is reasonable. For reasons similar to those given above, these limits of integration may be extended to $\pm\infty$. Moreover, we find that $\mathrm{e}^{-\Delta/T} \leq 0.02$ for $T \leq T_\lambda$, so it is a fair approximation to keep only the first term in the Taylor expansion of the logarithm. With the change in integration variable $x = \hbar(q - q_{\mathrm{R}})/\sqrt{2m_{\mathrm{R}} kT}$, we then have

$$\begin{aligned} F^{\mathrm{rot}} &= -\frac{kTVq_{\mathrm{R}}^3}{2\pi^2} \mathrm{e}^{-\Delta/T} \left(\frac{T}{\theta} \right)^{3/2} \int_{-\infty}^{\infty} \mathrm{d}x \left(x + \sqrt{\frac{\theta}{T}} \right)^2 \mathrm{e}^{-x^2} \\ &= -\frac{kTVq_{\mathrm{R}}^3}{2\pi^{3/2}} \mathrm{e}^{-\Delta/T} \left(\frac{T}{\theta} \right)^{3/2} \left(\frac{1}{2} + \frac{\theta}{T} \right) \end{aligned}$$

where $\theta = \hbar^2 q_{\mathrm{R}}^2 / 2m_{\mathrm{R}} k \simeq 137$ K. Clearly, for $T < T_\lambda$, the first term of $(1/2 + \theta/T)$ is much smaller than the second and can safely be neglected.

With this approximation, we estimate the roton contribution to the specific heat as

$$\begin{aligned} C_V^{\text{rot}} &= -\frac{T}{V}\left(\frac{\partial^2 F^{\text{rot}}}{\partial T^2}\right)_V \\ &= k\frac{q_{\text{R}}^3}{2\pi^{3/2}}\left(\frac{T}{\theta}\right)^{1/2} \mathrm{e}^{-\Delta/T}\left(\frac{3}{4}+\frac{\Delta}{T}+\frac{\Delta^2}{T^2}\right) \\ &\simeq (7.3\times 10^5 \text{ J K}^{-3/2}\text{ m}^{-3})T^{1/2}\,\mathrm{e}^{-\Delta/T}\left(\frac{3}{4}+\frac{\Delta}{T}+\frac{\Delta^2}{T^2}\right). \end{aligned}$$

(d) At this point, our estimate of the specific heat per unit volume is

$$C_V(T) \simeq C_V^{\text{ph}}(T) + C_V^{\text{rot}}(T).$$

This estimate excludes the contributions of modes which do not lie on the linear (phonon) part of the dispersion curve nor near the roton minimum (that is, modes with $q_0 < q < q_1$ or $q > q_2$). By extending limits of integration to $\pm\infty$, we have also, in effect, included contributions from modes which do not exist. At sufficiently low temperatures, the contributions of both the neglected modes and the fictitious modes will be negligible and, indeed, we have argued that the approximations should be reasonable when $T < T_\lambda = 2.18$ K. A numerical investigation reveals that $C_V^{\text{ph}}(T) = C_V^{\text{rot}}(T)$ when $T \simeq 0.78$ K. Below this temperature, C_V^{rot} is smaller than C_V^{ph}, mainly on account of the factor $\mathrm{e}^{-\Delta/T}$. At somewhat higher temperatures, C_V^{rot} makes the larger contribution. At temperatures above about 9 K, C_V^{ph} again becomes larger than C_V^{rot}, but at such temperatures our approximations are unreliable. In any case, the model makes no physical sense above T_λ.

• Problem 4.17

(a) As with the Bose gas treated in previous problems, an ideal Fermi gas is most conveniently treated using the grand canonical ensemble. The partition function is

$$\mathcal{Z} = \prod_{p,s}\sum_{n_{p,s}=0}^{1} \mathrm{e}^{-\beta(\varepsilon_{p,s}-\mu)n_{p,s}} = \prod_{p,s}(1+\mathrm{e}^{\beta(\varepsilon_{p,s}-\mu)}).$$

The index $s = \pm\frac{1}{2}$ labels spin states. If the single-particle energy is independent of spin (and here we shall take $\varepsilon = |\boldsymbol{p}|^2/2m$ for particles of mass m), then for a gas in a container of macroscopic size the pressure and number density can be written as

$$P = \frac{kT}{V}\ln\mathcal{Z} = \frac{kT}{V}\int_0^\infty \mathrm{d}\varepsilon\, g(\varepsilon)\ln(1+\mathrm{e}^{-\beta(\varepsilon-\mu)})$$

$$n = \frac{\langle N \rangle}{V} = \frac{1}{V}\left(\frac{\partial(\ln \mathcal{Z})}{\partial(\beta\mu)}\right)_{\beta,V} = \frac{1}{V}\int_0^\infty \mathrm{d}\varepsilon\, g(\varepsilon)\frac{1}{\mathrm{e}^{\beta(\varepsilon-\mu)}+1}.$$

Recall now that in d dimensions the density of states for free non-relativistic particles is $g(\varepsilon) = cV\varepsilon^{d/2-1}$ (see, for example, problem 4.2) where c is a constant, whose exact value does not yet concern us. At zero temperature the occupation numbers are $n(\varepsilon) = (\mathrm{e}^{\beta(\varepsilon-\mu)}+1)^{-1} = \theta(\varepsilon_\mathrm{F} - \varepsilon)$, where $\theta(\cdot)$ is the step function and $\varepsilon_\mathrm{F} = \mu(T=0)$ is the Fermi energy. We therefore obtain

$$n = \frac{c}{d/2}\varepsilon_\mathrm{F}^{d/2}$$

$$P = \frac{2}{d}\frac{1}{V}\int_0^\infty \mathrm{d}\varepsilon\, g(\varepsilon)\varepsilon n(\varepsilon) = \frac{c}{(d/2)(d/2+1)}\varepsilon_\mathrm{F}^{d/2+1} = \frac{n\varepsilon_\mathrm{F}}{d/2+1}.$$

This integral expression for P is obtained from the previous one by an integration by parts, and the result agrees with the general relation $P = (2/d)(U/V)$ obtained in problem 4.2. The isothermal compressibility can now be found from

$$\kappa_T(0)^{-1} = -V\left(\frac{\partial P}{\partial V}\right)_{N,T=0} = n\frac{\mathrm{d}P}{\mathrm{d}n} = \frac{2}{d}\left(\frac{d}{2}+1\right)P = \frac{2}{d}\varepsilon_\mathrm{F} n$$

or $\kappa_T(0) = d/2n\varepsilon_\mathrm{F}$. In order to calculate $n\,\mathrm{d}P/\mathrm{d}n$ (bearing in mind that ε_F depends on n), we have used the two previous equations to write

$$P = \frac{c}{(d/2)(d/2+1)}\left(\frac{d}{2c}n\right)^{(2/d)(d/2+1)}.$$

Alternatively, the same result may be obtained by making use of the thermodynamic relation $(\partial\mu/\partial N)_{T,V} = V/N^2\kappa_T$ (see problem 2.5) to write

$$\kappa_T(0) = \frac{V}{N^2}\left(\frac{\partial N}{\partial\mu}\right)_{T=0,V} = \frac{1}{n^2}\frac{\mathrm{d}n}{\mathrm{d}\varepsilon_\mathrm{F}} = \frac{d}{2n\varepsilon_\mathrm{F}}.$$

(b) In three dimensions, the calculation of problem 4.2 gives the constant c as

$$c = \frac{4}{\pi^2}\left(\frac{m}{2\hbar^2}\right)^{3/2}$$

if we include a factor of 2 for the spin degeneracy. We then find that

$$T_\mathrm{F} = \frac{\varepsilon_\mathrm{F}}{k} = \frac{\hbar^2}{2mk}(3\pi^2 n)^{2/3}$$

where $m = 9.11 \times 10^{-31}$ kg is the mass of an electron. The number of electrons per unit volume is the same as the number of atoms per unit volume, namely $n = (10^3\rho/M)N_\mathrm{A} \simeq 8.5 \times 10^{28}\ \mathrm{m}^{-3}$, where ρ is the density, M is the atomic weight and $N_A = 6.02 \times 10^{23}$ is the Avogadro number. With these numerical values, we estimate the Fermi temperature as $T_F \simeq 82\,000$ K.

Since the Fermi temperature is much greater than room temperature (about 300 K), the electrons in metallic copper can be considered as a highly degenerate Fermi gas, in the sense that the approximation $T \ll T_F$ is always justified. Note, however, that these electrons interact significantly with copper ions. Consequently, although the gas is degenerate, it is not free, and its Fermi surface is not spherical. Our calculation of T_F is therefore not exact.

• **Problem 4.18** We start from the usual grand canonical expression for the mean number of particles in terms of temperature and chemical potential, namely

$$N = \int_0^\infty d\varepsilon\, g(\varepsilon) n(\varepsilon) = \int_0^\infty d\varepsilon\, g(\varepsilon) \frac{1}{e^{\beta(\varepsilon-\mu)} + 1}$$

where $g(\varepsilon)$ is the density of states. Usually, this integral cannot be evaluated in closed form, which makes it impossible to solve this equation explicitly for the chemical potential μ. The two-dimensional gas is, however, a special case, because $g(\varepsilon)$ is simply a constant. In fact, from the general expression derived in problem 4.2, with an extra factor of 2 to account for spin degeneracy, we find that $g(\varepsilon) = mA/\pi\hbar^2$, and hence

$$\frac{\pi\hbar^2}{m}\frac{N}{A} = \int_0^\infty d\varepsilon \frac{1}{e^{\beta(\varepsilon-\mu)} + 1} = -\frac{1}{\beta} \ln(e^{-\beta\varepsilon} + e^{-\beta\mu})|_0^\infty$$
$$= kT \ln(e^{\mu/kT} + 1).$$

On taking the limit $T \to 0$ (figure 4.9), we obtain the Fermi energy as[2]

$$\varepsilon_F = \mu(T = 0) = \frac{\pi\hbar^2}{m}\frac{N}{A}.$$

Our previous result can then be written as $e^{\varepsilon_F/kT} = e^{\mu/kT} + 1$, which is easily solved to give

$$\mu = kT \ln(e^{\varepsilon_F/kT} - 1).$$

We note that, in the classical limit $T \to \infty$, the chemical potential becomes infinitely negative. However, at low temperatures ($kT \ll \varepsilon_F$), we have $\mu \simeq \varepsilon_F - kT\, e^{-\varepsilon_F/kT}$ and μ varies very little with temperature.

• **Problem 4.19** Using the grand canonical ensemble, the average numbers of particles having their spins parallel or antiparallel to the magnetic field can be expressed as

$$\langle N_\pm \rangle = \int_0^\infty d\varepsilon\, g(\varepsilon) \frac{1}{e^{\beta(\varepsilon-\mu\pm\mu_B H)} + 1}$$

[2] We are defining the Fermi energy to have the constant value $\varepsilon_F = \mu(T = 0)$. Some authors adopt the definition $n(\varepsilon_F) = \frac{1}{2}$, which implies that $\varepsilon_F = \mu$ at all temperatures.

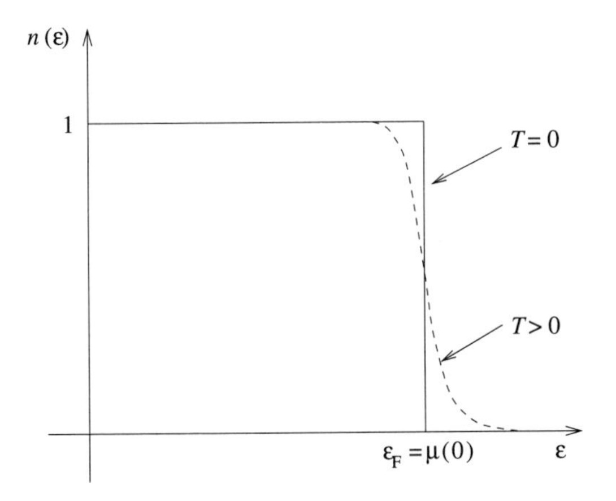

Figure 4.9. Fermi distribution function for zero temperature and low temperatures.

where $g(\varepsilon) = (2/\sqrt{\pi})(2\pi m/h^2)^{3/2} V \varepsilon^{1/2}$ is the density of states associated with translational motion in a container of volume V. (The minus sign corresponds to the lower-energy state in which the magnetic moment is parallel to the field.) To a good approximation, the magnetic moment of each electron is μ_{B}, so the average magnetization is

$$\begin{aligned}\langle M \rangle &= \mu_{\mathrm{B}}(\langle N_- \rangle - \langle N_+ \rangle) \\ &= \mu_{\mathrm{B}} A V [I_{1/2}(x_+) - I_{1/2}(x_-)]\end{aligned}$$

where $A = (2/\sqrt{\pi})(2\pi mkT/h^2)^{3/2}$, $x_\pm = \beta(\mu \pm \mu_{\mathrm{B}} H)$ and $I_{1/2}(x) = \int_0^\infty \mathrm{d}u\, u^{1/2}/(\mathrm{e}^{u-x} + 1)$. At low temperatures such that $x_\pm \gg 1$, we can use the approximation

$$I_{1/2}(x) = \tfrac{2}{3} x^{3/2} + \frac{\pi^2}{12} x^{-1/2} + \cdots$$

to get

$$\frac{\langle M \rangle}{\mu_{\mathrm{B}} A V} \simeq \tfrac{2}{3}(x_+^{3/2} - x_-^{3/2}) + \frac{\pi^2}{12}(x_+^{-1/2} - x_-^{-1/2})$$

and the paramagnetic susceptibility $\chi = \lim_{H \to 0}(\partial \langle M \rangle / \partial H)$ is

$$\chi = \frac{\mu_{\mathrm{B}}^2 A V}{kT}\left[2\left(\frac{\mu}{kT}\right)^{1/2} - \frac{\pi^2}{12}\left(\frac{\mu}{kT}\right)^{-3/2}\right].$$

For a fixed number N of electrons in the volume V, we determine the chemical potential (at $H = 0$) from the constraint equation

$$N = \langle N_+ \rangle + \langle N_- \rangle = 2AVI_{1/2}(\beta\mu) = \frac{4AV}{3}(\beta\mu)^{3/2}\left[1 + \frac{\pi^2}{8}\left(\frac{kT}{\mu}\right)^2 + \cdots\right]$$

whose solution for μ is

$$\mu = \varepsilon_{\mathrm{F}}\left[1 - \frac{\pi^2}{12}\left(\frac{kT}{\varepsilon_{\mathrm{F}}}\right)^2 + \cdots\right]$$

where $\varepsilon_{\mathrm{F}} = \mu(T = 0) = (h^2/8\pi^2 m)(3\pi^2 N/V)^{2/3}$ is the Fermi energy. Substituting this in our previous expression for χ finally yields

$$\chi = \chi_0\left[1 - \frac{\pi^2}{12}\left(\frac{kT}{\varepsilon_{\mathrm{F}}}\right)^2 + \cdots\right]$$

where $\chi_0 = 3N\mu_{\mathrm{B}}^2/2\varepsilon_{\mathrm{F}}$ is the zero-temperature susceptibility (a result first obtained by E C Stoner).

• Problem 4.20

(a) In equilibrium, both gases must be at the same temperature and pressure and their volumes are constrained by $V_{\mathrm{A}} + V_{\mathrm{B}} = V$. By the usual grand canonical methods, we can find expressions for the numbers of particles per unit volume and the pressure in terms of the chemical potential μ. Of course, the pressure is given by $PV = kT \ln \mathcal{Z}$, but it is convenient to use the fact that the internal energy of any non-relativistic monatomic ideal gas is related to its pressure by $U = 3PV/2$ (see problem 4.21). We thus have (for a = A, B)

$$N = \gamma_{\mathrm{a}} \int_0^\infty g_{\mathrm{a}}(\varepsilon) n_{\mathrm{a}}(\varepsilon)\, \mathrm{d}\varepsilon = c\gamma_{\mathrm{a}} V_{\mathrm{a}} (kT)^{3/2} \int_0^\infty \frac{u^{1/2}}{\mathrm{e}^{u - x_{\mathrm{a}}} + 1}\, \mathrm{d}u$$

$$U_{\mathrm{a}} = \gamma_{\mathrm{a}} \int_0^\infty \varepsilon\, g_{\mathrm{a}}(\varepsilon) n_{\mathrm{a}}(\varepsilon)\, \mathrm{d}\varepsilon = c\gamma_{\mathrm{a}} V_{\mathrm{a}} (kT)^{5/2} \int_0^\infty \frac{u^{3/2}}{\mathrm{e}^{u - x_{\mathrm{a}}} + 1}\, \mathrm{d}u.$$

In these expressions, $\gamma_{\mathrm{a}} = 2s_{\mathrm{a}} + 1$ denotes the spin degeneracy, $n_{\mathrm{a}}(\varepsilon) = (\mathrm{e}^{\beta(\varepsilon - \mu_{\mathrm{a}})} + 1)^{-1}$ is the Fermi distribution function, $g_{\mathrm{a}}(\varepsilon) = cV_{\mathrm{a}}\varepsilon^{1/2}$ with $c = (2/\sqrt{\pi})(2\pi m/h^2)^{3/2}$ is the density of states and $x_{\mathrm{a}} = \mu_{\mathrm{a}}/kT$. As we shall see, the two gases have different chemical potentials and therefore different distribution functions.

At very high temperatures, with a finite number of particles per unit volume, the integral in the last expression for N must be very small. This implies that $x_{\mathrm{a}} \to -\infty$ and thus

$$N \simeq c\gamma_{\mathrm{a}} V_{\mathrm{a}} (kT)^{3/2}\, \mathrm{e}^{x_{\mathrm{a}}} \int_0^\infty u^{1/2}\, \mathrm{e}^{-u}\, \mathrm{d}u = \gamma_{\mathrm{a}}\, \mathrm{e}^{\mu_{\mathrm{a}}/kT} \left(\frac{2\pi mkT}{h^2}\right)^{3/2}$$

$$U_{\mathrm{a}} \simeq c\gamma_{\mathrm{a}} V_{\mathrm{a}} (kT)^{5/2}\, \mathrm{e}^{x_{\mathrm{a}}} \int_0^\infty u^{3/2}\, \mathrm{e}^{-u}\, \mathrm{d}u = \tfrac{3}{2} NkT.$$

In this limit, of course, the gases behave classically. Their internal energies are independent of the spin s_{a} and, since $U_{\mathrm{a}} = 3PV_{\mathrm{a}}/2$ is the same for both gases, we easily find that $V_{\mathrm{A}} = V_{\mathrm{B}}$.

At very low temperatures, the chemical potentials approach the corresponding Fermi energies $\mu_a(0) = \varepsilon_{F,a}$ and we can use the approximations to the integrals given in appendix A to write

$$\frac{N}{V_a} \simeq \frac{2c}{3}\gamma_a \mu_a^{3/2}\left[1 + \frac{\pi^2}{8}\left(\frac{kT}{\mu_a}\right)^2 + \cdots\right]$$
$$\frac{U_a}{V_a} \simeq \frac{2c}{5}\gamma_a \mu_a^{5/2}\left[1 + \frac{5\pi^2}{8}\left(\frac{kT}{\mu_a}\right)^2 + \cdots\right].$$

From the first of these equations, evaluated at $T = 0$, we obtain the Fermi energies

$$\varepsilon_{F,a} = \left(\frac{3N}{2c\gamma_a V_a}\right)^{2/3}.$$

At a low but non-zero temperature, the two conditions $N_A = N_B = N$ and $U_A/V_A = U_B/V_B$ can be rearranged to give

$$r \equiv \frac{V_A}{V_B} = \left(\frac{\gamma_B}{\gamma_A}\right)^{2/5}\left[1 - \frac{\pi^2}{4}(kT)^2\left(\frac{1}{\mu_B^2} - \frac{1}{\mu_A^2}\right) + \cdots\right].$$

We immediately find the zero-temperature result $r_0 = (\gamma_B/\gamma_A)^{2/5} = 2^{2/5}$, where $r_0 \equiv r(T = 0)$. Note that both this result and the high-temperature limit could be obtained without using the constraint $V_A + V_B = V$. To find the leading correction to r_0, however, we do need this constraint, which we use to write $V_A = rV/(1 + r)$ and $V_B = V/(1 + r)$. In the correction term proportional to T^2, it is sufficient to set $\mu_a = \varepsilon_{F,a}$ and (since $\varepsilon_{F,a}$ depends on V_a) $r = r_0$. In this way, we finally obtain

$$r = r_0\left[1 - \kappa\frac{\pi^2}{4}\left(\frac{2\pi mkT}{h^2}\right)^2\left(\frac{4}{3\sqrt{\pi}}\frac{V}{N}\right)^{4/3} + \cdots\right]$$

where

$$\kappa = \left(\frac{\gamma_B}{1 + r_0}\right)^{4/3} - \left(\frac{\gamma_A r_0}{1 + r_0}\right)^{4/3} \simeq 0.88.$$

(b) Since the gas expands freely into empty space and is thermally isolated, its total energy remains unchanged. A classical gas at $T = 0$ has, of course, no energy and after a free adiabatic expansion will still be at $T = 0$. However, the Fermi gas at $T = 0$ has an internal energy $U = \frac{3}{5}N\varepsilon_F(V)$ where, as found above, the Fermi energy $\varepsilon_F(V) = (3N/4cV)^{2/3}$ is a decreasing function of volume. Thus, the gas in question has a fixed energy $U = \frac{3}{5}N\varepsilon_F(V_0)$ which, after the expansion, is greater than the energy $U' = \frac{3}{5}N\varepsilon_F(V_0 + \Delta V)$ that it would have if its temperature were still zero. Assuming that it eventually reaches a state of thermal equilibrium, the excess energy corresponds to a non-zero temperature. If the fractional change in volume is small, the final temperature will also be small, and we can use the approximations described in (a). The equation for N/V can be approximately solved for the chemical potential, giving

$$\mu = \varepsilon_{\mathrm{F}}\left[1 - \frac{\pi^2}{12}\left(\frac{kT}{\varepsilon_{\mathrm{F}}}\right)^2 + \cdots\right]$$

and this may be used to express the internal energy as

$$U = \tfrac{3}{5}N\varepsilon_{\mathrm{F}}\left[1 + \frac{5\pi^2}{12}\left(\frac{kT}{\varepsilon_{\mathrm{F}}}\right)^2 + \cdots\right].$$

The fact that this internal energy is unchanged is expressed by

$$\tfrac{3}{5}N\varepsilon_{\mathrm{F}}(V_0 + \Delta V)\left[1 + \frac{5\pi^2}{12}\left(\frac{kT}{\varepsilon_{\mathrm{F}}(V_0 + \Delta V)}\right)^2 + \cdots\right] = \tfrac{3}{5}N\varepsilon_{\mathrm{F}}(V_0)$$

and on using the expansion $\varepsilon_{\mathrm{F}}(V_0 + \Delta V) = \varepsilon_{\mathrm{F}}(1 - 2\,\Delta V/3V_0 + \cdots)$, with $\varepsilon_{\mathrm{F}} \equiv \varepsilon_{\mathrm{F}}(V_0)$, this becomes

$$\tfrac{3}{5}N\varepsilon_{\mathrm{F}}\left[1 + \frac{5\pi^2}{12}\left(\frac{kT}{\varepsilon_{\mathrm{F}}}\right)^2 - \frac{2}{3}\frac{\Delta V}{V_0} + \cdots\right] = \tfrac{3}{5}N\varepsilon_{\mathrm{F}}.$$

Clearly, the final temperature is

$$T = \frac{\varepsilon_{\mathrm{F}}}{k}\left(\frac{8}{5\pi^2}\frac{\Delta V}{V_0}\right)^{1/2} + \cdots$$

where the ellipsis indicate higher powers of $\Delta V/V_0$.

• Problem 4.21

(a) The grand canonical partition function for an ideal gas can be cast as a single expression applying to all three kinds of statistics, namely

$$\mathcal{Z}_a = \prod_i (1 + a\,\mathrm{e}^{-\beta(\varepsilon_i - \mu)})^{1/a}$$

where $a = 1$ corresponds to fermions and $a = -1$ to bosons. Since $\lim_{a\to 0}[(1 + ax)^{1/a}] = \mathrm{e}^x$, this limit describes Maxwell–Boltzmann particles. In all three cases, therefore, the pressure and internal energy are given by

$$\beta PV = \ln \mathcal{Z}_a = \frac{1}{a}\int_0^\infty \mathrm{d}\varepsilon\, g(\varepsilon)\ln(1 + a\,\mathrm{e}^{-\beta(\varepsilon-\mu)})$$

$$U = -\left(\frac{\partial(\ln \mathcal{Z}_a)}{\partial\beta}\right)_{\beta\mu} = \int_0^\infty \mathrm{d}\varepsilon\, g(\varepsilon)\varepsilon\,\frac{1}{\mathrm{e}^{\beta(\varepsilon-\mu)} + a}$$

where $g(\varepsilon)$ is the density of states. As discussed in detail in problem 4.12, this density of states has the form $g(\varepsilon) = \gamma \varepsilon^{d/s-1}$, where γ is a constant (depending on d and s). Consequently, using an integration by parts, we find that

$$\begin{aligned}\beta PV &= \frac{1}{a}\gamma \int_0^\infty \mathrm{d}\varepsilon\, \varepsilon^{d/s-1} \ln(1 + a\,\mathrm{e}^{-\beta(\varepsilon-\mu)}) \\ &= \beta\gamma \frac{s}{d} \int_0^\infty \mathrm{d}\varepsilon\, \varepsilon^{d/s} \frac{1}{\mathrm{e}^{\beta(\varepsilon-\mu)} + a} \\ &= \sigma\beta U\end{aligned}$$

where $\sigma = s/d$ is independent both of temperature and of statistics. For $s = 2$ and $d = 3$, we recover the usual relation $PV = 2U/3$ for a non-relativistic gas in 3 dimensions. For radiation (or, indeed, for any highly relativistic particles) we have $s = 1$, and the radiation pressure is given by $PV = U/d$.

(b) For fermions (with $a = 1$) the total number of particles is given in terms of the chemical potential by

$$\begin{aligned}N &= \gamma \int_0^\infty \mathrm{d}\varepsilon \frac{\varepsilon^{d/s-1}}{\mathrm{e}^{\beta(\varepsilon-\mu)} + 1} \\ &= \gamma (kT)^{d/s} I_{d/s-1}(x) \\ &= \frac{s}{d}\gamma\mu^{d/s}\left[1 + \frac{d}{s}\left(\frac{d}{s} - 1\right)\frac{\pi^2}{6}x^{-2} + \cdots\right]\end{aligned}$$

where $x = \mu/kT$ and $I_n(x) = \int_0^\infty \mathrm{d}u\, u^n(\mathrm{e}^{u-x} + 1)^{-1}$ can be approximated for large x by the expression given in appendix A. Solving this equation for μ, we find that

$$\mu = \varepsilon_\mathrm{F}\left[1 - \left(\frac{d}{s} - 1\right)\frac{\pi^2}{6}\left(\frac{kT}{\varepsilon_\mathrm{F}}\right)^2 + \cdots\right]$$

where the Fermi energy is $\varepsilon_\mathrm{F} = (dN/\gamma s)^{s/d}$. With the same approximations, the internal energy obtained in (a) becomes

$$U = \frac{s}{d+s}\gamma\mu^{d/s+1}\left[1 + \frac{d}{s}\left(\frac{d}{s} + 1\right)\frac{\pi^2}{6}x^{-2} + \cdots\right]$$

and so

$$\begin{aligned}\frac{U}{N} &= \frac{d}{d+s}\mu\left(1 + \frac{d}{s}\frac{\pi^2}{3}x^{-2} + \cdots\right) \\ &= \frac{d}{d+s}\varepsilon_\mathrm{F}\left(1 + \left(\frac{d}{s} + 1\right)\frac{\pi^2}{6}\left(\frac{kT}{\varepsilon_\mathrm{F}}\right)^2 + \cdots\right).\end{aligned}$$

• **Problem 4.22** In general, particles will flow from a region of higher chemical potential to a region of lower chemical potential. We therefore need to find out in which region the chemical potential is higher, and we do this by considering the grand canonical expression for the number of particles per unit volume. In the presence of a magnetic field, the single-particle energy is $\varepsilon \pm \tau H$, where ε is the kinetic energy, depending on whether the magnetic moment is parallel or antiparallel to the field. The total number of particles is then given by

$$N = \int_0^\infty \mathrm{d}\varepsilon\, g(\varepsilon) \frac{1}{\mathrm{e}^{\beta(\varepsilon-\mu-\tau H)}+1} + \int_0^\infty \mathrm{d}\varepsilon\, g(\varepsilon) \frac{1}{\mathrm{e}^{\beta(\varepsilon-\mu+\tau H)}+1}.$$

For non-relativistic particles in a d-dimensional volume V, the density of states is $g(\varepsilon) = \gamma V \varepsilon^{d/2-1}$, where γ is a constant. At $T = 0$, the Fermi distribution function is

$$\lim_{\beta\to\infty} \left(\frac{1}{\mathrm{e}^{\beta(\varepsilon-\mu\pm\tau H)}+1} \right) = \theta(\mu \mp \tau H - \varepsilon)$$

where $\theta(\cdot)$ is the step function, so the integrals are easily evaluated with the result

$$\frac{N}{V} = \frac{2\gamma}{d}[(\mu+\tau H)^{d/2} + (\mu-\tau H)^{d/2}].$$

At the moment that the wall is removed, N/V is the same in regions A and B; so (with $H = 0$ in region B) we have

$$(\mu_\mathrm{A}+\tau H)^{d/2} + (\mu_\mathrm{A}-\tau H)^{d/2} = 2\mu_\mathrm{B}^{d/2}.$$

For small fields, we can make use of the Taylor expansion

$$(1\pm x)^{d/2} = 1 \pm \frac{d}{2}x + \frac{d}{4}\left(\frac{d}{2}-1\right)x^2 + \cdots$$

to obtain

$$\left(\frac{\mu_\mathrm{B}}{\mu_\mathrm{A}}\right)^{d/2} = 1 + \frac{d(d-2)}{8}\left(\frac{\tau H}{\mu_\mathrm{A}}\right)^2 + \cdots.$$

We see that, for $d = 2$, the chemical potentials are equal, so there is no flow of particles. For $d > 2$, we have $\mu_\mathrm{B} > \mu_\mathrm{A}$ and so particles flow towards the magnetic field in region A while, for $d < 2$, the opposite is true. We can prove that the same result holds for any magnetic field strength as follows. For compactness, we write $\lambda = \tau H$. Since our basic equation $(\mu_\mathrm{A}+\lambda)^{d/2} + (\mu_\mathrm{A}-\lambda)^{d/2} = 2\mu_\mathrm{B}^{d/2}$ is unchanged if we change λ to $-\lambda$, we can take $\lambda > 0$ without loss of generality. Bearing in mind that μ_B is fixed, we calculate $\mathrm{d}\mu_\mathrm{A}/\mathrm{d}\lambda$ as

$$\frac{\mathrm{d}\mu_\mathrm{A}}{\mathrm{d}\lambda} = \frac{(\mu_\mathrm{A}-\lambda)^{d/2-1} - (\mu_\mathrm{A}+\lambda)^{d/2-1}}{(\mu_\mathrm{A}-\lambda)^{d/2-1} + (\mu_\mathrm{A}+\lambda)^{d/2-1}}.$$

Since $\mu_\mathrm{A}+\lambda > \mu_\mathrm{A}-\lambda$, we have $(\mu_\mathrm{A}+\lambda)^{d/2-1} > (\mu_\mathrm{A}-\lambda)^{d/2-1}$ if $d > 2$ and vice versa. Therefore, if $d > 2$, then $\mathrm{d}\mu_\mathrm{A}/\mathrm{d}\lambda$ is negative and, as the field is increased, μ_A decreases from its zero-field value μ_B and is always smaller than μ_B. Conversely, if $d < 2$, then μ_A is always greater than μ_B. For $d = 2$, we have $\mu_A = \mu_B$ independent of the field.

• **Problem 4.23**

(a) Classically, an electron in a magnetic field oriented in the z direction describes a helical path whose projection onto the x–y plane is a circle. Quantum-mechanically, the z component of momentum is a conserved quantity (it commutes with the Hamiltonian) and so p_z is a good quantum number. However, the motion in the x–y plane is essentially equivalent to that of a harmonic oscillator, giving rise to the energies $(\hbar eB/mc)(j + \frac{1}{2})$. The degeneracy $g = 2V^{2/3}eB/2\pi\hbar c$, which is the same for all p_z and j, corresponds to two possibilities for the spin polarization and, roughly speaking, to the possible locations of a circular orbit in a plane of area $V^{2/3}$. (The precise value of g can be checked by requiring that the partition function that we are about to compute reduce to the correct ideal gas form in the limit $B \to 0$, where the sum over j can be replaced by a momentum integral with $p_x^2 + p_y^2 = 2eB\hbar j/c$).

The grand canonical partition function is

$$\mathcal{Z} = \prod_{p_z, j} (1 + z\,\mathrm{e}^{-\beta\varepsilon(p_z, j)})^g.$$

For a system of length $L = V^{1/3}$ in the z direction, with periodic boundary conditions applied to single-particle wavefunctions, the allowed values of p_z are $2\pi\hbar n/L$, where n is a positive or negative integer. The logarithm of $\mathcal{Z}$ is, as usual, well approximated if we replace the sum over p_z by an integral, with the result

$$\ln \mathcal{Z} = \frac{gV^{1/3}}{2\pi\hbar} \int_{-\infty}^{\infty} \mathrm{d}p_z \sum_{j=0}^{\infty} \ln(1 + z\,\mathrm{e}^{-\beta\varepsilon(p_z, j)}).$$

(b) For a gas with a fixed number of particles per unit volume N/V, the fugacity is obtained in the usual way from the condition that the grand canonical mean particle number be equal to N:

$$N = z\left(\frac{\partial(\ln \mathcal{Z})}{\partial z}\right)_{T,V,B} = \frac{gV^{1/3}}{2\pi\hbar} \int_{-\infty}^{\infty} \mathrm{d}p_z \sum_{j=0}^{\infty} \frac{1}{z^{-1}\,\mathrm{e}^{\beta\varepsilon(p_z, j)} + 1}.$$

When $T \to \infty$, we can make the change of variable $p_z = \sqrt{2mkT}u$ and replace $\sum_j \to (mckT/\hbar eB)\int \mathrm{d}v$, with $v = (\hbar eB/mckT)j$ to obtain

$$N \simeq \frac{2}{\sqrt{\pi}} V \left(\frac{2\pi mkT}{h^2}\right)^{3/2} \int_{-\infty}^{\infty} \mathrm{d}u \int_0^{\infty} \mathrm{d}v\, \frac{1}{z^{-1}\,\mathrm{e}^{u^2+v} + 1}.$$

In order that N/V remain fixed, z must become very small (which means that μ tends to $-\infty$ much faster than T tends to $+\infty$). We can therefore expand in powers of z to find that

$$\frac{N}{V} \simeq \frac{2}{\sqrt{\pi}} \left(\frac{2\pi mkT}{h^2}\right)^{3/2} z \int_{-\infty}^{\infty} du\, e^{-u^2} \int_0^{\infty} dv\, e^{-v} = 2z \left(\frac{2\pi mkT}{h^2}\right)^{3/2}.$$

This is precisely the same result as for a classical gas (apart from the spin degeneracy factor $2s + 1 = 2$) and leads to

$$z = \frac{1}{2} \frac{N}{V} \left(\frac{h^2}{2\pi mkT}\right)^{3/2}.$$

(c) The diamagnetic susceptibility is $\chi(T) = kT[\partial^2(\ln \mathcal{Z})/\partial B^2]_{T,V,z}$. In the high-temperature limit studied above, where the gas is essentially classical, we have $N \simeq \ln \mathcal{Z}$. This quantity is seen to be independent of B, so that $\chi = 0$. This corresponds to the fact that the energy of a classical particle is unaffected by a magnetic field. Thus, any two circular orbits having the same speed but opposite senses are equally likely, and the gas does not develop any net magnetization (van Leeuwen's theorem). Quantum-mechanically, however, the single-particle energies are field dependent, leading to a non-zero magnetization and susceptibility. To evaluate the susceptibility, we retain the small-z approximation but carry out the sum over Landau levels exactly:

$$\begin{aligned}
\ln \mathcal{Z} &\simeq z \frac{gV^{1/3}}{2\pi\hbar} \int_{-\infty}^{\infty} dp_z\, e^{-p_z^2/2mkT} \sum_{j=0}^{\infty} e^{-(2j+1)x} \\
&\simeq 4Vz \left(\frac{2\pi mkT}{h^2}\right)^{3/2} \frac{x\, e^{-x}}{1 - e^{-2x}} \\
&\simeq 2Vz \left(\frac{2\pi mkT}{h^2}\right)^{3/2} \left(1 - \frac{x^2}{6} + \cdots\right)
\end{aligned}$$

where $x = \hbar eB/2mckT$. The expansion in x is legitimate, because we want to find the zero-field susceptibility. On differentiating, and substituting our previous result for z, we obtain

$$\chi(T) \simeq -\frac{N}{3} \left(\frac{e\hbar}{2mc}\right)^2 \frac{1}{kT}.$$

The $1/T$ dependence is reminiscent of Curie's law for the paramagnetic susceptibility of a collection of permanent magnetic dipole moments, although the physical mechanism is rather different. Indeed, the diamagnetic susceptibility is negative (which can be thought of as a manifestation of Lenz's law) whereas the paramagnetic susceptibility is positive. We see from the factor $\hbar^2$ that the diamagnetism of free electrons is indeed a quantum effect. The susceptibility vanishes both in the limit $T \to \infty$, where a quantum gas behaves like a classical gas, and at all temperatures in the limit $\hbar \to 0$ of a purely classical system.

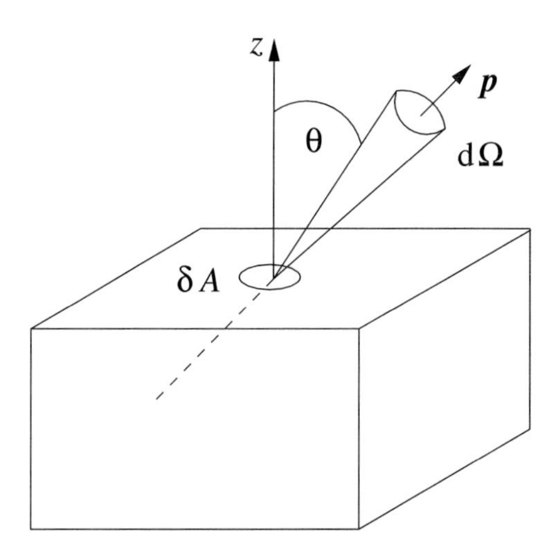

Figure 4.10. Electrons with energy $E > \mathcal{V}$ escaping through a hole of a cavity.

• **Problem 4.24** Our main task is to find the number of particles per unit time escaping from the cavity. To this end, we first suppose that the cavity is a cube of side L and that single-particle wavefunctions satisfy periodic boundary conditions at its walls. Then the allowed momentum eigenvalues are $p_i = hn_i/L$ ($i = 1, \ldots, 3$), where the n_i are positive or negative integers. Then (allowing for two spin polarizations) the number of states with momentum in the range d^3p is $(2V/h^3)\,\mathrm{d}^3p$, where $V = L^3$ is the volume. (To an excellent approximation, this result is independent of the shape of the cavity.) Multiplying by the grand canonical occupation numbers for these states, we find that the number of particles per unit volume with momentum in the range d^3p near $\boldsymbol{p}$ is $n(p)\,\mathrm{d}^3p$, where

$$n(p) = \frac{2}{h^3}\frac{1}{\mathrm{e}^{\beta[\varepsilon(p)-\mu]}+1}$$

with $\varepsilon(p) = |\boldsymbol{p}|^2/2m$. Now we consider, in particular, a small volume enclosing the hole in the cavity wall, and adopt a polar coordinate system with the usual angles θ and ϕ, such that the axis $\theta = 0$ (the z axis, say) is the outward normal to the hole (figure 4.10).

The number of electrons per unit volume whose momentum has a magnitude between p and $p + \mathrm{d}p$ and is directed into a solid angle $\mathrm{d}\Omega = \sin\theta\,\mathrm{d}\theta\,\mathrm{d}\phi$ surrounding the direction (θ, ϕ) is

$$n(p)p^2 \sin\theta\,\mathrm{d}p\,\mathrm{d}\theta\,\mathrm{d}\phi$$

and, since these electrons have a speed p/m, the number per unit time crossing a unit area normal to the (θ, ϕ) direction is

$$\frac{1}{m} n(p) p^3 \sin\theta \, \mathrm{d}p \, \mathrm{d}\theta \, \mathrm{d}\phi.$$

The hole subtends an area $\delta A \cos\theta$ normal to this direction, so the number of electrons per unit time passing through the hole with momentum between p and $p + \mathrm{d}p$ into the solid angle $\mathrm{d}\Omega$ is

$$\frac{\delta A}{m} n(p) p^3 \sin\theta \cos\theta \, \mathrm{d}p \, \mathrm{d}\theta \, \mathrm{d}\phi.$$

It is useful to check this for the case of a classical gas, with $n(p) = n(2\pi mkT)^{-3/2} \mathrm{e}^{-p^2/2mkT}$, where n is the total number density, and with $\mathcal{V} = 0$. Bearing in mind that only those particles escape for which $0 \leq \theta < \pi/2$, we then find that the total number of particles escaping per unit time is

$$\frac{\delta A \, n}{m(2\pi mkT)^{3/2}} \int_0^\infty \mathrm{d}p \int_0^{\pi/2} \mathrm{d}\theta \int_0^{2\pi} \mathrm{d}\phi \; p^3 \mathrm{e}^{-p^2/2mkT} \sin\theta \cos\theta = \tfrac{1}{4} n \bar{c} \, \delta A$$

where $\bar{c} = \sqrt{8kT/\pi m}$ is the mean speed. This standard result can be obtained by several methods in elementary kinetic theory.

For the case in hand, the number of electrons escaping per unit time is

$$\begin{aligned}
\frac{\mathrm{d}N}{\mathrm{d}t} &= \frac{\delta A}{m} \int_{\sqrt{2m\mathcal{V}}}^\infty \mathrm{d}p \int_0^{\pi/2} \mathrm{d}\theta \int_0^{2\pi} \mathrm{d}\phi \; p^3 n(p) \sin\theta \cos\theta \\
&= \frac{2\pi \, \delta A}{mh^3} \int_{\sqrt{2m\mathcal{V}}}^\infty \frac{p^3}{\mathrm{e}^{[\varepsilon(p)-\mu]/kT} + 1} \mathrm{d}p.
\end{aligned}$$

If $\mathcal{V} - \mu \gg kT$, then $[\varepsilon(p) - \mu]/kT$ is large for all values of p in the range of integration, and we can use the approximation

$$\begin{aligned}
\frac{\mathrm{d}N}{\mathrm{d}t} &\simeq \frac{2\pi \, \delta A}{mh^3} \int_{\sqrt{2m\mathcal{V}}}^\infty p^3 \mathrm{e}^{-[\varepsilon(p)-\mu]/kT} \, \mathrm{d}p \\
&\simeq \frac{\pi \, \delta A}{mh^3} (2mkT)^2 \mathrm{e}^{-(\mathcal{V}-\mu)/kT} \left(1 + \frac{\mathcal{V}}{kT}\right).
\end{aligned}$$

Since μ is positive for a Fermi gas at low temperatures, we also have $\mathcal{V} \gg kT$, so the 1 in $1 + \mathcal{V}/kT$ can be neglected. Finally, therefore, on multiplying by the charge $-e$ of each electron, we estimate the current as

$$I = -\left(\frac{4\pi me}{h^3}\right) \delta A \, \mathcal{V} kT \, \mathrm{e}^{-(\mathcal{V}-\mu)/kT}.$$

If the temperature is low enough that $kT \ll \varepsilon_{\mathrm{F}}$, then μ can be replaced by the Fermi energy $\varepsilon_{\mathrm{F}} = (h^2/8m)(3n/\pi)^{2/3}$, where n is the total number of particles per unit volume (see problem 4.17). Of course, the current that we have calculated is the charge per unit time emerging from the hole. The number of electrons per unit solid angle at an angle θ to the normal (z direction) is proportional to $\cos\theta$, so, although no electrons emerge tangentially to the wall ($\theta = \pi/2$), not all of them travel in the z direction.

• **Problem 4.25** As in problem 3.25, we apply the grand canonical ensemble to this problem by regarding the gas of free particles and each absorbing site as separately in equilibrium with a heat and particle reservoir. Each of these systems therefore has the temperature T and the chemical potential μ of the reservoir and the chemical potential is determined by the constraint $N_\mathrm{a}(T,\mu) + N_\mathrm{f}(T,\mu) = N$. The average number of electrons absorbed at each site is $(1 + \mathrm{e}^{-\beta(\mu+\varepsilon_0)})^{-1}$, so the average number of absorbed electrons is

$$N_\mathrm{a} = \frac{N_0}{1 + \mathrm{e}^{-\beta(\mu+\varepsilon_0)}}.$$

The number of free electrons is (see, for example, problem 4.17 for details)

$$\begin{aligned} N_\mathrm{f} &= cV \int_0^\infty \mathrm{d}\varepsilon \, \frac{\varepsilon^{1/2}}{\mathrm{e}^{\beta(\varepsilon-\mu)} + 1} \\ &= cV(kT)^{3/2} \int_0^\infty \mathrm{d}x \, \frac{x^{1/2}}{z^{-1}\,\mathrm{e}^x + 1} \\ &\equiv cV(kT)^{3/2} I_{1/2}\left(\frac{\mu}{kT}\right) \end{aligned}$$

where $c = (4/\pi^2)(m/2\hbar^2)^{3/2}$, the integration variable in the second expression is $x = \varepsilon/kT$ and the fugacity is $z = \mathrm{e}^{\mu/kT}$.

(a) If $N > N_0$, then N_f is always finite and non-zero. As for a Fermi gas in a non-absorbent container, this requires that $z \to \infty$ when $T \to 0$ and $z \to 0$ when $T \to \infty$. In the low-temperature limit, therefore, we find that $N_\mathrm{a} \to N_0$ and $N_\mathrm{f} \to N - N_0$ while, in the high-temperature limit, $N_\mathrm{a} \to 0$ and $N_\mathrm{f} \to N$. As might be expected, all the electrons are in their lowest-energy state (absorbed) at $T = 0$, while very few are in this state at high temperatures.

(b) For $N = N_0$, the constraint $N = N_\mathrm{a} + N_\mathrm{f}$ becomes $N_0 - N_\mathrm{a} = N_\mathrm{f}$ or, explicitly,

$$\frac{N_0/V}{\mathrm{e}^{(\mu+\varepsilon_0)/kT} + 1} = c(kT)^{3/2} I_{1/2}\left(\frac{\mu}{kT}\right).$$

In order to find an appropriate approximation for the integral $I_{1/2}(\mu/kT)$, we must determine how μ/kT behaves as $T \to 0$. Suppose first that μ/kT approaches some finite value. Then the left-hand side of the constraint varies as $\mathrm{e}^{-\varepsilon_0/kT}$, while the right-hand side varies as $(kT)^{3/2}$, which is inconsistent. Similarly, suppose that $\mu/kT \to +\infty$. The left-hand side of the constraint behaves as $e^{-(\mu+\varepsilon_0)/kT}$. From the approximation given in appendix A, we find that $I_{1/2}(\mu/kT) \sim (\mu/kT)^{3/2}$, so the right-hand side of the constraint is approximately $c\mu^{3/2}$ which is also inconsistent. We conclude that $\mu/kT \to -\infty$ and $z \to 0$. Consequently, the integral may be expanded in powers of z:

$$I_{1/2}\left(\frac{\mu}{kT}\right) = z\int_0^\infty dx\, x^{1/2}\, e^{-x}[1 + O(z)] = \frac{\sqrt{\pi}}{2}z + \cdots$$

and the constraint reads approximately

$$\frac{N_0/V}{e^{(\mu+\varepsilon_0)/kT} + 1} \simeq \frac{\sqrt{\pi}c}{2}(kT)^{3/2}z$$

or

$$e^{\varepsilon_0/kT}\, z^2 + z - \kappa(kT)^{-3/2} \simeq 0$$

where $\kappa = 2N_0/\sqrt{\pi}cV$. Note that the term proportional to z^2 has a large coefficient $e^{\varepsilon_0/kT}$ and must be retained, even though a term proportional to z^2 (but with a coefficient of order 1) was neglected in approximating the integral. When $T \to 0$, the solution to this quadratic equation is approximately

$$z \simeq \sqrt{\kappa}(kT)^{-3/4}\, e^{-\varepsilon_0/2kT}$$

and the logarithm of this expression yields the chemical potential as

$$\mu \simeq \tfrac{1}{2}\left[kT \ln\left(\frac{\kappa}{(kT)^{3/2}}\right) - \varepsilon_0\right].$$

Substituting this into our earlier expression for N_f, we obtain

$$N_f \simeq \left(\frac{\sqrt{\pi}cN_0V}{2}\right)^{1/2}(kT)^{3/4}\, e^{-\varepsilon_0/2kT}$$

and, of course, $N_a = N_0 - N_f$. We see that $\mu(T = 0)$ has the value $-\varepsilon_0/2$—the same value that we would have obtained for a classical gas. This result may seem surprising at first sight, since quantum effects are generally important at low temperatures. It arises from the fact that at $T = 0$ all the particles are located in the absorbing sites and, since there is only one particle per site, the quantum statistics become irrelevant.

• Problem 4.26

(a) For fermions, a given state can be unoccupied, say with probabilty $\mathcal{P}_0(\varepsilon, \mu)$, or occupied by one particle, with probability $\mathcal{P}_1(\varepsilon, \mu)$. The mean occupation number is therefore

$$n_p(\varepsilon, \mu) = 0 \times \mathcal{P}_0(\varepsilon, \mu) + 1 \times \mathcal{P}_1(\varepsilon, \mu) = \mathcal{P}_1(\varepsilon, \mu).$$

So the probability of finding the state occupied by a particle is just the Fermi–Dirac occupation number $n_{\mathrm{p}}(\varepsilon, \mu) = (\mathrm{e}^{\beta(\varepsilon-\mu)} + 1)^{-1}$. For reasons which will shortly become apparent, we introduce the subscript p to distinguish particles from holes. The probability of finding the state unoccupied is

$$\mathcal{P}_0(\varepsilon, \mu) = 1 - \mathcal{P}_1(\varepsilon, \mu) = (\mathrm{e}^{-\beta(\varepsilon-\mu)} + 1)^{-1}$$

and it is simple to see that $\mathcal{P}_0(\mu - \alpha, \mu) = \mathcal{P}_1(\mu + \alpha, \mu) = (\mathrm{e}^{\beta\alpha} + 1)^{-1}$.

More generally, we see that $\mathcal{P}_0(\varepsilon, \mu)$ has the form of the Fermi–Dirac occupation number for a state of energy $-\varepsilon$ and chemical potential $-\mu$. At least in this respect, it is apparently possible to treat the system as a gas of two particle species, the particles and holes, both having a positive energy. From this point of view, the particles have a density of states $g_{\mathrm{p}}(\varepsilon) = A(\varepsilon - \varepsilon_0)^{1/2}$ and occupation numbers $n_{\mathrm{p}}(\varepsilon, \mu)$, while the holes have a density of states $g_{\mathrm{h}}(\varepsilon) = B\varepsilon^{1/2}$ and occupation numbers $n_{\mathrm{h}}(\varepsilon, \mu) = \mathcal{P}_0(-\varepsilon, \mu) = (\mathrm{e}^{\beta(\varepsilon+\mu)} + 1)^{-1}$. If the number of positive-energy particles is N_{p} and the number of holes is N_{h}, then the chemical potential μ is conjugate to the net particle number $N = N_{\mathrm{p}} - N_{\mathrm{h}}$. This net particle number is a conserved quantity, since the promotion of a particle from a negative energy to a positive energy creates one particle and one hole. This interpretation can, in fact, be applied quite consistently. If the Dirac sea has a particle of momentum $\boldsymbol{p}$ missing, then the remaining particles carry a net momentum $-\boldsymbol{p}$, which can be attributed to the hole. If the particles are electrons, say, with a negative charge, then the filled Dirac sea gives rise to a uniform background of negative charge against which a hole constitutes a single positive charge. In the context of semiconductor physics, this description is a convenient fiction. The energy of the Dirac sea is finite, because the energy spectrum does not really extend to $-\infty$, and its charge just serves to cancel that of the positive ions in the host material. In relativistic quantum theory, the holes correspond to antiparticles. Here, the infinite negative energy and charge of the Dirac sea are an embarrassment. However, the methods of second quantization permit a fully consistent treatment of particles and antiparticles, in which the notion of a Dirac sea is not needed. This is fortunate, because bosons also have antiparticles but do not obey the Pauli exclusion principle, so for them the Dirac sea construction is meaningless.

(b) If, at $T = 0$, all the negative-energy states are filled and all the positive-energy states are empty, then the conserved particle number $N = N_{\mathrm{p}} - N_{\mathrm{h}}$ is zero. At all temperatures, therefore, we have $N_{\mathrm{p}} = N_{\mathrm{h}}$ and this gives an integral equation which, in principle, can be solved for μ:

$$A \int_{\varepsilon_0}^{\infty} \frac{(\varepsilon - \varepsilon_0)^{1/2}}{\mathrm{e}^{\beta(\varepsilon-\mu)} + 1} \,\mathrm{d}\varepsilon = B \int_0^{\infty} \frac{\varepsilon^{1/2}}{\mathrm{e}^{\beta(\varepsilon+\mu)} + 1} \,\mathrm{d}\varepsilon.$$

By making the change of variable $\varepsilon = \varepsilon_0 + kTu$ in the first integral and $\varepsilon = kTu$

in the second, we obtain

$$\frac{A}{B} = \frac{I_{1/2}(-\mu/kT)}{I_{1/2}[(\mu-\varepsilon_0)/kT]}$$

where, as in previous problems, $I_{1/2}(x) = \int_0^\infty u^{1/2}(\mathrm{e}^{u-x}+1)^{-1}\,\mathrm{d}u$. In general, it is impossible to obtain a closed-form solution to this equation. In the symmetrical case $A = B$, however, the solution is clearly $\mu = \varepsilon_0/2$ and at low temperatures we find that

$$\begin{aligned} N_\mathrm{p} = N_\mathrm{h} &= A(kT)^{3/2}\int_0^\infty \frac{u^{1/2}}{\mathrm{e}^{u+\varepsilon_0/2kT}+1}\,\mathrm{d}u \\ &\simeq A(kT)^{3/2}\,\mathrm{e}^{-\varepsilon_0/2kT}\int_0^\infty u^{1/2}\,\mathrm{e}^{-u}\,\mathrm{d}u \\ &= \frac{\sqrt{\pi}A}{2}(kT)^{3/2}\,\mathrm{e}^{-\varepsilon_0/2kT}. \end{aligned}$$

(c) If, at $T = 0$, all the negative-energy states are filled and, in addition, N_0 particles occupy positive-energy states, then the conserved particle number is $N = N_\mathrm{p} - N_\mathrm{h} = N_0$, so the constraint equation for μ is

$$N_0 = (kT)^{3/2}\left[AI_{1/2}\left(\frac{\mu-\varepsilon_0}{kT}\right) - BI_{1/2}\left(-\frac{\mu}{kT}\right)\right].$$

Since at low temperatures $N_\mathrm{p} \simeq N_0$ and $N_\mathrm{h} \simeq 0$, we expect $(\mu-\varepsilon_0)/kT \to +\infty$ and $-\mu/kT \to -\infty$ as $T \to 0$ (see, for example, problem 4.25 for details). In the two terms of the constraint equation, we therefore use the two limits

$$I_{1/2}(x) \simeq \begin{cases} \dfrac{2}{3}x^{3/2}\left[1+\dfrac{\pi^2}{8}x^{-2}+\cdots\right] & \text{if } x \to +\infty \\ \dfrac{\sqrt{\pi}}{2}\,\mathrm{e}^x & \text{if } x \to -\infty \end{cases}$$

to obtain

$$N_0 \simeq (kT)^{3/2}\left\{\frac{2A}{3}\left(\frac{\mu-\varepsilon_0}{kT}\right)^{3/2}\left[1+\frac{\pi^2}{8}\left(\frac{kT}{\mu-\varepsilon_0}\right)^2\right] - \frac{\sqrt{\pi}B}{2}\,\mathrm{e}^{-\mu/kT}\right\}.$$

To solve approximately for $\mu(T)$, we first set $T = 0$ to find $\mu(T=0) \equiv \varepsilon_\mathrm{F} = \varepsilon_0 + (3N_0/2A)^{2/3}$. Then, on setting $\mu = \varepsilon_\mathrm{F}$ in the non-leading terms, the remaining equation can be solved to obtain

$$\bar{\mu} = \bar{\varepsilon}_\mathrm{F}\left[1 - \frac{\pi^2}{12}\left(\frac{kT}{\bar{\varepsilon}_\mathrm{F}}\right)^2 + \frac{B}{A}\frac{\sqrt{\pi}}{2}\left(\frac{kT}{\bar{\varepsilon}_\mathrm{F}}\right)^{3/2}\mathrm{e}^{-\varepsilon_\mathrm{F}/kT} + \cdots\right]$$

where $\bar{\mu} = \mu - \varepsilon_0$ and $\bar{\varepsilon}_\mathrm{F} = \varepsilon_\mathrm{F} - \varepsilon_0$. The first two terms in the square brackets are precisely the expression obtained in several earlier problems for the chemical potential of an ideal Fermi gas in the absence of a Dirac sea (with ε_F and μ defined relative to the lower edge of the conduction band and thus corresponding to $\bar{\varepsilon}_F$ and $\bar{\mu}$ here). The last term, proportional to B/A, can therefore be identified as the correction due to the Dirac sea, and we note the characteristic Boltzmann factor $e^{-\varepsilon_F/kT}$ arising from the fact that a particle in the sea must gain at least an energy ε_F to be promoted to an unoccupied positive-energy state. We can recover the standard result, without the Dirac-sea correction, by taking the limit $\varepsilon_0 \to \infty$ of an infinite band gap. In that limit, $\varepsilon_F \to \infty$, but $\bar{\varepsilon}_F$ remains finite.

• Problem 4.27

(a) The energy of a highly relativistic particle is $\varepsilon(\boldsymbol{p}) = c|\boldsymbol{p}|$, where c is the speed of light, so, as found in problem 4.12(a) (with $\alpha = c$, $s = 1$ and $d = 3$), the density of translational states is $g(\varepsilon) = (4\pi V/h^3c^3)\varepsilon^2$. For a single species, the grand potential is therefore

$$\begin{aligned}\Omega(T, V, \mu) = kT \ln \mathcal{Z} &= \pm kT \frac{4\pi V}{(hc)^3} f^{\text{spin}} \int_0^\infty \mathrm{d}\varepsilon\, \varepsilon^2 \ln(1 \pm e^{-\beta(\varepsilon-\mu)}) \\ &= \frac{4\pi V}{3(hc)^3} f^{\text{spin}} \int_0^\infty \mathrm{d}\varepsilon\, \frac{\varepsilon^3}{e^{\beta(\varepsilon-\mu)} \pm 1}\end{aligned}$$

where the upper signs refer to fermions while the lower signs refer to bosons, and the second integral is obtained from an integration by parts. The spin degeneracy factor is $f^{\text{spin}} = 1$ for particles of spin $s = 0$. It is $f^{\text{spin}} = 2$ for both spin-$\frac{1}{2}$ and spin-1 particles, whose helicity is restricted to be ± 1 in the relativistic limit. The particle number density n, the energy density u and the entropy density σ are given by

$$\begin{aligned}n &= \frac{1}{V}\left(\frac{\partial \Omega}{\partial \mu}\right)_{T,V} = \frac{4\pi f^{\text{spin}}}{(hc)^3} \int_0^\infty \mathrm{d}\varepsilon\, \frac{\varepsilon^2}{e^{\beta(\varepsilon-\mu)} \pm 1} \\ u &= -\frac{1}{V}\left(\frac{\partial(\ln \mathcal{Z})}{\partial \beta}\right)_{\beta\mu,V} = \frac{4\pi f^{\text{spin}}}{(hc)^3} \int_0^\infty \mathrm{d}\varepsilon\, \frac{\varepsilon^3}{e^{\beta(\varepsilon-\mu)} \pm 1} \\ \sigma &= \frac{1}{V}\left(\frac{\partial \Omega}{\partial T}\right)_{\mu,V} = \frac{4\pi f^{\text{spin}}}{3(hc)^3 T} \int_0^\infty \mathrm{d}\varepsilon\, \frac{\varepsilon^2(4\varepsilon - 3\mu)}{e^{\beta(\varepsilon-\mu)} \pm 1}.\end{aligned}$$

Under the stated conditions, the chemical potentials for all particle species in the gas are zero. This is because the stationary probability density (which is summed and integrated to get $\mathcal{Z}$) can contain chemical potentials only in a combination of the form $\exp(\sum_i \mu_i \mathcal{N}_i)$, where the particle numbers $\mathcal{N}_i$ are conserved quantities. For particles which are their own antiparticles and here assumed to be created and annihilated in arbitrary numbers, there is no conserved particle number, so the corresponding chemical potential must vanish. For species which are created and annihilated in particle–antiparticle pairs, the difference $\mathcal{N} = N - \bar{N}$ in the numbers of particles and antiparticles is conserved, so a term of the form

$\exp[\mu(N-\bar{N})]$ is permitted, corresponding to chemical potentials μ and $\bar{\mu} = -\mu$ for particles and antiparticles. If equal numbers of particles and antiparticles are to be present, we must then have $\mu = \bar{\mu} = 0$. (Note that, depending on the nature of the interactions between various particle species, a variety of constraints on the chemical potentials different from those considered here are possible.)

If all the chemical potentials are zero, then the change of variable $\varepsilon = kTx$ in each of the above integrals leads to

$$n = \frac{4\pi f^{\text{spin}}}{(hc)^3}(kT)^3 \int_0^\infty \mathrm{d}x \, \frac{x^2}{\mathrm{e}^x \pm 1}$$

$$u = \frac{4\pi f^{\text{spin}}}{(hc)^3}(kT)^4 \int_0^\infty \mathrm{d}x \, \frac{x^3}{\mathrm{e}^x \pm 1}$$

$$\sigma = k\frac{16\pi f^{\text{spin}}}{3(hc)^3}(kT)^3 \int_0^\infty \mathrm{d}x \, \frac{x^3}{\mathrm{e}^x \pm 1} = \frac{4u}{3T}.$$

We evidently require bosonic and fermionic integrals of the form

$$I_n^{(\mathrm{b})} = \int_0^\infty \mathrm{d}x \, \frac{x^n}{\mathrm{e}^x - 1} = \Gamma(n+1)\zeta(n+1) \qquad I_n^{(\mathrm{f})} = \int_0^\infty \mathrm{d}x \, \frac{x^n}{\mathrm{e}^x + 1}.$$

The bosonic integrals that we actually need are $I_2^{(\mathrm{b})} = 2\zeta(3)$, with $\zeta(3) = 1.202\ldots$ and $I_3^{(\mathrm{b})} = \pi^4/15$. The fermionic integrals can be related to these by writing

$$I_n^{(\mathrm{b})} = \int_0^\infty \mathrm{d}x \, \frac{x^n}{(\mathrm{e}^{x/2} - 1)(\mathrm{e}^{x/2} + 1)} = 2^n(I_n^{(\mathrm{b})} - I_n^{(\mathrm{f})})$$

or $I_n^{(\mathrm{f})} = (1 - 2^{-n})I_n^{(\mathrm{b})}$. Summing over all particle species, we get

$$n = \sum_i n_i = \sum_i f_i^{\text{spin}} f_i^{\text{stat}} \bar{f}_i \frac{8\pi\zeta(3)}{(hc)^3}(kT)^3$$

$$u = \sum_i u_i = \sum_i f_i^{\text{spin}} g_i^{\text{stat}} \bar{f}_i \frac{4\pi^5}{15(hc)^3}(kT)^4$$

and $\sigma = 4u/3T$, where f_i^{stat} is equal to 1 for bosons and $\frac{3}{4}$ for fermions, while g_i^{stat} is 1 for bosons and $\frac{7}{8}$ for fermions. The index i labels particle species and we have taken particles and antiparticles to belong to the same species, defining $\bar{f}_i$ to be 1 if particles of the ith species are their own antiparticles or 2 if the particles and antiparticles are distinct.

(b) The case of electrons and positrons which are created and annihilated in pairs satisfies the conditions described above, so, when they are present in equal numbers, their chemical potentials are zero. When there is an excess of electrons, say, n_0 per unit volume, their chemical potential is determined by

$$n_0 = \frac{N - \bar{N}}{V} = 8\pi\left(\frac{kT}{hc}\right)^3 \int_0^\infty \mathrm{d}x \, x^2 \left(\frac{1}{\mathrm{e}^{x-\mu/kT} + 1} - \frac{1}{\mathrm{e}^{x+\mu/kT} + 1}\right)$$

where we have used $f^{\rm spin} = 2$ for electrons. At high temperatures, the number density of thermally produced electrons should be much greater than n_0. Then μ should be small, so we can expand in powers of μ/kT to obtain

$$n_0 \simeq 16\pi \left(\frac{kT}{hc}\right)^3 \frac{\mu}{kT} A$$

where

$$A = \int_0^\infty \mathrm{d}x \, \frac{x^2 \, \mathrm{e}^x}{(\mathrm{e}^x + 1)^2}.$$

The energy density of electrons and positrons is

$$u_{e^+e^-} = \frac{8\pi}{(hc)^3}(kT)^4 \int_0^\infty \mathrm{d}x \, x^3 \left(\frac{1}{\mathrm{e}^{x-\mu/kT}+1} + \frac{1}{\mathrm{e}^{x+\mu/kT}+1}\right)$$

and this can also be expanded in powers of μ/kT to give the leading correction

$$\Delta u = \frac{8\pi}{(hc)^3}(kT)^4 B \left(\frac{\mu}{kT}\right)^2 = \frac{B}{32\pi A^2}\left(\frac{hc}{kT}\right)^3 n_0^2 \, kT$$

with

$$B = \int_0^\infty \mathrm{d}x \, \frac{x^3 \, \mathrm{e}^x \, (\mathrm{e}^x - 1)}{(\mathrm{e}^x + 1)^3}.$$

Numerical evaluation of the integrals gives $A \simeq 1.645$ and $B \simeq 4.935$.

• **Problem 4.28** The internal energies for bosons of spin $s_{\rm b}$ and fermions of spin $s_{\rm f}$ can be expressed as

$$U^{\rm BE}(T, V, \mu) = (2s_{\rm b}+1)\frac{(2m)^{3/2}V}{4\pi^2\hbar^3}(kT)^{5/2} \, B\left(\frac{\mu}{kT}\right)$$

$$U^{\rm FD}(T, V, \mu) = (2s_{\rm f}+1)\frac{(2m)^{3/2}V}{4\pi^2\hbar^3}(kT)^{5/2} \, F\left(\frac{\mu}{kT}\right)$$

where

$$B(\alpha) = \int_0^\infty \frac{x^{3/2}\,\mathrm{d}x}{\mathrm{e}^{x-\alpha}-1} \qquad F(\alpha) = \int_0^\infty \frac{x^{3/2}\,\mathrm{d}x}{\mathrm{e}^{x-\alpha}+1}.$$

We readily verify that these integrals are related by

$$B\left(\frac{\alpha}{2}\right) - F\left(\frac{\alpha}{2}\right) = 2\int_0^\infty \frac{x^{3/2}\,\mathrm{d}x}{\mathrm{e}^{2x-\alpha}-1} = 2^{-3/2}\,B(\alpha)$$

which can be rewritten as

$$F(\alpha) = B(\alpha) - 2^{-3/2}B(2\alpha).$$

Since pressure is related to internal energy by $PV = 2U/3$, this immediately implies the relation

$$P^{\mathrm{FD}}(T, V, \mu) = \frac{2s_{\mathrm{f}} + 1}{2s_{\mathrm{b}} + 1}[P^{\mathrm{BE}}(T, V, \mu) - 2^{-3/2}P^{\mathrm{BE}}(T, V, 2\mu)].$$

We wish, however, to invert this relation, expressing P^{BE} in terms of P^{FD}. To this end, we guess that $B(\alpha)$ can be expressed as

$$B(\alpha) = \sum_{k=0}^{\infty} b_k F(2^k\alpha).$$

Substituting this into our original relation, we find that

$$\begin{aligned} F(\alpha) &= \sum_{k=0}^{\infty} b_k[F(2^k\alpha) - 2^{-3/2}F(2^{k+1}\alpha)] \\ &= b_0 F(\alpha) + \sum_{k=1}^{\infty}[b_k - 2^{-3/2}b_{k-1}]F(2^k\alpha). \end{aligned}$$

For the two sides of this equation to match, we clearly need $b_0 = 1$ and $b_k = 2^{-3/2}b_{k-1}$, which implies that $b_k = 2^{-3k/2}$. Proceeding as above, we finally obtain

$$P^{\mathrm{BE}}(T, V, \mu) = \frac{2s_{\mathrm{b}} + 1}{2s_{\mathrm{f}} + 1}\sum_{k=0}^{\infty} 2^{-3k/2}P^{\mathrm{FD}}(T, V, 2^k\mu)$$

which has the required form, with $a_k = 2^{-3k/2}(2s_{\mathrm{b}}+1)/(2s_{\mathrm{f}}+1)$ and $\mu_k = 2^k\mu$.

• **Problem 4.29** When the average interparticle separation $v^{1/3}$ is much larger than the thermal wavelength $\lambda_T = \sqrt{h^2/2\pi mkT}$ (or $\lambda_T^3/v \ll 1$), we expect quantum effects to be negligible. This limit, which corresponds to high temperatures and low densities, is the classical limit. In order to study quantum corrections to the classical behaviour, we first express the constraint equation, which determines the chemical potential (or the fugacity $z = \mathrm{e}^{\beta\mu}$) in terms of the number density $N/V = 1/v$ for both Bose–Einstein and Fermi–Dirac gases as (see, for example, problem 4.17)

$$\text{Fermi–Dirac gas:}\quad \frac{\lambda_T^3}{v} = (2s_{\mathrm{f}} + 1)\frac{4}{\sqrt{\pi}}\int_0^{\infty} \mathrm{d}x\, \frac{x^2}{z^{-1}\,\mathrm{e}^{x^2} + 1} \equiv (2s_{\mathrm{f}} + 1)\, f_{3/2}(z)$$

$$\text{Bose–Einstein gas:}\quad \frac{\lambda_T^3}{v} = (2s_{\mathrm{b}} + 1)\frac{4}{\sqrt{\pi}}\int_0^{\infty} \mathrm{d}x\, \frac{x^2}{z^{-1}\,\mathrm{e}^{x^2} - 1} \equiv (2s_{\mathrm{b}} + 1)g_{3/2}(z)$$

where s_{f} and s_{b} are the spins of fermions and bosons respectively. In the equation for the Bose–Einstein gas, we neglect the ground-state contribution since we

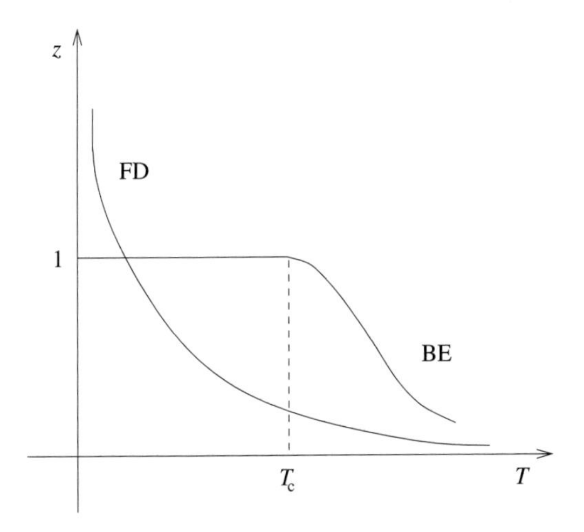

Figure 4.11. Temperature dependences of the fugacity for Fermi–Dirac (FD) and Bose–Einstein (BE) ideal gases.

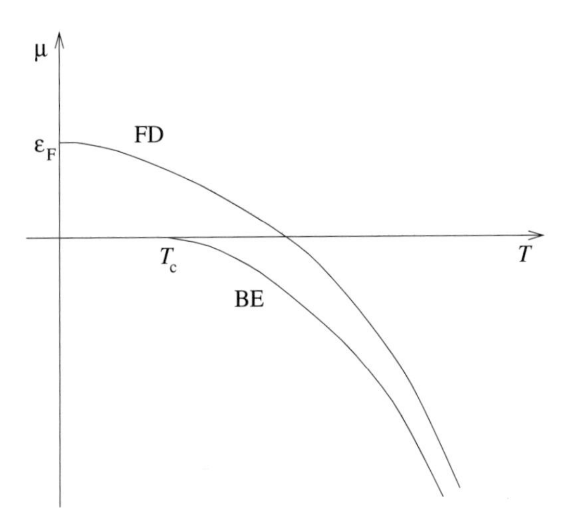

Figure 4.12. Temperature dependences of the chemical potential for Fermi–Dirac (FD) and Bose–Einstein (BE) ideal gases.

are interested in temperatures much larger than the condensation temperature ($T \gg T_c$). When $\lambda_T^3/v \ll 1$, z must be very small, so we expand the integrals in powers of z, using

$$
\begin{aligned}
z\int_0^\infty \mathrm{d}x\,\frac{x^2}{\mathrm{e}^{x^2}\pm z} &= \int_0^\infty \mathrm{d}x\,x^2(z\,\mathrm{e}^{-x^2}\mp z^2\,\mathrm{e}^{-2x^2}+z^3\,\mathrm{e}^{-3x^2}\mp\cdots) \\
&= \int_0^\infty \mathrm{d}x\,x^2\,\mathrm{e}^{-x^2}\left(z\mp\frac{z^2}{2^{3/2}}+\frac{z^3}{3^{3/2}}\mp\cdots\right) \\
&= \frac{\sqrt{\pi}}{4}\left(z\mp\frac{z^2}{2^{3/2}}+\frac{z^3}{3^{3/2}}\mp\cdots\right)
\end{aligned}
$$

which gives

$$
\text{Fermi–Dirac gas:}\quad \frac{\lambda_T^3}{v} = (2s_\mathrm{f}+1)\left(z-\frac{z^2}{2^{3/2}}+\frac{z^3}{3^{3/2}}-\cdots\right)
$$

$$
\text{Bose–Einstein gas:}\quad \frac{\lambda_T^3}{v} = (2s_\mathrm{b}+1)\left(z+\frac{z^2}{2^{3/2}}+\frac{z^3}{3^{3/2}}+\cdots\right).
$$

Inverting these equations, we find that

$$
\text{Fermi–Dirac gas:}\quad z = \frac{\lambda_T^3}{(2s_\mathrm{f}+1)v}+\frac{1}{2^{3/2}}\left(\frac{\lambda_T^3}{(2s_\mathrm{f}+1)v}\right)^2+\cdots
$$

$$
\text{Bose–Einstein gas:}\quad z = \frac{\lambda_T^3}{(2s_\mathrm{b}+1)v}-\frac{1}{2^{3/2}}\left(\frac{\lambda_T^3}{(2s_\mathrm{b}+1)v}\right)^2+\cdots.
$$

The equation of state follows from calculating the pressure. Using the same method as above, we obtain

$$
\begin{aligned}
\text{Fermi–Dirac gas:}\quad \beta P &= \frac{8(2s_\mathrm{f}+1)}{3\sqrt{\pi}\lambda_T^3}\int_0^\infty \mathrm{d}x\,\frac{x^4}{z^{-1}\,\mathrm{e}^{x^2}+1} \\
&= \frac{(2s_\mathrm{f}+1)}{\lambda_T^3}\left(z-\frac{z^2}{2^{5/2}}+\cdots\right) \\
\text{Bose–Einstein gas:}\quad \beta P &= \frac{8(2s_\mathrm{b}+1)}{3\sqrt{\pi}\lambda_T^3}\int_0^\infty \mathrm{d}x\,\frac{x^4}{z^{-1}\,\mathrm{e}^{x^2}-1} \\
&= \frac{(2s_\mathrm{b}+1)}{\lambda_T^3}\left(z+\frac{z^2}{2^{5/2}}+\cdots\right)
\end{aligned}
$$

and, on inserting our expansion for z, these become

$$
\text{Fermi–Dirac gas:}\quad \frac{PV}{kT} = N\left(1+\frac{1}{2^{5/2}}\frac{\lambda_T^3}{(2s_\mathrm{f}+1)v}+\cdots\right)
$$

$$
\text{Bose–Einstein gas:}\quad \frac{PV}{kT} = N\left(1-\frac{1}{2^{5/2}}\frac{\lambda_T^3}{(2s_\mathrm{b}+1)v}+\cdots\right).
$$

These expansions are similar to virial expansions. They arise not from interactions between the particles (as is the case for cluster expansions of classical gases), but from quantum effects. The leading terms correspond to the classical ideal gas and we see that Fermi–Dirac and Bose–Einstein statistics give the same result in that limit, the fugacity $z \sim \lambda_T^3/v$ being very small. The temperature dependences of z and μ respectively are sketched for Fermi–Dirac and Bose–Einstein gases in figures 4.11 and 4.12. The first quantum correction in the equation for the pressure is positive for fermions and negative for bosons. We see that, for given T, V and N, the pressure of a Fermi gas is enhanced relative to that of a classical gas, while that of a Bose gas is depressed. The increased pressure for fermions can be understood in terms of the Pauli exclusion principle, which forbids two fermions from occupying the same quantum state and can be thought of as resulting in an effective repulsion at short distances. Bosons, on the other hand, have an enhanced probability of occupying the same state, which can loosely be interpreted as resulting in an attractive force and a decrease in pressure.

5

INTERACTING SYSTEMS

In contrast with the preceding chapters, this chapter deals with systems in which interactions between microscopic constituents have important manifestations at the macroscopic level. By and large, exact solutions can be found only for low-dimensional models, but we illustrate some of the approximation schemes that are available for more complicated systems. Problems 5.1–5.9 deal with interactions in classical fluids arising from interparticle potentials; virial coefficients and critical exponents associated with second-order phase transitions are calculated. The remaining problems of the chapter focus on lattice models. Problem 5.11 studies the one-dimensional Potts model, while problems 5.12 and 5.13 illustrate how the Ising model for magnetism can be related to systems whose physical natures are quite different. In problems 5.14–5.18, exact treatments of several variants of the one-dimensional Ising model are given, while problem 5.19 is an exact computation for the one-dimensional Heisenberg model. Correlation functions in the Ising model will be found in problems 5.20 and 5.21. Problems 5.22–5.26 illustrate the mean-field approximation, mainly for magnetic systems and the last problem, 5.28, introduces the renormalization-group approach to critical phenomena in the context of the one-dimensional Ising model.

5.1 Questions

• **Problem 5.1** Consider a gas of N classical point particles of mass m, which interact through a two-body potential of the form

$$\Phi(r_{ij}) = ar_{ij}^{-\nu}$$

where $r_{ij} = |\boldsymbol{r}_i - \boldsymbol{r}_j|$, a is a positive constant and ν is a positive index. If this gas occupies a volume V at temperature T, show that:

(a) its canonical partition function $Z(T, V, N)$ is a homogeneous function, in the sense that

$$Z(\alpha T, \alpha^{-3/\nu} V, N) = \alpha^{3N(1/2-1/\nu)}\, Z(T, V, N)$$

where α is an arbitrary scaling factor,

(b) its Helmholtz free energy $F(T, V, N)$ obeys the relation

$$T\left(\frac{\partial F}{\partial T}\right)_V - \frac{3}{\nu}V\left(\frac{\partial F}{\partial V}\right)_T = F - 3\left(\frac{1}{2} - \frac{1}{\nu}\right)NkT$$

and

(c) its internal energy U is related to its pressure P by $U = yPV + xNkT$, where y and x are functions of ν, which you are invited to determine.

• **Problem 5.2** For a monatomic gas of N atoms with number density n at temperature T, assume that the equation of state can be expressed as a virial expansion $P = kT\sum_{j=1}^{\infty} B_j(T)n^j$.

(a) Use thermodynamic reasoning to express the entropy difference $S - S^{\text{ideal}}$, and the specific heat difference $C_V - C_V^{\text{ideal}}$ in terms of the virial coefficients $B_j(T)$, where S^{ideal} and C_V^{ideal} are the entropy and specific heat respectively of an ideal gas with the same number density and temperature. Note any additional assumptions that you need.

(b) Obtain a virial expansion for the Helmholtz free energy from first principles and hence confirm the results of (a).

• **Problem 5.3** A gas in d dimensions has a pairwise interaction potential $\Phi(|\boldsymbol{r}_i - \boldsymbol{r}_j|)$ between its constituent particles given by

$$\Phi(r) = \begin{cases} \infty & 0 < r < a \\ -\varepsilon & a < r < b \\ 0 & b < r < \infty. \end{cases}$$

(a) Compute the second virial coefficient $B_2(T)$ and interpret its behaviour in the limits $T \to \infty$ and $T \to 0$, both for $\varepsilon > 0$ and for $\varepsilon < 0$.

(b) Using this $\Phi(r)$ as a model for the potential between two argon atoms in three dimensions, find suitable values of ε and of the ratio a/b, given that argon has a Boyle temperature of 410 K and a maximum Joule–Kelvin inversion temperature of 720 K.

• **Problem 5.4**

(a) Compute the second virial coefficient $B_2(T)$ for a gas in three dimensions, when the potential between two gas particles separated by a distance r is $\Phi(r) = (\alpha/r)^\nu$ with $\alpha > 0$ and $\nu > 3$.

(b) For a monatomic gas in d dimensions with the same interparticle potential (now with $\nu > d$), and whose particles each have a kinetic energy of the form

$\varepsilon \sim |\boldsymbol{p}|^s$, show that the leading corrections to the internal energy U and the pressure P at low densities are related by

$$\frac{U - U_{\text{ideal}}}{U_{\text{ideal}}} \simeq \gamma \frac{P - P_{\text{ideal}}}{P_{\text{ideal}}}$$

and find the constant γ.

• **Problem 5.5** Let $\Phi(r)$ be the pair potential of a gas of interacting molecules at temperature T such that $\Phi(r) = \infty$ for $r < r_0$ while $\Phi(r)/kT \ll 1$ for $r > r_0$.

(a) Evaluate the difference between the internal energy of this gas and that of an ideal gas, up to first order in the virial expansion.

(b) Show, in the same approximation, that the entropy of this gas is smaller than that of the ideal gas. Interpret this result.

• **Problem 5.6** Show that the second and third virial coefficients of a system of hard spheres are related by $B_3(T) = \frac{5}{8}B_2(T)^2$.

• **Problem 5.7**
(a) Consider a mixture of two molecular species A and B which interact via pair potentials $\Phi_{\text{AA}}(r)$, $\Phi_{\text{AB}}(r)$ and $\Phi_{\text{BB}}(r)$ (that is, $\Phi_{\text{ab}}(r)$ is the potential energy of a molecule of type a and a molecule of type b, separated by a distance r). Find an expression for the second virial coefficient of this mixture at temperature T, when a fraction x of the molecules present are of type A while the remaining fraction $1 - x$ are of type B.

(b) Find the equation of state, up to second order of the virial expansion, of a hard-sphere gas consisting of N_{A} balls of radius R_{A} and N_{B} balls of radius R_{B}.

• **Problem 5.8** For a fluid which has the van der Waals equation of state, determine the critical exponents δ, γ and β which are defined as follows.

(a) Along the critical isotherm $T = T_{\text{c}}$, pressure and volume are related by

$$P - P_{\text{c}} \sim (V - V_{\text{c}})^{\delta} \quad \text{as } P \to P_{\text{c}} \text{ and } V \to V_{\text{c}}$$

where P_{c} and V_{c} are the critical pressure and volume, respectively.

(b) When $V = V_{\text{c}}$ and $T > T_{\text{c}}$, the isothermal compressibility varies with temperature as

$$\kappa_T \sim (T - T_{\text{c}})^{-\gamma} \text{ as } T \to T_{\text{c}}.$$

(c) Along the coexistence curve $P = P_0(T)$, where $P_0(T)$ is the vapour pressure and $T < T_{\text{c}}$,

$$(v_{\text{g}} - v_{\ell}) \sim (T_{\text{c}} - T)^{\beta} \text{ as } T \to T_{\text{c}}$$

where v_g and v_ℓ are the specific volumes of the coexisting gas and liquid phases respectively.

• **Problem 5.9** The Gibbs free energy per particle of a fluid near its critical point at $T = T_c$, $P = P_c$ and $V = V_c$ can be written, to an adequate approximation, as

$$G(T, V) \simeq G_{\text{reg}}(t, p) + G_{\text{sing}}(t, p) = G_{\text{reg}}(t, p) + t^{2-\alpha}\mathcal{G}\left(\frac{p}{t^{\Delta}}\right)$$

for $T > T_c$. The scaling fields are $t = (T - T_c)/T_c$ and $p = [P - P_0(T)]/P_c$, where $P = P_0(T)$ is the equation of the coexistence curve (which, we assume, can be analytically continued to temperatures above T_c). The regular part $G_{\text{reg}}(t, p)$ has a well-defined Taylor expansion in powers of t and p when these variables are small. The singular part $G_{\text{sing}}(t, p)$ takes the indicated scaling form, in which the scaling function $\mathcal{G}(x)$ has a well-defined Taylor expansion in powers of $x = p/t^{\Delta}$ when x is small. The measured values of the exponents are $\alpha \simeq 0.1$ and $\Delta \simeq 1.6$.

(a) Show that the critical isochore (the curve in the P–T plane along which $V = V_c$) is given, sufficiently close to the critical point, by $p = x_0 t^{\Delta}$, where x_0 is the value of x for which $\mathrm{d}\mathcal{G}(x)/\mathrm{d}x = 0$.

(b) As the critical point is approached along the critical isochore, show that the isothermal compressibility diverges as $\kappa_T \propto t^{-\gamma}$, and find the exponent γ in terms of α and Δ. Show also that the specific heat at constant volume diverges as $C_V \propto t^{-\alpha}$.

(c) Investigate the dependence on volume of the pressure and the isothermal compressibility as the critical point is approached along the critical isotherm $T = T_c$.

• **Problem 5.10** In the shell model of large atomic nuclei, a spin–orbit interaction causes single-particle states with a given radial quantum number to split into shells characterized by definite values of the angular momentum j. Within a shell, there are $2J = 2j + 1$ degenerate states, which can be labelled by the azimuthal quantum number m ($m = -j, \ldots, j$), and J is an integer because j (being composed of orbital angular momentum and spin) is a half-odd-integer. Within a single shell, interparticle interactions appear to give rise to a pairing force, in such a way that, if the state with quantum number m is occupied, then the state with quantum number $-m$ is likely to be occupied also. A reasonable approximation gives the energy of the *whole shell* as $E = -\varepsilon P[1 + (P - N + 1)/J]$, where P is the number of pairs (so there are $2P$ particles in paired states), N is the total number of particles in the shell (assumed to be identical, so that protons are not distinguished from neutrons) and ε is a positive energy characterizing the strength of the pairing force.

Supposing that the N particles in one such shell can be treated as an isolated system in thermal equilibrium, find the temperature of this system as a function of P, assuming that $J > N \gg 1$. What fraction of the particles are in paired states when $T = 0$ and when $T \to \infty$?

• **Problem 5.11** A set of N idealized atoms is arranged in a closed one-dimensional chain. Each atom can be in any of p states, labelled by an integer ν ($\nu = 1, 2, \ldots, p$) and two neighbouring atoms on the chain interact so as to have a potential energy equal to $-J$ (with $J > 0$) if they are in the same state and 0 otherwise. The system can thus be described by the Hamiltonian

$$\mathcal{H} = -J \sum_{i=1}^{N} \delta_{\nu_i, \nu_{i+1}}$$

where $\delta_{\nu,\nu'}$ is the Kronecker symbol, provided that ν_{N+1} is taken to refer to the same atom as ν_1.

(a) Obtain the canonical partition function for this system.
(b) Compute the internal energy $U(T)$ for the system in the thermodynamic limit $N \to \infty$ and find its low- and high-temperature limits.

• **Problem 5.12** In the lattice gas model for the liquid–gas phase transition, gas particles are allowed to exist only at the sites of a regular spatial lattice, each of which can accommodate at most one particle. The particles interact so as to give a potential energy of $-\varepsilon$ for each pair of nearest-neighbour sites that are simultaneously occupied. Show that the grand canonical partition function for this model is equivalent to the canonical partition function for the Ising model of a ferromagnet.

• **Problem 5.13** It is often possible to estimate the thermodynamic properties of a quantum-mechanical system in d dimensions by establishing a correspondence between this system and a classical system in $d + 1$ dimensions. To illustrate how such a correspondence may come about (although in this case it is not particularly useful), consider a zero-dimensional quantum system, consisting of a single two-state object whose Hamiltonian can be expressed in matrix form as

$$\mathbf{H} = \begin{pmatrix} \varepsilon & -\Delta \\ -\Delta & -\varepsilon \end{pmatrix}.$$

By using the fact that $\exp(-\beta\mathbf{H}) = \lim_{N\to\infty}[(1 - N^{-1}\beta\mathbf{H})^N]$, show that the partition function $Z = \mathrm{Tr}\,[\exp(-\beta\mathbf{H})]$ can be expressed in terms of the partition function of an Ising model on a one-dimensional chain of N sites.

• Problem 5.14

(a) Consider a one-dimensional Ising chain of N sites with nearest-neighbour interactions and free ends (that is, the spin variables at sites 1 and N are independent, and there is no interaction between these two sites). When there is no magnetic field, use a suitable set of variables, describing the states of links of the chain rather than of sites, to relate the partition function of this system to that of another Ising chain in which there is a magnetic field, but no interactions between the sites.

(b) Adding a next-nearest-neighbour interaction, to give a total Hamiltonian of the form

$$\mathcal{H} = -J_1 \sum_{i=1}^{N-1} S_i S_{i+1} - J_2 \sum_{i=1}^{N-2} S_i S_{i+2}$$

(where, of course, each S_i takes the values ± 1), obtain the partition function in the limit $N \to \infty$. Find the correlation function $\langle S_i S_{i+1} \rangle$, also in the limit $N \to \infty$.

• Problem 5.15 A lattice of $N + 1$ sites has spins $S_i = \pm 1$ at each site, all of which are acted on by a magnetic field. There are interactions of equal strength between one of the spins, S_0, and each of the others. Thus, the Hamiltonian is

$$\mathcal{H} = -h \sum_{i=0}^{N} S_i - J \sum_{i=1}^{N} S_i S_0.$$

(a) Find the canonical partition function $Z(T, N)$, the average energy $\langle E \rangle$ and, for $i \neq 0$, the statistical averages $\langle S_i \rangle$ and $\langle S_0 S_i \rangle$. Compute the limits of these averages when $h \to 0$ with $J \neq 0$ and when $J \to 0$ with $h \neq 0$.

(b) When $h = 0$, show, for $i, j \neq 0$ and $i \neq j$, that $\langle S_i S_j \rangle = \langle S_0 S_i \rangle \langle S_0 S_j \rangle$.

• Problem 5.16 An Ising model in one dimension, with position-dependent couplings between nearest-neighbour spins, has the Hamiltonian

$$\mathcal{H} = -\sum_{i=1}^{N} J_i \, S_i S_{i+1}$$

in which, as usual, each of the N spins takes the values $S_i = \pm 1$. The chain is closed, so that S_{N+1} refers to the same site as S_1, and, for simplicity, we take $N \gg 1$.

(a) Obtain an expression for the correlation function $\langle S_i S_{i+r} \rangle$.

(b) What is the probability $\mathcal{P}_i(1)$ of finding the spin at site i to have the value $+1$? What is the probability $\mathcal{P}(S_i = S_j)$ of finding the spins at sites i and j to have the same value?

(c) For the special cases of position-independent ferromagnetic coupling ($J_i = J > 0$) and antiferromagnetic coupling ($J_i = J < 0$), investigate the low-temperature behaviour of spin correlations by expressing $\mathcal{P}(S_i = S_{i+r})$ as a function of $r/\xi(T)$, where $\xi(T)$ is a suitably defined correlation length.

(d) Determine the ground-state configuration of the system for the case that $J_i = (-1)^i J$, with $J > 0$.

• **Problem 5.17** Suppose that the coupling J_{ij} between two arbitrary sites in the Ising model for ferromagnetism has the same value for any pair of sites i and j, that is $J_{ij} = g/N$, where g is a constant and N is the number of lattice sites. This model yields a phase transition in all dimensions. Find the critical temperature in the thermodynamic limit $N \to \infty$.

• **Problem 5.18** Spins on a one-dimensional lattice have a three-level Ising Hamiltonian, which in the absence of any external field is

$$\mathcal{H} = -J \sum_{i=1}^{N} S_i S_{i+1} \qquad S_i = 1\,, 0\,, -1 \qquad J > 0\,.$$

Obtain the exact partition function $Z(T, N)$ in terms of temperature and the number of sites. Analyse the low-temperature limit of the internal energy for $N \gg 1$.

• **Problem 5.19** Consider the Heisenberg Hamiltonian for spins on a one-dimensional lattice of N sites with a nearest-neighbour interaction. When no external field is present it is

$$\mathcal{H} = -J \sum_{i}^{N} \boldsymbol{S}_i \cdot \boldsymbol{S}_{i+1}$$

where $\boldsymbol{S}_i$ is a three-dimensional unit vector, J is a positive constant and periodic boundary conditions are applied. Show that the partition function can be expressed as

$$Z(T, N) = (2\pi)^N \sum_{l=0}^{\infty} (2l+1) \left(\int_{-1}^{1} \mathrm{e}^{\beta J x}\, P_l(x)\, \mathrm{d}x \right)^N$$

where $P_l(x)$ is the lth Legendre polynomial.

• **Problem 5.20** For the Ising system, express the magnetic susceptibility in terms of the correlation functions $G_{ij} = \langle S_i S_j \rangle - \langle S_i \rangle \langle S_j \rangle$, where $\langle \cdots \rangle$ denotes the ensemble average. Show that, in the thermodynamic limit, the magnetic susceptibility is proportional to the zero Fourier mode of the correlation function.

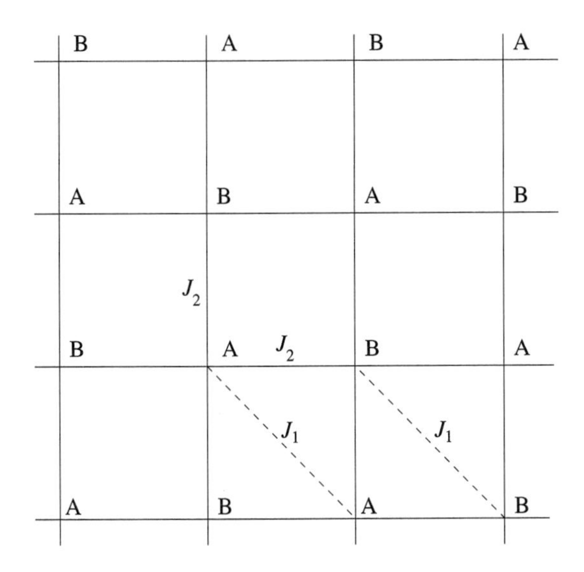

Figure 5.1. Two kinds of site interspersed in a two-dimensional square lattice.

• **Problem 5.21** For a one-dimensional Ising model of N spins, with periodic boundary conditions, evaluate the correlation function for two nearest-neighbour spins:

$$G(j, j+1) = \langle S_j S_{j+1} \rangle - \langle S_j \rangle \langle S_{j+1} \rangle$$

in the limit $N \to \infty$. Investigate the special cases $J \to 0$ (in which the spins do not interact with each other) and $H \to 0$ (in which there is no applied magnetic field).

• **Problem 5.22** Consider a two-dimensional lattice with two kinds of site, A and B, as shown in figure 5.1. Ising spins ($S_i = \pm 1$) are on all the sites and interact in the following way: there is an interaction energy $-J_1 S_i^{\mathrm{A}} S_j^{\mathrm{A}}$, with $J_1 > 0$, between nearest neighbours i and j of the A sublattice (which are next-nearest neighbours on the whole lattice) and nearest neighbours of the B sublattice interact in the same way; there is also an interaction energy $+J_2 S_i^{\mathrm{A}} S_j^{\mathrm{B}}$, with $0 < J_2 < J_1$, between nearest-neighbour sites, when site i is on the A sublattice and j is on the B sublattice.

(a) Write down the Hamiltonian for the system, allowing for an external magnetic field H, which acts on all the spins in the same way (via magnetic moments $\mu_{\mathrm{A}} = \mu_{\mathrm{B}} = \mu$), in terms of S_i^{A} and S_i^{B}.

(b) Using the mean-field approximation, calculate the effective field seen by spins on the A sublattice and that seen by spins on the B sublattice. Find equations that can be solved for $\langle S^{\mathrm{A}} \rangle$ and $\langle S^{\mathrm{B}} \rangle$ (which are independent of the sites within each sublattice on which the spins live).

(c) For $H = 0$, find the temperature T_{f}, below which a ferromagnetic state with $\langle S^{\mathrm{A}} \rangle = \langle S^{\mathrm{B}} \rangle \neq 0$ is possible, and the temperature T_{a} below which an antiferromagnetic state with $\langle S^{\mathrm{A}} \rangle = -\langle S^{\mathrm{B}} \rangle \neq 0$ is possible. Over what range of temperature, if any, is each of these states thermodynamically stable?

(d) Show that at high temperatures the zero-field magnetic susceptibility follows the Curie–Weiss law $\chi \sim (T - T_{\mathrm{f}})^{-1}$, but that it has a finite value at $T = T_{\mathrm{f}}$.

• **Problem 5.23** The Heisenberg model for ferromagnetism is described by

$$\mathcal{H} = -J \sum_{\langle i,j \rangle} \boldsymbol{S}_i \cdot \boldsymbol{S}_j - \mu \sum_i \boldsymbol{H} \cdot \boldsymbol{S}_i$$

where $\boldsymbol{S}_i$ is a three-dimensional unit vector and the lattice is d dimensional. Consider $J > 0$, let the external field be uniform, and let $h = \mu |\boldsymbol{H}|$. Using mean-field theory,

(a) find the critical temperature T_{c} and

(b) find the critical exponents δ, β, γ and γ' by studying the magnetization m and susceptibility χ, which behave near the critical point as

$$\begin{aligned} h &\sim m^{\delta} & T &= T_{\mathrm{c}} \\ m &\sim (T_{\mathrm{c}} - T)^{\beta} & h &= 0, T < T_{\mathrm{c}} \\ \chi &\sim (T - T_{\mathrm{c}})^{-\gamma} & h &= 0,\ T > T_{\mathrm{c}} \\ \chi &\sim (T_{\mathrm{c}} - T)^{-\gamma'} & h &= 0,\ T < T_{\mathrm{c}}. \end{aligned}$$

• **Problem 5.24** Consider an Ising model on a lattice with coordination number z and with ferromagnetic nearest-neighbour couplings of strength J in the mean-field approximation.

(a) Obtain the critical exponents β and γ, which describe the behaviour of the spontaneous magnetization and magnetic susceptibility near the critical temperature through the relations $m \sim (T_{\mathrm{c}} - T)^{\beta}$ and $\chi \sim |T - T_{\mathrm{c}}|^{-\gamma}$.

(b) Show that the specific heat at constant magnetic field C_H is discontinuous at the phase transition and determine the discontinuity.

• **Problem 5.25** Consider a two-dimensional Ising system (with magnetic field H and coupling $J > 0$) on a square lattice and investigate improvements to the mean-field approximation in the following way.

(a) Find the effective field seen by a *pair* of nearest-neighbour spins, when the remaining spins are replaced by their mean values. Use this effective field to obtain a self-consistent equation for the mean value of each spin and calculate the quantity $\beta_c J$, where $T_c = 1/k\beta_c$ is the critical temperature. Compare this with the value obtained from the usual mean-field approximation.

(b) Now find the effective field seen by the spins in an entire row of lattice sites. For a one-dimensional Ising model, the exact mean value of each spin is $\sigma = \sinh(\beta\mu H)\,[\sinh^2(\beta\mu H) + \mathrm{e}^{-4\beta J}]^{-1/2}$. Use this fact to obtain an improved estimate of $\beta_c J$.

(c) Compare the approximations of $\beta_c J$ found in (a) and (b) with the exact value obtained by Onsager.

• **Problem 5.26** A generalized Ising model, representing some (but not all) of the properties of a collection of magnetic spin-s ions, has the usual Hamiltonian

$$\mathcal{H} = -J \sum_{\langle ij \rangle} S_i S_j - h \sum_{i=1}^{N} S_i$$

but each spin variable takes the values $S_i = -s, -s+1, \ldots, s-1, s$, where s may be either an integer or a half-odd integer. Using the mean-field approximation to this model,

(a) find the critical temperature of the system and

(b) using the result in (a) recover the critical temperatures corresponding to the standard two-state Ising model and to the Heisenberg model.

• **Problem 5.27** A collection of N oscillators with frequencies ω_i interact in such a way that (ignoring zero-point energies) the energy of the whole system can be represented as

$$E = \sum_{i=1}^{N} \varepsilon_i n_i + \frac{\lambda}{2} \sum_{i=1}^{N} \sum_{j=1}^{N} V_{ij} n_i n_j .$$

Here, $\varepsilon_i = \hbar\omega_i$, V_{ij} is a symmetric matrix with diagonal elements $V_{ii} = 0$, $n_i = 0, \ldots, \infty$ is the occupation number for the ith oscillator and λ is a small parameter measuring the strength of interactions. We attempt to approximate this interacting system as a system of non-interacting oscillators, with effective temperature-dependent energy-level spacings $\varepsilon_i + \lambda\varepsilon_i^{(1)}(T)$.

The canonical partition functions of the interacting and non-interacting systems can be calculated as perturbative expansions in powers of λ. Show that these expansions agree to first order in λ, if the correction $\varepsilon_i^{(1)}(T)$ is appropriately chosen.

• **Problem 5.28** The reduced Hamiltonian $\tilde{\mathcal{H}} = \mathcal{H}/kT$ of an Ising model with nearest-neighbour ferromagnetic coupling J and magnetic field H on an infinite one-dimensional lattice can be written as

$$\tilde{\mathcal{H}} = -b\sum_i S_i - K\sum_i S_i S_{i+1} \qquad S_i = \pm 1$$

where $K = J/kT$ and $b = \mu H/kT$.

(a) By carrying out the sums over spins S_i on all the odd-numbered lattice sites, show that the partition function obeys a relation of the form

$$Z(K, b) = Z_0(K, b)Z(\bar{K}, \bar{b})$$

where $\bar{K}(K, b)$ and $\bar{b}(K, b)$ are effective interaction parameters for the system that consists of just the even-numbered spins. (This is an example of a *renormalization-group transformation.*) If $\tau = \mathrm{e}^{-4K}$ and $\eta = \mathrm{e}^{-2b}$, find the corresponding renormalized parameters $\bar{\tau}$ and $\bar{\eta}$ in terms of τ and η.

(b) For a truly infinite lattice, this transformation can be repeated any number of times, yielding a sequence of renormalized parameters τ_n and η_n. A *fixed point* of the transformation is a pair of values (τ^*, η^*) such that $\bar{\tau}(\tau^*, \eta^*) = \tau^*$ and $\bar{\eta}(\tau^*, \eta^*) = \eta^*$. Suppose that the initial parameters are $\tau = \tau^* + \delta\tau$ and $\eta = \eta^* + \delta\eta$, where $\delta\tau$ and $\delta\eta$ are small. Then the fixed point is said to be *stable* with respect to perturbations in τ if $\delta\tau_n \to 0$ as $n \to \infty$ and *unstable* if $\delta\tau_n$ increases in magnitude as the transformation is repeated. Its stability with respect to perturbations in η is defined in the same way. Show that $(\tau^*, \eta^*) = (0, 1)$ is a fixed point that is unstable to perturbations in both τ and η.

(c) Use the renormalization-group transformation to show that, near $T = 0$ and $H = 0$, the correlation length and magnetization per spin have the scaling forms

$$\xi(T, H) \simeq \tau^{-\nu} X(b\tau^{-\Delta}) \qquad m(T, H) \simeq \mathcal{M}(b\tau^{-\Delta})$$

and deduce the values of the exponents ν and Δ.

5.2 Answers

• Problem 5.1

(a) The canonical partition function of this system is given (up to a factor of $1/h^{3N}N!$ which is irrelevant for our present purposes) by

$$Z(T,V,N) = \int \mathrm{d}^{3N}p \int_V \mathrm{d}^{3N}r \exp\left(-\beta \sum_{k=1}^{3N} \frac{p_k^2}{2m} - \beta \sum_{\langle ij \rangle} \frac{a}{r_{ij}^\nu} \right)$$

$$= (2\pi mkT)^{3N/2} \int_V \mathrm{d}^{3N}r \, \exp\left(-\frac{1}{kT} \sum_{\langle ij \rangle} \frac{a}{r_{ij}^\nu} \right)$$

where $\langle ij \rangle$ indicates a sum over pairs of particles and, in the second expression, the integration over momenta has been performed, yielding the initial factor. On making the change of variable $\boldsymbol{r}_i = \alpha^{1/\nu}\boldsymbol{\rho}_i$ and, correspondingly, $r_{ij} = \alpha^{1/\nu}\rho_{ij}$, we obtain

$$Z(T,V,N) = (2\pi mkT)^{3N/2}\alpha^{3N/\nu} \int_{\alpha^{-3/\nu}V} \mathrm{d}^{3N}\rho \, \exp\left(-\frac{1}{\alpha kT} \sum_{\langle ij \rangle} \frac{a}{\rho_{ij}^\nu} \right)$$

$$= \alpha^{3N/\nu - 3N/2} Z(\alpha T, \alpha^{-3/\nu}V, N)$$

which is the result that we wanted.

(b) It is easy to see that, for any function $f(x,y)$,

$$\alpha \frac{\mathrm{d}}{\mathrm{d}\alpha}[f(\alpha^a x, \alpha^b y)]\bigg|_{\alpha=1} = ax\left(\frac{\partial f(x,y)}{\partial x}\right)_y + by\left(\frac{\partial f(x,y)}{\partial y}\right)_x .$$

Applying this to the result of (a), we obtain

$$T\left(\frac{\partial Z}{\partial T}\right)_V - \frac{3}{\nu}V\left(\frac{\partial Z}{\partial V}\right)_T = 3N\left(\frac{1}{2} - \frac{1}{\nu}\right)Z$$

and then substituting $Z = \mathrm{e}^{-F/kT}$ gives

$$T\left(\frac{\partial F}{\partial T}\right)_V - \frac{3}{\nu}V\left(\frac{\partial F}{\partial V}\right)_T = F - 3\left(\frac{1}{2} - \frac{1}{\nu}\right)NkT.$$

(c) Using the fact that $(\partial F/\partial T)_V = -S$ and $(\partial F/\partial V)_T = -P$, we can write the relation obtained in (b) as $-TS + (3/\nu)PV = F - 3(1/2 - 1/\nu)NkT$ and, since $F = U - TS$, this is equivalent to

$$U = 3\left(\frac{1}{2} - \frac{1}{\nu}\right)NkT + \frac{3}{\nu}PV.$$

It is interesting to note that we recover the ideal gas result $U = 3NkT/2$ in the limit $\nu \to \infty$. To make this limit meaningful, we must take the constant a in the potential as $a = \varepsilon r_0^\nu$, where ε is a constant energy and r_0 a characteristic distance. In the limit $\nu \to \infty$, we then have $\Phi(r) = 0$ if $r > r_0$ and $\Phi(r) = \infty$ if $r < r_0$ (a hard-sphere gas). The particles of such a gas have only their kinetic energy, so it is reasonable that U should be the same as for an ideal gas. Mathematically, we see that the configurational integral contributes only a temperature-independent factor to Z in this limit, so $U = -\partial(\ln Z)/\partial\beta$ depends, as for an ideal gas, only on the factor $(2\pi mkT)^{3N/2}$.

• Problem 5.2

(a) In order to find the entropy, we first obtain an expression for the Helmholtz free energy $F(T, V, N)$ from the relation $P = -(\partial F/\partial V)_{T,N}$. In terms of the number density $n = N/V$, this is $P = N^{-1}n^2(\partial F/\partial n)_{T,N}$, so we can integrate to find

$$F = NkT\left(\sum_{j=1}^{\infty} B_j(T) \int n^{j-2}\,\mathrm{d}n + h(T)\right)$$

$$= NkT\left(\ln n + \sum_{j=2}^{\infty} \frac{n^{j-1}}{j-1} B_j(T) + h(T)\right)$$

where $h(T)$ is an undetermined function. In writing this result, we have assumed that in the limit $n \to 0$ the gas behaves like an ideal gas, with $P = nkT$, so that $B_1(T) = 1$. If we also assume that the free energy is extensive, then $h(T)$ is indeed a function of T only. This being so, the quantity $NkT\,[\ln n + h(T)]$ ought to be the free energy of an ideal gas, since the remaining sum vanishes when $n \to 0$. We therefore have

$$\Delta F \equiv F - F^{\text{ideal}} = NkT \sum_{j=2}^{\infty} \frac{n^{j-1}}{j-1} B_j(T) = NkT \sum_{j=1}^{\infty} \frac{n^j}{j} B_{j+1}(T)$$

and, on differentiating,

$$\Delta S = -\left(\frac{\partial \Delta F}{\partial T}\right)_{N,V} = -Nk \sum_{j=1}^{\infty} \frac{n^j}{j}\left(B_{j+1}(T) + T\frac{\mathrm{d}B_{j+1}(T)}{\mathrm{d}T}\right)$$

$$\Delta C_V = T\left(\frac{\partial(\Delta S)}{\partial T}\right)_{N,V} = -Nk \sum_{j=1}^{\infty} \frac{n^j}{j}\left(2T\frac{\mathrm{d}B_{j+1}(T)}{\mathrm{d}T} + T^2\frac{\mathrm{d}^2 B_{j+1}(T)}{\mathrm{d}T^2}\right)$$

where $\Delta S = S - S^{\text{ideal}}$ and $\Delta C_V = C_V - C_V^{\text{ideal}}$.

(b) At first sight, these results may seem surprising, since in general the equation of state does not contain all the information needed to obtain the entropy. The crucial feature of the above calculation is, of course, the assumption (albeit reasonably well justified) that the undetermined function $h(T)$ is just that associated with an ideal gas. By obtaining the Helmholtz free energy from first principles, we hope to find $h(T)$ explicitly, thereby confirming that the corrections to ideal gas behaviour of *all* thermodynamic quantities can be expressed in terms of the virial coefficients of the equation of state.

The virial expansion is useful when the number density n is very small (that is for a dilute gas). Since particles flow towards a region of low chemical potential μ and also, other things being equal, towards a region of low particle concentration, we expect that $\mu \to -\infty$ as $n \to 0$ for a given temperature, and this has been confirmed by explicit calculations in earlier chapters. Consequently, the fugacity $z = e^{\beta\mu}$ should be small when n is small, and we might expect to be able to convert the expression

$$\mathcal{Z} = \sum_{N=0}^{\infty} z^N Z(T, V, N)$$

for the grand canonical partition function into an expansion in powers of n. For a classical gas, the canonical partition function is

$$Z(T, V, N) = \frac{\tilde{Z}(T, V, N)}{N!\lambda^{3N}}$$

where $\lambda = \sqrt{h^2/2\pi mkT}$ is the thermal wavelength and $\tilde{Z}(T, V, N)$ is the configurational partition function, which involves an integral only over the positions of the N particles. In the thermodynamic limit $V, N \to \infty$ with $n = N/V$ fixed, $\tilde{Z}$ will be large (in particular, we have $\tilde{Z}(T, V, N) = V^N$ for an ideal gas), so it is not correct simply to truncate the sum which defines $\mathcal{Z}$. Instead, we deal with the grand potential $\Omega(T, V, \mu)$, which can also be written as a power series in z:

$$\Omega(T, V, \mu) = kT \ln \mathcal{Z} = kT \sum_{l=1}^{\infty} y^l \Omega_l(T, V)$$

where $y = z/\lambda^3$. In this expansion, the coefficient Ω_l involves all the $\tilde{Z}(T, V, N)$ with $N \leq l$, so l is not directly related to the number of particles. For an ideal gas, we have $\mathcal{Z} = \exp(zV/\lambda^3)$, which gives $\Omega_1 = V$ and $\Omega_l = 0$ for $l \geq 2$. Even when particles interact, we have $\Omega_1 = \tilde{Z}(T, V, 1) = V$, since there are obviously no interaction effects for a single particle. We now assume that $\Omega(T, V, \mu)$ is extensive in the thermodynamic limit, which will normally be true if the interparticle potential decays rapidly enough at large distances. Then we can write $\Omega_l(T, V) = V\omega_l(T)$, where $\omega_l(T)$ is a function only of temperature (and of the interaction potential) and $\omega_1(T) = 1$. We thus obtain a basic pair of equations for the number density n and the pressure P as power series in y, namely

$$\frac{P}{kT} = \frac{\Omega}{kTV} = \sum_{l=1}^{\infty} \omega_l(T) y^l$$

$$n = z\left[\frac{\partial}{\partial z}\left(\frac{\Omega}{kTV}\right)\right]_{T,V} = \sum_{l=1}^{\infty} l\omega_l(T)y^l$$

and these determine all the thermodynamic properties of the gas.

To obtain the standard virial expansion, we wish to convert these expansions in y to expansions in n. Our basic variables will then be (T, N, n) or, equivalently, (T, N, V), so our central task is to obtain the Helmholtz free energy, which we shall express in the form $F(T, V, N) = NkT\tilde{F}(T, n)$. For an ideal gas, we easily find that $y = n$, $P/kT = n$ and $\tilde{F} = (\mu N - \Omega)/NkT = \ln z - P/nkT = \ln(\lambda^3 n) - 1$. According to the usual description, the effects of interactions are encoded in the virial coefficients of the equation of state, so our two sets of variables are related through the quantity $\Pi \equiv P/kT$, which can be expressed either as a function of y or as a function of n:

$$\Pi = \sum_{l=1}^{\infty} \omega_l(T)y^l = \sum_{j=1}^{\infty} B_j(T)n^j.$$

We need to express the quantity

$$\Delta\tilde{F} \equiv \tilde{F} - \tilde{F}^{\text{ideal}} = \left(\ln z - \frac{P}{nkT}\right) - [\ln(\lambda^3 n) - 1] = \ln\left(\frac{y}{n}\right) + 1 - \frac{\Pi}{n}$$

in terms of the $B_j(T)$. To do this, we first note that in the limit $n \to 0$ we have $y/n \to 1$ and $\Delta\tilde{F} \to 0$. Then, by differentiating our first expansion for Π (always keeping T fixed), we find that

$$\frac{\mathrm{d}\Pi}{\mathrm{d}n} = \sum_{l=1}^{\infty} l\omega_l y^{l-1}\frac{\mathrm{d}y}{\mathrm{d}n} = \frac{n}{y}\frac{\mathrm{d}y}{\mathrm{d}n}$$

$$\frac{\mathrm{d}(\Delta\tilde{F})}{\mathrm{d}n} = -\frac{1}{n} + \frac{1}{y}\frac{\mathrm{d}y}{\mathrm{d}n} + \frac{\Pi}{n^2} - \frac{1}{n}\frac{\mathrm{d}\Pi}{\mathrm{d}n} = -\frac{1}{n} + \frac{\Pi}{n^2} = \sum_{j=1}^{\infty} B_{j+1}(T)n^{j-1}.$$

On integrating this equation, with the constant of integration determined by $\Delta\tilde{F}(T, 0) = 0$, we easily discover that

$$\begin{aligned} F(T, V, N) &= NkT(\tilde{F}^{\text{ideal}} + \Delta\tilde{F}) \\ &= NkT\left(\ln(\lambda^3 n) - 1 + \sum_{j=1}^{\infty}\frac{1}{j}B_{j+1}(T)n^j\right). \end{aligned}$$

This is, of course, the same expression that we found in (a), except that we have now derived the function $h(T) = -1 + \ln(\lambda^3)$ explicitly. (Strictly, of course, only the combination $\ln(\lambda^3 n)$ is well defined, having a dimensionless argument for the logarithm!) The results obtained in (a) for ΔS and ΔC_V will naturally be reproduced by our complete expression for $F(T, V, N)$.

• **Problem 5.3**

(a) Following the derivation of the virial expansion given in problem 5.2, we find from the series for Ω that $\omega_2(T) = \lim_{V\to\infty}\{[\tilde{Z}(T,V,2) - \tilde{Z}(T,V,1)^2]/2V\}$, from the series for the number density n that $y = n - 2\omega_2(T)n^2 + O(n^3)$ and from the equality of the two expansions of Π in powers of y and n that $B_2(T) = -\omega_2(T)$. Thus, we have

$$\begin{aligned}
B_2(T) &= -\lim_{V\to\infty}\left[\frac{1}{2V}\left(\int \mathrm{d}^d r_1 \int \mathrm{d}^d r_2\, \mathrm{e}^{-\beta\Phi(|\boldsymbol{r}_1-\boldsymbol{r}_2|)} - V^2\right)\right] \\
&= -\tfrac{1}{2}\int \mathrm{d}^d r\,(\mathrm{e}^{-\beta\Phi(|\boldsymbol{r}|)} - 1) \\
&= -\frac{\pi^{d/2}}{\Gamma(d/2)}\int_0^\infty \mathrm{d}r\, r^{d-1}(\mathrm{e}^{-\beta\Phi(r)} - 1) \\
&= \frac{\pi^{d/2}}{d\Gamma(d/2)}[b^d - \mathrm{e}^{\varepsilon/kT}(b^d - a^d)].
\end{aligned}$$

At high temperatures, $B_2(T)$ approaches a finite constant. In fact, it is the same as that of a gas of hard spheres of radius a, regardless of the sign of ε. This must clearly be so, because the particles have a high kinetic energy and are not significantly affected by the potential well. At low temperatures and for $\varepsilon > 0$, $B_2(T)$ becomes large and negative. This reflects the fact that particles are likely to become bound together by the potential well and therefore exert a lower pressure than an ideal gas with the same density and temperature. Qualitatively, at least, these two limits are typical of the behaviour of real gases. For $\varepsilon < 0$, the low-temperature limit of $B_2(T)$ is that of a gas of hard spheres of radius b, because the step in the potential acts like an infinite step for particles whose kinetic energy is less than ε.

(b) The Boyle temperature T_B is that for which Boyle's law (which asserts that PV is independent of P at fixed temperature) holds at low pressures. More precisely, if $v = 1/n$ is the specific volume, then Pv can be expanded in powers of P as $Pv = kT + B_2(T)P + O(P^2)$, and the definition $\lim_{P\to 0}\{[\partial(Pv)/\partial P]_{T=T_\mathrm{B}}\} = 0$ clearly implies that $B_2(T_\mathrm{B}) = 0$. For our model potential, we therefore find that

$$\mathrm{e}^{\varepsilon/kT_\mathrm{B}} = \left(1 - \frac{a^3}{b^3}\right)^{-1}.$$

The Joule–Kelvin process is an expansion at constant enthalpy. According to problem 2.5(c), the temperature may either increase or decrease during this process and is in fact unchanged if $(\partial v/\partial T)_P = v/T$. The locus in the T–P plane along which this condition holds is called the inversion curve and the maximum temperature on this curve, T_i, occurs in the limit $P \to 0$. By differentiating the equation $Pv = kT + B_2(T)P + \mathrm{O}(P^2)$, we easily find that

$$\left.\frac{\mathrm{d}B_2(T)}{\mathrm{d}T}\right|_{T=T_\mathrm{i}} = \frac{B_2(T_\mathrm{i})}{T_\mathrm{i}}$$

or, for our model,

$$\left(\frac{\varepsilon}{kT_\mathrm{i}} + 1\right)\mathrm{e}^{\varepsilon/kT_\mathrm{i}} = \left(1 - \frac{a^3}{b^3}\right)^{-1}.$$

We now have two equations to solve for the values of ε and a/b. On writing $x = \varepsilon/kT_\mathrm{i}$, we may combine them to obtain

$$x + 1 = \mathrm{e}^{\alpha x}$$

where $\alpha = T_\mathrm{i}/T_\mathrm{B} - 1 \simeq 0.76$ for argon. This equation clearly has $x = 0$ as one solution. This corresponds to $\varepsilon = 0$ and $a = 0$, representing an ideal gas. This trivial solution is expected, since $B_2 \equiv 0$ for an ideal gas, and our two conditions hold at any temperature. There is, however, a second solution, which can be obtained numerically (for example, by plotting out the function $1 + x - \mathrm{e}^{\alpha x}$). It is given by $x \simeq 0.69$, implying that $\varepsilon \simeq 0.69\,kT_\mathrm{i} \simeq 0.04\,\mathrm{eV}$. Finally, we calculate that

$$\frac{a}{b} = (1 - \mathrm{e}^{-\varepsilon/kT_\mathrm{B}})^{1/3} \simeq (1 - \mathrm{e}^{-0.69T_\mathrm{i}/T_\mathrm{B}})^{1/3} \simeq 0.89.$$

Although our model for the interatomic potential is rather crude, this calculation illustrates how measurements of the bulk properties of gases can be used to obtain information about the properties of individual atoms.

• Problem 5.4

(a) As shown in problem 5.3, the second virial coefficient is given by the integral

$$B_2(T) = -\tfrac{1}{2}\int \mathrm{d}^3r\,(\mathrm{e}^{-\beta\Phi(|\boldsymbol{r}|)} - 1).$$

Since the potential depends only on the distance $r = |\boldsymbol{r}|$, we use spherical polar coordinates to express this as

$$B_2(T) = 2\pi\int_0^\infty \mathrm{d}r\,r^2(1 - \mathrm{e}^{-\beta\Phi(r)}).$$

Integration by parts yields

$$B_2(T) = \left.\frac{2\pi}{3}r^3(1 - \mathrm{e}^{-\beta\alpha^\nu r^{-\nu}})\right|_0^\infty + \frac{2\nu\pi\alpha^\nu}{3kT}\int_0^\infty \mathrm{d}r\,\frac{\mathrm{e}^{-\beta\alpha^\nu r^{-\nu}}}{r^{\nu-2}}.$$

The first term vanishes if $\alpha > 0$ and $\nu > 3$ and in the second term we make a change in integration variable, $x = \beta\alpha^\nu r^{-\nu}$, getting

$$B_2(T) = \frac{2\pi}{3}\frac{\alpha^3}{(kT)^{3/\nu}}\int_0^\infty \mathrm{d}x\,\frac{\mathrm{e}^{-x}}{x^{3/\nu}} = \frac{2\pi}{3}\frac{\alpha^3\Gamma(1 - 3/\nu)}{(kT)^{3/\nu}}$$

in which $\Gamma(1 - 3/\nu)$ has a finite value if $\nu > 3$.

(b) In d dimensions, the above calculation can clearly be repeated (if $\nu > d$), giving a temperature dependence of the form

$$B_2(T) \sim \frac{1}{(kT)^{d/\nu}}.$$

From the expressions found in problem 5.2 for the pressure and Helmholtz free energy, we can write

$$\begin{aligned}
P &\simeq nkT + n^2kTB_2(T) = P_{\mathrm{ideal}}[1 + nB_2(T)] \\
U &= F + TS \\
&= F - T\left(\frac{\partial F}{\partial T}\right)_{V,N} \\
&\simeq U_{\mathrm{ideal}} - NnkT^2\frac{\mathrm{d}B_2(T)}{\mathrm{d}T} \\
&\simeq U_{\mathrm{ideal}}\left(1 + \frac{s}{\nu}nB_2(T)\right)
\end{aligned}$$

where we have used the fact (see, for example, problem 3.10) that $U_{\mathrm{ideal}} = (d/s)NkT$ for a monatomic gas. We therefore find that

$$\frac{U}{U_{\mathrm{ideal}}} - 1 \simeq \frac{s}{\nu}\left(\frac{P}{P_{\mathrm{ideal}}} - 1\right)$$

which is the expected relation, with $\gamma = s/\nu$ independent of dimensionality.

• Problem 5.5

(a) From the general results obtained in problem 5.2, we see that the virial expansion for the internal energy of the system is given by

$$U = U_{\mathrm{ideal}} - NkT\sum_{j=1}^{\infty}\frac{1}{j}T\frac{\mathrm{d}B_{j+1}(T)}{\mathrm{d}T}\frac{1}{v^j}$$

where $v = V/N$ is the specific volume and $B_j(T)$ is the jth virial coefficient. To evaluate the first-order correction to the ideal system, we simply need the coefficient (derived in problem 5.3)

$$B_2(T) = -\tfrac{1}{2}\int \mathrm{d}^3r\,(\mathrm{e}^{-\beta\Phi(|\boldsymbol{r}|)} - 1) = 2\pi\int_0^{\infty}\mathrm{d}r\,r^2\,(1 - \mathrm{e}^{-\beta\Phi(r)}).$$

Under the given conditions, the integrand is equal to 1 for $r < r_0$ and approximately equal to $\beta\Phi(r)$ for $r > r_0$; so we obtain

$$B_2(T) \simeq \frac{2\pi r_0^3}{3} + \frac{2\pi}{kT}\int_{r_0}^{\infty} r^2\Phi(r)\,\mathrm{d}r.$$

Therefore, the difference between the internal energies of the real gas and the ideal gas is

$$U - U_{\text{ideal}} \simeq \frac{N^2}{2}\left(\frac{1}{V}\int_{r_0}^{\infty} 4\pi r^2 \Phi(r)\,\mathrm{d}r\right) \simeq N_{\text{pairs}}\langle\Phi\rangle$$

where $N_{\text{pairs}} = N(N-1)/2 \simeq N^2/2$ is the number of pairs of molecules. That is, the leading correction is just the total potential energy between pairs of molecules, averaged over the relative positions of each pair. This simple result is valid only at temperatures which are high enough for the molecular configurations to be unaffected by the potential.

(b) In the same way, the first-order correction to the entropy is given by

$$S - S_{\text{ideal}} \simeq -\frac{Nk}{v}\frac{\mathrm{d}}{\mathrm{d}T}[T B_2(T)] = -k\frac{N^2}{2}\frac{V_0}{V} \simeq -N_{\text{pairs}}k\frac{V_0}{V}$$

where $V_0 = 4\pi r_0^3/3$ is the excluded volume for a pair of molecules. We see that the entropy is reduced by the hard-core part of the interaction. This can be interpreted as reflecting a reduction in the number of microstates available to the gas, owing to the smaller volume in which the molecules are allowed to move.

• **Problem 5.6** The virial coefficients $B_2(T)$ and $B_3(T)$ can be obtained straightforwardly from first principles using the method developed in problems 5.2 and 5.3 or from the graphical rules summarized in chapter 1. They are given by

$$B_2(T) = -\frac{1}{2V}$$

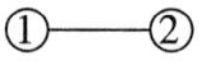

$$= -\frac{1}{2}\int \mathrm{d}^3x\, f(|\boldsymbol{x}|)$$

$$B_3(T) = -\frac{1}{3V}$$

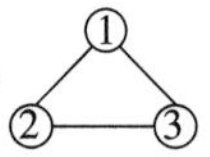

$$= -\frac{1}{3}\int \mathrm{d}^3x\,\mathrm{d}^3y\, f(|\boldsymbol{x}|)f(|\boldsymbol{y}|)f(|\boldsymbol{x}-\boldsymbol{y}|)$$

where $f(r) = \mathrm{e}^{-\beta\Phi(r)} - 1$. For a gas of hard spheres, with

$$\Phi(r) = \begin{cases} \infty & \text{if } r < r_0 \\ 0 & \text{if } r > r_0 \end{cases}$$

$f(|\boldsymbol{r}|)$ is equal to -1 when $\boldsymbol{r}$ lies inside a sphere of radius r_0 and equal to zero when $\boldsymbol{r}$ is outside this sphere, so we immediately obtain $B_2(T) = 2\pi r_0^3/3$. In general, the third virial coefficient is quite complicated to evaluate, but for the hard-sphere potential we can calculate it rather easily. Consider, indeed, the subintegral

$$W(\boldsymbol{x}) = \int \mathrm{d}^3y\, f(|\boldsymbol{y}|)f(|\boldsymbol{y}-\boldsymbol{x}|).$$

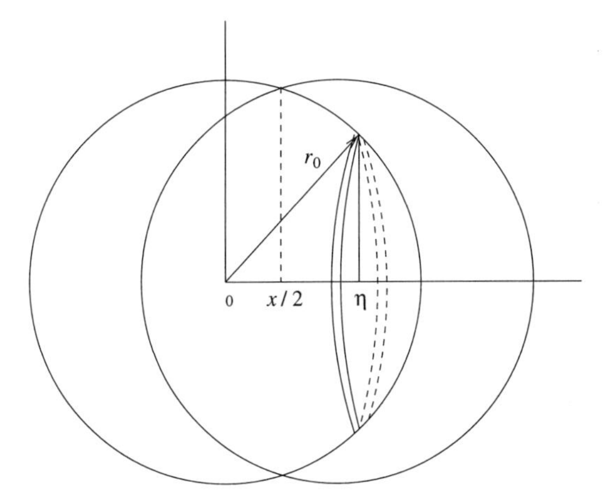

Figure 5.2. Volume $W(x)$ defined by the intersection of two overlapping spheres.

The integrand is equal to 1 if $\boldsymbol{y}$ lies both inside a sphere of radius r_0 centred at $\boldsymbol{y} = \mathbf{0}$ and inside a similar sphere centred at $\boldsymbol{y} = \boldsymbol{x}$ and is equal to zero otherwise. Therefore $W(\boldsymbol{x})$ is equal to the volume of space in which these two spheres overlap, as depicted in figure 5.2.

In this figure, let η be a coordinate measured from the centre of the left-hand sphere in the direction of $\boldsymbol{x}$. One half of the overlap region is that part of the left-hand sphere for which $|\boldsymbol{x}|/2 < \eta < r_0$. Considering this spherical cap as a stack of discs with radii $\sqrt{r_0^2 - \eta^2}$, we compute

$$W(\boldsymbol{x}) = 2\int_{x/2}^{r_0} \pi(r_0^2 - \eta^2)\,\mathrm{d}\eta = \pi\left(\tfrac{4}{3}r_0^3 - r_0^2 x + \frac{x^3}{12}\right)$$

where $x = |\boldsymbol{x}|$. Finally, $B_3(T)$ is given by

$$B_3(T) = -\tfrac{1}{3}\int \mathrm{d}^3x\, f(x)W(x) = \tfrac{1}{3}4\pi\int_0^{r_0} \mathrm{d}x\, x^2 W(x) = \frac{5\pi^2 r_0^6}{18} = \tfrac{5}{8}B_2^2(T)$$

which is what we hoped to prove.

• Problem 5.7

(a) It is quite straightforward to generalize the derivation of the virial coefficients given in problems 5.2 and 5.3 to the case of a mixture of two gases. The main steps may be outlined as follows. First, the grand canonical partition function is a sum of terms corresponding to M molecules of type A and N molecules of type B:

$$\mathcal{Z} = \sum_{M,N=0}^{\infty} \frac{1}{M!N!} y_{\mathrm{A}}^M y_{\mathrm{B}}^N \tilde{Z}(T, V, M, N)$$

where, for molecules of type a, $y_a = e^{\beta\mu_a}(2\pi m_a kT/h^2)^{3/2}$. In the thermodynamic limit, the grand potential has the form

$$\Omega(T, V, \mu_A, \mu_B) = kTV \sum_{k=0}^{\infty}\sum_{l=0}^{\infty} y_A^k y_B^l \omega_{kl}(T)$$

where, with the abbreviation $\tilde{Z}_{MN} = \tilde{Z}(T, V, M, N)$, the first few ω_{kl} are

$$\omega_{00} = 0 \qquad \omega_{10} = \omega_{01} = 1$$

$$\omega_{11} = \lim_{V\to\infty}\left(\frac{1}{V}(\tilde{Z}_{11} - V^2)\right)$$

$$\omega_{20} = \lim_{V\to\infty}\left(\frac{1}{2V}(\tilde{Z}_{20} - V^2)\right) \qquad \omega_{02} = \lim_{V\to\infty}\left(\frac{1}{2V}(\tilde{Z}_{02} - V^2)\right)$$

and so on. Next, the number densities for the two particle species are given by

$$xn = n_A = \sum_{k=0}^{\infty}\sum_{l=0}^{\infty} k y_A^k y_B^l \omega_{kl}(T)$$

$$(1-x)n = n_B = \sum_{k=0}^{\infty}\sum_{l=0}^{\infty} l y_A^k y_B^l \omega_{kl}(T)$$

and these equations can be reverted to obtain y_A and y_B as power series in the overall number density $n = n_A + n_B$ using the formulae for reversion of power series given in appendix A. Finally, since the pressure is $P = \Omega/V$, we obtain the virial expansion for the equation of state by substituting these series into the above expression for Ω. After carrying out the algebra in detail, we find that

$$\frac{P}{kT} = n - [x^2\omega_{20}(T) + x(1-x)\omega_{11}(T) + (1-x)^2\omega_{02}(T)]n^2 + O(n^3)$$

and, from the second term, we obtain

$$B_2(T) = -\tfrac{1}{2}\int d^3r\, [x^2 e^{-\beta\Phi_{AA}(r)} + 2x(1-x)e^{-\beta\Phi_{AB}(r)} + (1-x)^2 e^{-\beta\Phi_{BB}(r)} - 1].$$

The coefficients x^2, $2x(1-x)$ and $(1-x)^2$ have a straightforward interpretation as the relative numbers of A–A, A–B and B–B interactions respectively. We observe that the second virial coefficient for a single species is recovered (as it must be) either when $x = 1$ or $x = 0$, so that only one species is present, or when $\Phi_{AA} = \Phi_{AB} = \Phi_{BB}$, so that both species interact in the same way.

(b) When both molecular species are hard spheres, the potentials are

$$\Phi_{AA}(r) = \begin{cases} \infty & r < 2R_A \\ 0 & r > 2R_A \end{cases}$$

$$\Phi_{BB}(r) = \begin{cases} \infty & r < 2R_B \\ 0 & r > 2R_B \end{cases}$$

$$\Phi_{AB}(r) = \begin{cases} \infty & r < R_A + R_B \\ 0 & r > R_A + R_B. \end{cases}$$

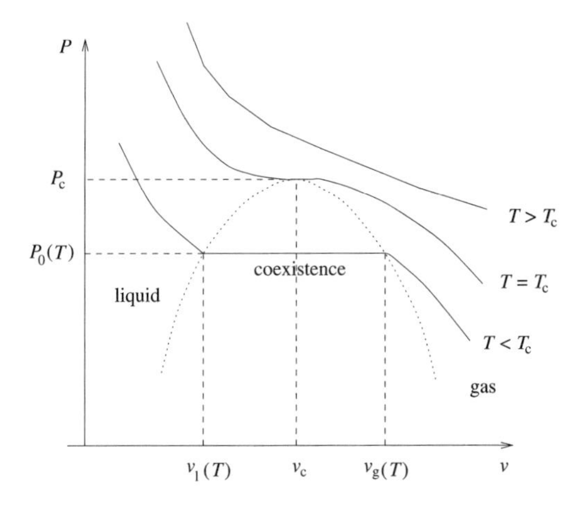

Figure 5.3. P–v diagram for a van der Waals gas.

We see that $B_2(T)$ is a sum of terms, each involving an integral of the form

$$\tfrac{1}{2}\int \mathrm{d}^3r\,(1 - \mathrm{e}^{-\beta\Phi(r)}) = \frac{1}{2}\frac{4\pi(2R)^3}{3}$$

the integrand being equal to one inside a sphere of radius $2R$ and zero outside. With a total of $N = N_\mathrm{A} + N_\mathrm{B}$ molecules, we have $x = N_\mathrm{A}/N$ and $(1-x) = N_\mathrm{B}/N$ and the equation of state can be written as $P(V - V_0) \simeq NkT$, where the excluded volume is

$$V_0 = NB_2 = \frac{16\pi}{3N}\left[N_\mathrm{A}^2R_\mathrm{A}^3 + N_\mathrm{B}^2R_\mathrm{B}^3 + 2N_\mathrm{A}N_\mathrm{B}\left(\frac{R_\mathrm{A}+R_\mathrm{B}}{2}\right)^3\right].$$

• **Problem 5.8** The van der Waals equation of state can be written as

$$\left(P + \frac{a}{v^2}\right)(v - b) = kT$$

where v is the specific volume and a and b are positive constants. The critical isotherm has a point of inflection, namely the critical point (figure 5.3), at the critical pressure and volume, that is $P = P_\mathrm{c}$ and $v = v_\mathrm{c}$, which are therefore determined by

$$\left(\frac{\partial P}{\partial v}\right)_T = 0 \qquad \left(\frac{\partial^2 P}{\partial v^2}\right)_T = 0.$$

It is straightforward, if a little tedious, to solve these equations, but a slightly less direct method will prove to be convenient. We define $\phi = v/v_\mathrm{c} - 1$ and, since ϕ is small in the region of interest, express P as a power series in ϕ. To obtain a point of inflection in the critical isotherm $P = P(T_c, v)$, we must retain terms of order ϕ^3, but higher powers can be neglected. Using the abbreviations $p = kTv_c/a$ and $q = v_c/(v_c - b)$, we obtain

$$P \simeq \frac{a}{v_c^2}[(pq-1) - (pq^2-2)\phi + (pq^3-3)\phi^2 - (pq^4-4)\phi^3].$$

The critical point is then given by $pq^2 = 2$ and $pq^3 = 3$ which implies that $q = \frac{3}{2}$ and $p = \frac{8}{9}$, or $v_c = 3b$ and $kT_c = 8a/27b$. On setting $\phi = 0$, we obtain $P_c = (pq-1)a/v_c^2 = a/27b^2$.

To simplify matters, we define the reduced pressure and temperature by $\pi = P/P_c - 1$ and $t = T/T_c - 1$, in terms of which the equation of state reads

$$\pi \simeq 4t - 6t\phi + 9t\phi^2 - \tfrac{3}{2}(1+9t)\phi^3.$$

Since we are concerned with a region where ϕ and t are both very small, $t\phi^2$ is much smaller than $t\phi$ and can be neglected, while $1 + 9t$ can be replaced by 1, giving

$$\pi \simeq 4t - 6t\phi - \tfrac{3}{2}\phi^3.$$

Clearly, the limiting form of the critical isotherm is $\pi \propto \phi^3$, so the first critical exponent is $\delta = 3$.

The isothermal compressibility is given by

$$\frac{1}{\kappa_T} = -V\left(\frac{\partial P}{\partial V}\right)_T = -P_c(1+\phi)\left(\frac{\partial \pi}{\partial \phi}\right)_t \simeq P_c(1+\phi)(6t - \tfrac{9}{2}\phi^2)$$

and, when $\phi = 0$, this reduces to $\kappa_T = (6P_c t)^{-1}$, so our second exponent is $\gamma = 1$.

In order to determine the third exponent β, we have to find the reduced specific volumes $\phi_g(t)$ and $\phi_\ell(t)$ of the gas and liquid phases which coexist along the vapour pressure curve $\pi = \pi_0(t)$, using the fact that the Gibbs free energies of the two coexisting phases are equal. The Gibbs free energy $G(T, P)$ (we shall ignore its dependence on the number N of particles, which is fixed throughout) is obtained up to an unknown but irrelevant function of temperature by integrating the relation $(\partial G/\partial P)_T = V$. In terms of our reduced variables, this is

$$\left(\frac{\partial g}{\partial \pi}\right)_t = 1 + \phi$$

where $g(t, \pi) = G(T, P)/P_c V_c$. One way of proceeding would be to find two solutions, $\phi_g(t, \pi)$ and $\phi_\ell(t, \pi)$, of the equation of state, which we write as

$$\pi = F(\phi, t) \equiv 4t - 6t\phi - \tfrac{3}{2}\phi^3 + \cdots$$

which could then be integrated to find $g_g(t, \pi)$ and $g_\ell(t, \pi)$. Here, however, we use a less direct method, which in the end is algebraically simpler and is of rather general use in the study of phase transitions. We introduce an auxiliary function

$$\Gamma(x, t, \pi) = \int_0^x [\pi - F(x', t)]\,\mathrm{d}x' \simeq (\pi - 4t)x + 3tx^2 + \tfrac{3}{8}x^4.$$

This function is usually called the Landau free energy but, since it depends on an arbitrary variable x, it is clearly not one of the thermodynamic potentials. However, the extrema of this function, satisfying $(\partial\Gamma/\partial x)_{t,\pi} = 0$, are clearly the values of ϕ which solve the equation of state. Moreover, if $x = \phi(t, \pi)$ is one of these extrema, then the Gibbs free energy of the corresponding state is given by

$$g(t, \pi) = g_0(t) + \pi + \Gamma(\phi, t, \pi)$$

where $g_0(t)$ is an undetermined function of t only, because then

$$\left(\frac{\partial g}{\partial \pi}\right)_t = 1 + \left(\frac{\partial \Gamma}{\partial \pi}\right)_t = 1 + \phi + [\pi - F(\phi, t)]\left(\frac{\partial \phi}{\partial \pi}\right)_t = 1 + \phi.$$

Since the Gibbs free energy is a minimum in thermal equilibrium (see problem 2.1), the extrema of Γ corresponding to stable thermodynamic states are its minima. At a given pressure and temperature, two states with the same Gibbs free energy also have the same value of Γ, from which we deduce that

$$\Gamma(\phi_{\mathrm{g}}) - \Gamma(\phi_\ell) = \pi(\phi_{\mathrm{g}} - \phi_\ell) - \int_{\phi_\ell}^{\phi_{\mathrm{g}}} F(x, t)\,\mathrm{d}x = 0.$$

This relation gives rise to the Maxwell equal-areas construction, and we could obtain $\phi_{\mathrm{g}}(t)$, $\phi_\ell(t)$ and $\pi_0(t)$ by solving this relation, together with $\pi_0(t) = F(\phi_{\mathrm{g}}, t) = F(\phi_\ell, t)$. It is much easier, however, to observe that $\Gamma(x)$ is a quartic, with leading coefficient $\frac{3}{8}$, that it has minima at $x = \phi_{\mathrm{g}}$ and $x = \phi_\ell$, and that it has the same value, say, Γ_0, at both minima. It must therefore have the form

$$\begin{aligned}\Gamma(x) &= \tfrac{3}{8}(x - \phi_{\mathrm{g}})^2(x - \phi_\ell)^2 + \Gamma_0 \\ &= \tfrac{3}{8}[x^4 - 2(\phi_{\mathrm{g}} + \phi_\ell)x^3 + (\phi_{\mathrm{g}}^2 + \phi_\ell^2 + 4\phi_{\mathrm{g}}\phi_\ell)x^2 \\ &\quad - 2\phi_{\mathrm{g}}\phi_\ell(\phi_{\mathrm{g}} + \phi_\ell)x + \phi_{\mathrm{g}}^2\phi_\ell^2] + \Gamma_0\end{aligned}$$

and we can identify the coefficients as

$$\begin{aligned}\phi_{\mathrm{g}} + \phi_\ell &= 0 \\ \phi_{\mathrm{g}}^2 + \phi_\ell^2 + 4\phi_{\mathrm{g}}\phi_\ell &= 8t \\ -\phi_{\mathrm{g}}\phi_\ell(\phi_{\mathrm{g}} + \phi_\ell) &= \tfrac{4}{3}(\pi - 4t).\end{aligned}$$

It is now a simple matter to solve these equations and find that $\phi_{\mathrm{g}}(t) = -\phi_\ell(t) = 2\sqrt{-t}$, so that our third critical exponent is $\beta = \frac{1}{2}$, while the limiting form of the coexistence curve is $\pi_0(t) = 4t$. Finally, we note that the equation of state in the neighbourhood of the critical point has the scaling form characteristic of all critical phenomena, that is

$$\pi - \pi_0(t) = \phi^\delta f(t|\phi|^{-1/\beta})$$

where, in this case, the scaling function is $f(x) = -\frac{3}{2}(1 + 4x)$.

• Problem 5.9

(a) In order to be sure of locating the critical isochore correctly, we need some knowledge of the behaviour of the scaling function $\mathcal{G}(x)$ when $x = p/t^\Delta$ is either very large or very small. The information that we need can be deduced from the fact that, for $T \geq T_c$, the only place in the phase diagram where the free energy can be singular is exactly at the critical point, where $P = P_c$ and $T = T_c$, or $p = t = 0$. At some fixed non-zero value of t, the free energy is a well-behaved function of p, even at $p = 0$, so it has a well-defined series expansion in p. Therefore, as stated in the question, $\mathcal{G}(x)$ has a well-defined expansion in powers of x. Conversely, at some fixed non-zero value of p, the free energy is a well-behaved function of t, even at $t = 0$, where $x = p/t^\Delta \to \infty$ and must have a well-defined expansion in powers of t. For this to be true, the scaling function must behave as

$$\mathcal{G}(x) = x^{(2-\alpha)/\Delta}\tilde{\mathcal{G}}(x^{-1/\Delta})$$

when x is large, where $\tilde{\mathcal{G}}(y)$ has a well-defined expansion in powers of $y = x^{-1/\Delta} = t/p^{1/\Delta}$. This ensures that $G_{\text{sing}} = p^{(2-\alpha)/\Delta}\tilde{\mathcal{G}}(t/p^{1/\Delta})$ is a well-behaved function of t.

With this in mind, we investigate the critical isochore, identifying the volume as

$$V = N\left(\frac{\partial G}{\partial P}\right)_T = \frac{N}{P_c}\left(\frac{\partial G_{\text{reg}}}{\partial p}\right)_t + \frac{N}{P_c}t^{2-\alpha-\Delta}\mathcal{G}'\left(\frac{p}{t^\Delta}\right).$$

Here, $\mathcal{G}'(x)$ means $\mathrm{d}\mathcal{G}(x)/\mathrm{d}x$ and the factor of N appears because we have denoted by $G(T, P)$ the free energy *per particle*. The critical isochore is, of course, a curve in the p–t plane that passes through the critical point $p = t = 0$, as sketched in figure 5.4, and we want to know what happens as the critical point is approached along this curve. As far as the scaling variable $x = p/t^\Delta$ is concerned, there are three possibilities, namely $x \to 0$, $x \to \infty$ or $x \to x_0$, where x_0 is a non-zero constant. In each case, since the exponent $2-\alpha-\Delta \simeq 0.3$ is positive, it is straightforward to see from our knowledge of $\mathcal{G}$ that the singular part of the expression for V vanishes at $p = t = 0$. Since $V = V_c$ at this point, we can write

$$\frac{P_c(V - V_c)}{N} = At + Bp + \cdots + t^{2-\alpha-\Delta}\mathcal{G}'\left(\frac{p}{t^\Delta}\right)$$

where A and B are constants, coming from the power series for G_{reg}. We must now consider in turn the three possible behaviours of x as $p \to 0$ and $t \to 0$ along the critical isochore $V = V_{\mathrm{c}}$. Suppose first that $x \to 0$. If $\mathcal{G}'(0)$ happens to vanish, then the curve $V = V_{\mathrm{c}}$ corresponds to $p = -(A/B)t$, or $x = p/t^{\Delta} = -(A/B)t^{1-\Delta} \sim t^{-0.6}$, which is not consistent with $x \to 0$. If $\mathcal{G}'(0)$ is a non-zero constant, then At can be neglected in comparison with $\mathcal{G}'(0)t^{2-\alpha-\Delta}$ and we find instead that $x = p/t^{\Delta} \sim t^{2-\alpha-2\Delta} \sim t^{-1.3}$ which is also not consistent with $x \to 0$. The possibility that $x \to 0$ on the critical isochore is therefore not self-consistent and can be discarded. In just the same way, readers may satisfy themselves that the possibility $x \to \infty$ is not self-consistent either. Finally, then, suppose that $x \to x_0$. In that case, both At and $Bp \sim Bt^{\Delta}$ vanish faster than $t^{2-\alpha-\Delta}$ as $t \to 0$. Consequently, the curve $V = V_{\mathrm{c}}$ must correspond to $\mathcal{G}'(x_0) = 0$, which is what we hoped to show.

(b) The isothermal susceptibility is given by

$$\begin{aligned}\kappa_T &= -\frac{1}{V}\left(\frac{\partial V}{\partial P}\right)_T \\ &= -\frac{1}{P_{\mathrm{c}}V}\left(\frac{\partial V}{\partial p}\right)_t \\ &= -\frac{N}{P_{\mathrm{c}}^2 V}\left[\left(\frac{\partial^2 G_{\mathrm{reg}}}{\partial p^2}\right)_t + t^{2-\alpha-2\Delta}\mathcal{G}''\left(\frac{p}{t^{\Delta}}\right)\right].\end{aligned}$$

In the last expression, V can be set equal to V_{c} near the critical point. The first term inside the square brackets, $(\partial^2 G_{\mathrm{reg}}/\partial p^2)_t$, is a power series in t and p and approaches a finite constant at the critical point. On the critical isochore, the second term becomes $t^{2-\alpha-2\Delta}\mathcal{G}''(x_0)$, and $\mathcal{G}''(x_0)$ is a finite non-zero constant. We see, therefore, that κ_T diverges as $\kappa_T \propto t^{-\gamma}$ as $t \to 0$, with the exponent $\gamma = 2\Delta + \alpha - 2 \simeq 1.3$.

The specific heat at constant volume is a little harder to find. We first obtain the entropy per particle, which is given by

$$S = -\left(\frac{\partial G}{\partial T}\right)_P = -\frac{1}{T_{\mathrm{c}}}\left(\frac{\partial G}{\partial t}\right)_p + \frac{1}{P_{\mathrm{c}}}\frac{\mathrm{d}P_0}{\mathrm{d}T}\left(\frac{\partial G}{\partial p}\right)_t$$

because $p = [P - P_0(T)]/P_{\mathrm{c}}$. Working this out, we get the somewhat forbidding result

$$\begin{aligned}S = &-\frac{1}{T_{\mathrm{c}}}\left[\left(\frac{\partial G_{\mathrm{reg}}}{\partial t}\right)_p + (2-\alpha)t^{1-\alpha}\mathcal{G}(x) - \Delta t^{1-\alpha}x\mathcal{G}'(x)\right] \\ &+ \frac{1}{P_{\mathrm{c}}}\frac{\mathrm{d}P_0}{\mathrm{d}T}\left[\left(\frac{\partial G_{\mathrm{reg}}}{\partial p}\right)_t + t^{2-\alpha-\Delta}\mathcal{G}'(x)\right]\end{aligned}$$

where, of course, $x = p/t^{\Delta}$. To find C_V on the critical isochore, we set $V = V_{\mathrm{c}}$ and then differentiate with respect to t. Now, although the derivatives of G_{reg}

are power series in t and p, we have $p \propto t^{\Delta}$ when $V = V_c$ and sufficiently many derivatives of a positive non-integer power of t produce a negative power. In this case, fortunately, we need only one derivative and, since $\Delta > 1$, the terms in C_V involving G_{reg} approach finite values. Of the remaining terms in S, those proportional to $\mathcal{G}'(x)$ vanish when $V = V_c$. In fact, the divergent part of C_V is

$$C_V = NT\left(\frac{\partial S}{\partial T}\right)_{V=V_c} \simeq N\left(\frac{\partial S}{\partial t}\right)_{x=x_0} \simeq -\frac{(1-\alpha)(2-\alpha)N\mathcal{G}(x_0)}{T_c}t^{-\alpha}$$

and this diverges with the expected power.

(c) On the critical isotherm $t = 0$, we have $x = p/t^{\Delta} \to \infty$. The discussion of (a) shows that in this limit $\mathcal{G}'(x) = x^{(2-\alpha-\Delta)/\Delta}\mathcal{V}(x^{-1/\Delta})$, where $\mathcal{V}(y)$ is a well-behaved function with an expansion in powers of y. We therefore find that

$$V - V_c = \frac{NB}{P_c}p + \mathrm{O}(p^2) + \frac{N\mathcal{V}(0)}{P_c}p^{1/\delta}$$

where $\delta = \Delta/(2-\alpha-\Delta) \simeq 5.3$. Now, of course, the scaling field p is just $p = [P - P_0(T_c)]/P_c = (P - P_c)/P_c$. Clearly, the regular part is negligible when $p \to 0$ and we obtain the limiting behaviour $p \propto (V - V_c)^{\delta}$.

The limiting behaviour of the isothermal compressibility is easily found:

$$\kappa_T \simeq \frac{1}{P_c V_c}\left(\frac{\partial V}{\partial p}\right)_{t=0} \propto (V - V_c)^{-\gamma/\beta}$$

where $\beta = 2 - \alpha - \Delta = \Delta/\delta \simeq 0.3$. This is in fact the same exponent that governs the temperature dependence of $V - V_c$ along the coexistence curve for $T < T_c$ (see problem 5.8). In this region, the fluid exhibits a similar scaling property, with

$$\frac{P_c(V - V_c)}{N} = \bar{A}t + \bar{B}p + \cdots + (-t)^{2-\alpha-\Delta}\bar{\mathcal{G}}'\left(\frac{p}{(-t)^{\Delta}}\right).$$

The coexistence curve is $p = 0$, and we see that there $V - V_c \propto (-t)^{2-\alpha-\Delta}$. Evidently, the scaling property is extremely powerful, yielding the values of a wide range of critical exponents in terms of any two of them. First advanced as a hypothesis by B Widom, this property is well verified in many fluid, magnetic and other systems having critical points, and in a variety of theoretical models.

• **Problem 5.10** This closed system is appropriately treated using the microcanonical ensemble, and we need to find the number $\Omega(E, N)$ of microstates available to it when its energy is E, although this energy is determined by the number P of pairs. To compute Ω, we first assign the $2P$ paired particles to pairs of states. There are a total of J pairs of states, so the

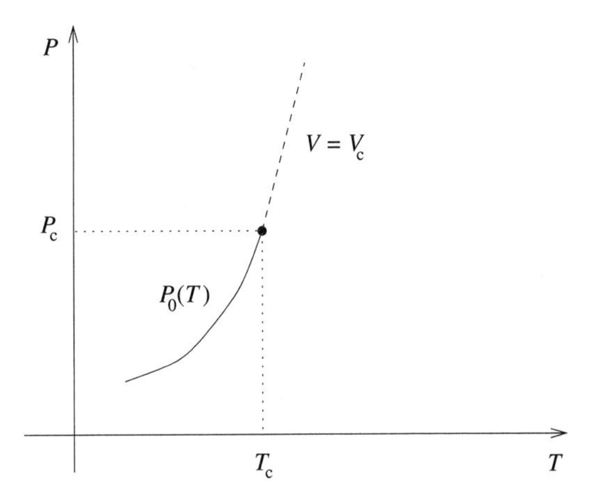

Figure 5.4. Coexistence curve and critical isochore for the liquid–gas phase transition.

number of ways of assigning P particle pairs to them is $\binom{J}{P}$. Now we have to assign the remaining $N-2P$ particles to the remaining $2(J-P)$ states without forming more pairs. Each particle must therefore be assigned to one of the $J-P$ pairs of states, which can be done in $\binom{J-P}{N-2P}$ ways. Having done this, we must finally assign each particle to one of the two states in its pair, and this can be done in 2^{N-2P} ways. Collecting these factors together, we obtain

$$\Omega(E, N) = \binom{J}{P}\binom{J-P}{N-2P}2^{N-2P} = \frac{2^{N-2P}J!}{P!(N-2P)!(J+P-N)!}$$

and use Stirling's approximation to find

$$\ln[\Omega(E, N)] \simeq (N-2P)\ln 2 + J\ln J - P\ln P - (N-2P)\ln(N-2P) - (J+P-N)\ln(J+P-N).$$

The temperature is then given by

$$\frac{1}{kT} = \left(\frac{\partial(\ln\Omega)}{\partial E}\right)_N = \frac{[\partial(\ln\Omega)/\partial P]_N}{(\partial E/\partial P)_N}$$

which yields

$$kT = \frac{\varepsilon[1+(2P-N+1)/J]}{\ln[4P(J+P-N)] - \ln[(N-2P)^2]}.$$

To get $T = 0$, the argument of one of the logarithms must vanish. With $J > N > 2P$, the argument of the first logarithm can vanish only when $P \to 0$.

This, however, corresponds to a negative (very high) temperature $1/kT \to -\infty$. The absolute zero temperature is obtained when the second logarithm diverges at $P = N/2$, at which point all the particles are paired, as we should expect. The limit $T \to \infty$ is obtained when the two logarithms are equal. This occurs when $P = N^2/4J$, and then the fraction of particles in paired states, $2P/N$, is equal to $N/2J$.

• **Problem 5.11** The model described in this question is known as the p-state Potts model, after R B Potts who invented and first solved it. It is one of a variety of idealized models whose interest derives largely from the fact that they can be solved exactly (or that one can make significant progress towards an exact solution), although in some cases they also capture significant features of real physical systems. The given expression for the energy is conventionally referred to as the 'Hamiltonian' but, since it does not contain conjugate coordinates and momenta, it could not be used to derive Hamilton's equations for a *bona fide* dynamical system. Rather, it is a model for the potential energy of an interacting system and the partition function associated with it is really only a model for the configurational factor in the partition function of a classical system (this being the part which is generally difficult to calculate).

(a) In this case, the canonical partition function is given by

$$Z(T, N) = \sum_{\{\nu_i\}} \exp\left(\beta J \sum_{i=1}^{N} \delta_{\nu_i,\nu_{i+1}}\right)$$

where each of the ν_i is summed over the values $\nu_i = 1, \ldots, p$ on the understanding that $\nu_{N+1} = \nu_1$. It can be calculated exactly by the transfer matrix method, which works as follows. Consider a chain of n atoms, and suppose that the sums over all the ν_i have been carried out, except for ν_1 and ν_{n+1} which, for now, we regard as being independent. In that case, we have an object $Z^{(n)}_{\nu_1,\nu_{n+1}}$, such that

$$Z(T, n) = \sum_{\nu=1}^{p} Z^{(n)}_{\nu,\nu} = \mathrm{Tr}(\mathbf{Z}^{(n)})$$

if we regard $Z^{(n)}_{\nu_1,\nu_{n+1}}$ as the elements of a $p \times p$ matrix $\mathbf{Z}^{(n)}$. Now we consider adding an extra atom to the chain, and ask how $\mathbf{Z}^{(n+1)}$ can be obtained from $\mathbf{Z}^{(n)}$. In terms of matrix elements, the answer is

$$Z^{(n+1)}_{\nu_1,\nu_{n+2}} = \sum_{\nu_{n+1}=1}^{p} Z^{(n)}_{\nu_1,\nu_{n+1}} \exp(\beta J \delta_{\nu_{n+1},\nu_{n+2}}).$$

This can be recognized as a product of two matrices, namely $\mathbf{Z}^{(n+1)} = \mathbf{Z}^{(n)}\mathbf{T}$, where $\mathbf{T}$ is the matrix whose elements are

$$T_{\nu,\nu'} = \exp(\beta J \delta_{\nu,\nu'}) = (\mathrm{e}^{\beta J} - 1)\delta_{\nu,\nu'} + 1.$$

That these two expressions are equivalent can easily be checked by considering the two cases $\nu = \nu'$ and $\nu \neq \nu'$. The matrix $\mathbf{T}$ is called the transfer matrix. It is easy to see that $\mathbf{Z}^{(1)} = \mathbf{T}$ and hence that

$$Z(T, N) = \mathrm{Tr}(\mathbf{T}^N).$$

Now $\mathbf{T}$, being a real symmetric matrix, can be diagonalized by an orthogonal transformation. That is, it can be expressed as $\mathbf{T} = \mathbf{U}^{-1}\bar{\mathbf{T}}\mathbf{U}$, where $\mathbf{U}$ is an orthogonal matrix and $\bar{\mathbf{T}}$ is a diagonal matrix whose diagonal elements are the eigenvalues λ_j of $\mathbf{T}$. We can therefore use the cyclic property of the trace operation to write

$$Z(T, N) = \mathrm{Tr}\,(\mathbf{U}^{-1}\bar{\mathbf{T}}\mathbf{U}\mathbf{U}^{-1}\bar{\mathbf{T}}\mathbf{U}\cdots\mathbf{U}^{-1}\bar{\mathbf{T}}\mathbf{U}) = \mathrm{Tr}\,(\bar{\mathbf{T}}^N) = \sum_{j=1}^{p} \lambda_j^N.$$

Naturally, this method is useful only if we can find the eigenvalues λ_j. In this case, we can find them as follows. We have to solve the eigenvalue equation $\mathbf{T}\boldsymbol{v} = \lambda\boldsymbol{v}$, where the column vector $\boldsymbol{v}$ is an eigenvector of $\mathbf{T}$. In terms of the components of $\boldsymbol{v}$, this is

$$(\lambda - \mathrm{e}^{\beta J} + 1)v_\nu = \sum_{\nu'=1}^{p} v_{\nu'}.$$

The right-hand side of this equation is independent of ν, so if $(\lambda - \mathrm{e}^{\beta J} + 1) \neq 0$, then all the v_ν are equal and we might as well set them equal to unity, since we are not interested in finding normalized eigenvectors. This gives us one eigenvalue as

$$\lambda_1 = \mathrm{e}^{\beta J} + p - 1$$

corresponding to a single eigenvector. The only other possibility is that

$$\lambda = \mathrm{e}^{\beta J} - 1 \equiv \lambda_2 \qquad \sum_{\nu'=1}^{p} v_{\nu'} = 0.$$

Since $\mathbf{T}$ has a total of p eigenvalues, $p - 1$ of these must be equal to λ_2. In fact, we can see that there are exactly $p - 1$ eigenvectors of the form

$$\boldsymbol{u} = \begin{pmatrix} 1 \\ -1 \\ 0 \\ 0 \\ 0 \\ \vdots \end{pmatrix} \begin{pmatrix} 1 \\ 1 \\ -2 \\ 0 \\ 0 \\ \vdots \end{pmatrix} \begin{pmatrix} 1 \\ 1 \\ 1 \\ -3 \\ 0 \\ \vdots \end{pmatrix} \cdots$$

which are mutually orthogonal and whose elements sum to zero, although we do not actually need these. The partition function is now given by

$$Z(T, N) = \lambda_1^N + (p-1)\lambda_2^N = (\mathrm{e}^{\beta J} + p - 1)^N + (p-1)(\mathrm{e}^{\beta J} - 1)^N.$$

(b) In the thermodynamic limit, the partition function is given by

$$Z(T,N) = \lambda_1^N\left[1+(p-1)\left(\frac{\lambda_2}{\lambda_1}\right)^N\right] \simeq \lambda_1^N = (e^{\beta J}+p-1)^N$$

because $\lambda_2 < \lambda_1$ and so $\lim_{N\to\infty}[(\lambda_2/\lambda_1)^N] = 0$. Evidently, the partition function in the thermodynamic limit will always involve only the largest eigenvalue of the transfer matrix. The internal energy is then

$$U(T) = -\left(\frac{\partial(\ln Z)}{\partial\beta}\right)_N = -\frac{NJ}{1+(p-1)\,e^{-\beta J}}.$$

At $T=0$, it is $U(0) = -NJ$, corresponding to all the atoms being in the same state. In the high-temperature limit, we have $U(T\to\infty) = -NJ/p$. This corresponds to a state of maximum entropy because, when the atomic states are randomly distributed, the probability of finding a pair of neighbouring atoms in the same state (and thus contributing an energy $-J$) is $1/p$.

• **Problem 5.12** If the total number of lattice sites is N and the number of particles at the ith site is n_i ($n_i = 0, 1$), then the grand canonical partition function is

$$\mathcal{Z} = \sum_{\{n_i=0,1\}} \exp\left(\beta\mu\sum_{i=1}^{N} n_i + \beta\varepsilon\sum_{\langle i,j\rangle} n_i n_j\right)$$

where μ is the chemical potential and $\langle i,j\rangle$ indicates a sum over pairs of neighbouring lattice sites. The potential energy of a given pair of neighbouring sites $\langle i,j\rangle$ is $-\varepsilon$ if $n_i = n_j = 1$ and zero otherwise. To relate this partition function to that of the Ising model, we write n_i in terms of a spin variable S_i, which takes the values ± 1, as $n_i = (S_i+1)/2$. Then an occupied site corresponds to spin up ($S_i = 1$) and an unoccupied site to spin down ($S_i = -1$). In terms of these variables, the partition function becomes

$$\mathcal{Z} = \sum_{\{S_i=\pm 1\}} \exp\left[\frac{\beta\mu}{2}\sum_{i=1}^{N}(S_i+1) + \frac{\beta\varepsilon}{4}\sum_{\langle i,j\rangle}(S_iS_j + S_i + S_j + 1)\right].$$

A little thought enables us to calculate sums over lattice sites as follows. First, it is obvious that $\sum_{i=1}^{N} 1 = N$. The sum $\sum_{\langle i,j\rangle} 1$ is the number of 'bonds' which might be drawn between neighbouring sites. If we assign half of each bond to each of the sites at its ends, then the total number of bonds belonging to each site is $z/2$, where the coordination number z is the number of nearest neighbours that each site has, and the total number of bonds in the lattice is $\sum_{\langle i,j\rangle} 1 = Nz/2$. (This assumes that N is large enough for the missing neighbours at the surfaces to be negligible compared with the total number of sites.) Finally, we see that

$$\sum_{\langle i,j\rangle} S_i = \sum_{\langle i,j\rangle} S_j = \frac{z}{2}\sum_{i=1}^{N} S_i$$

because we must recover the value $Nz/2$ when all the S_i are equal to 1. Using these results, we can write the partition function as

$$\mathcal{Z} = \mathcal{Z}_0 \sum_{\{S_i=\pm 1\}} \exp\left(\beta h \sum_{i=1}^{N} S_i + \beta J \sum_{\langle i,j \rangle} S_i S_j\right)$$

with

$$\mathcal{Z}_0 = \exp\left[\beta N\left(\frac{\mu}{2} + \frac{z\varepsilon}{8}\right)\right] \qquad h = \frac{\mu}{2} + \frac{z\varepsilon}{4} \qquad J = \frac{\varepsilon}{4}.$$

Apart from the factor $\mathcal{Z}_0$, this is the partition function for the Ising model, where h is the magnetic field multiplied by the magnetic moment of each atom and J is the interaction energy. Although $\mathcal{Z}_0$ contributes to the total free energy of the system, it is independent of the spin configuration and therefore irrelevant to a phase transition. In view of this correspondence, we expect close analogies between the critical properties of fluids and magnets, and these are well verified in practice.

• **Problem 5.13** First of all, we remark that the partition function of this system is very easy to calculate. The eigenvalues of $\mathbf{H}$ are $E_\pm = \pm\sqrt{\varepsilon^2 + \Delta^2}$, and so

$$Z = \exp(\beta E_+) + \exp(\beta E_-).$$

The following analysis is therefore intended only to illustrate a technique that may be useful for more complicated problems. We introduce a set of basis vectors $|u\rangle$ for the Hilbert space of the quantum object, labelled by a variable u, which takes the values ± 1:

$$|1\rangle \equiv \begin{pmatrix} 1 \\ 0 \end{pmatrix} \qquad |-1\rangle \equiv \begin{pmatrix} 0 \\ 1 \end{pmatrix}.$$

The completeness relation for these vectors is $\sum_u |u\rangle\langle u| = I$, where I is the identity operator, and we can use this to write

$$\begin{aligned} Z &= \mathrm{Tr}(\mathrm{e}^{-\beta \mathbf{H}}) \\ &= \lim_{N\to\infty}\left(\sum_{u_1} \langle u_1|(1 - N^{-1}\beta\mathbf{H})^N|u_1\rangle\right) \\ &= \lim_{N\to\infty}\Bigg(\sum_{\{u_1,\cdots,u_N\}} \langle u_1|(1 - N^{-1}\beta\mathbf{H})|u_2\rangle\langle u_2|(1 - N^{-1}\beta\mathbf{H})|u_3\rangle \cdots \\ &\qquad \times \cdots \langle u_N|(1 - N^{-1}\beta\mathbf{H})|u_1\rangle\Bigg). \end{aligned}$$

Now suppose that each of these matrix elements can be expressed in the form

$$T_{u,u'} \equiv \langle u|(1 - N^{-1}\beta\mathbf{H})|u'\rangle = A\exp(Kuu' + Hu + Lu').$$

Then we would have

$$Z = \lim_{N\to\infty}\left[A^N \sum_{\{u_i\}} \exp\left(\sum_{i=1}^{N}(Ku_iu_{i+1} + hu_i)\right)\right]$$

where $h = H + L$, and it is understood that u_{N+1} means u_1. Apart from the factor A^N, this is the partition function for the one-dimensional Ising model with nearest-neighbour interaction constant $J = kT_\mathrm{I}K$ in a magnetic field of strength $kT_\mathrm{I}h$, where T_I means the temperature of the equivalent Ising system, which is not necessarily $1/\beta k$. That the matrix elements can indeed be written in this way follows from evaluating the four elements explicitly:

$$\begin{aligned}
T_{1,1} &= A\,\mathrm{e}^{K+H+L} = 1 - N^{-1}\beta\varepsilon \\
T_{1,-1} &= A\,\mathrm{e}^{-K+H-L} = N^{-1}\beta\Delta \\
T_{-1,1} &= A\,\mathrm{e}^{-K-H+L} = N^{-1}\beta\Delta \\
T_{-1,-1} &= A\,\mathrm{e}^{K-H-L} = 1 + N^{-1}\beta\varepsilon.
\end{aligned}$$

These four equations can be solved to give

$$\begin{aligned}
A &= \left(\frac{\beta\Delta}{N}\right)^{1/2}\left(1 - \frac{\beta^2\varepsilon^2}{N^2}\right)^{1/4} \\
K &= \frac{1}{4}\ln\left(\frac{N^2 - \beta^2\varepsilon^2}{\beta^2\Delta^2}\right) \\
h = 2H = 2L &= \frac{1}{2}\ln\left(\frac{N - \beta\varepsilon}{N + \beta\varepsilon}\right).
\end{aligned}$$

It should be noted that only the parameters K and h of the corresponding Ising system are determined, its temperature, interaction strength and magnetic field cannot be unambiguously identified. Also, K diverges and h approaches zero when $N \to \infty$, whereas these quantities are independent of N in a standard Ising system. In order to obtain the correct partition function, one must therefore solve the model with a finite N and then take the limit $N \to \infty$. The method which naturally presents itself is the transfer matrix (discussed in problem 5.11). In fact, the transfer matrix for this problem is none other than the matrix $\mathbf{T}$ whose elements we have just calculated. After a little algebra, we find that its eigenvalues are

$$\lambda_\pm = 1 \pm \frac{\beta}{N}\sqrt{\varepsilon^2 + \Delta^2}$$

and the calculation

$$Z = \lim_{N\to\infty}(\lambda_+^N + \lambda_-^N) = \exp(\beta\sqrt{\varepsilon^2 + \Delta^2}) + \exp(-\beta\sqrt{\varepsilon^2 + \Delta^2})$$

reproduces our original result.

• Problem 5.14

(a) Consider the set of $(N-1)$ variables b_i $(i = 1, \ldots, N-1)$ such that $b_i = 1$ if $S_{i+1} = S_i$ and $b_i = -1$ if $S_{i+1} = -S_i$. Together with the first spin S_1, these give a set of N variables, which uniquely specify the state of the chain. Thus, knowing S_1 and b_1, we can deduce the value of S_2; then, knowing the value of b_2, we can deduce the value of S_3, and so on. Clearly, we can express these link variables as $b_i = S_i S_{i+1}$. The energy of the ith link is Jb_i and the total energy of the chain is independent of S_1. (This is because reversing S_1 with the b_i unchanged is equivalent to reversing all the spins, which does not change the energy.) The partition function is therefore

$$Z(T, N) = \sum_{S_1} \sum_{\{b_i\}} \exp\left(\beta J \sum_{i=1}^{N-1} b_i\right) = 2[2\cosh(\beta J)]^{N-1}.$$

Evidently, $Z(T, N)/2$ is equivalent to the partition function of a chain of $N-1$ sites without interactions, on which we reinterpret the b_i as spin variables and J as the magnetic moment of each spin multiplied by a magnetic field.

(b) With next-nearest-neighbour interactions, we can write $S_i S_{i+2} = (S_i S_{i+1})(S_{i+1} S_{i+2}) = b_i b_{i+1}$, since $S_{i+1}^2 = 1$. The partition function is therefore

$$Z(T, N) = \sum_{S_1} \sum_{\{b_i\}} \exp\left(\beta J_1 \sum_{i=1}^{N-1} b_i + \beta J_2 \sum_{i=1}^{N-2} b_i b_{i+1}\right).$$

Now, $Z(T, N)/2$ is equivalent to the partition function for a chain of $N-1$ sites, again with free ends, with spin variables b_i in a magnetic field proportional to J_1 and with *nearest*-neighbour interactions of strength J_2. This partition function can be calculated by the transfer matrix method (see problem 5.11). The transfer matrix is

$$\mathbf{T} = \begin{pmatrix} e^{\beta(J_1+J_2)} & e^{\beta(J_1-J_2)} \\ e^{\beta(-J_1-J_2)} & e^{\beta(-J_1+J_2)} \end{pmatrix}$$

but, with free ends, the partition function is not equal to $\operatorname{Tr} \mathbf{T}^{N-1}$. It should not be hard to see that the correct expression is

$$\tfrac{1}{2} Z(T, N) = (1\ 1)\mathbf{T}^{N-2} \begin{pmatrix} e^{\beta J_1} \\ e^{-\beta J_1} \end{pmatrix}.$$

The column on the right accounts for the single-site energy of the spin at one end of the chain; each factor of $\mathbf{T}$ adds a link and a spin, accounting for the energies of both, and multiplying by the row at the left completes the sum over states of the last spin. If the eigenvalues of $\mathbf{T}$ are $\lambda_\pm$, we can find a matrix $\mathbf{U}$ such that

$$\mathbf{T} = \mathbf{U}^{-1} \begin{pmatrix} \lambda_+ & 0 \\ 0 & \lambda_- \end{pmatrix} \mathbf{U}$$

although $\mathbf{U}$ is not orthogonal in this case, because $\mathbf{T}$ is not symmetric. Then the partition function is given by

$$\tfrac{1}{2}Z(T,N) = (1 \quad 1)\,\mathbf{U}^{-1}\begin{pmatrix} \lambda_+^{N-2} & 0 \\ 0 & \lambda_-^{N-2} \end{pmatrix}\mathbf{U}\begin{pmatrix} e^{\beta J_2} \\ e^{-\beta J_2} \end{pmatrix}.$$

It is, of course, possible to calculate $\mathbf{U}$ and hence the partition function exactly, but the algebra is rather lengthy. When N is very large, however, we easily find that $\ln[Z(T,N)] = N \ln \lambda_+ \ (1 + \cdots)$, with corrections of order $1/N$ and $(\lambda_-/\lambda_+)^N$, if λ_+ is the larger eigenvalue, so it is sufficient to use the approximation $Z(T,N) \simeq \lambda_+^N$. The result is

$$Z(T,N) \simeq \left[e^{\beta J_2}\cosh(\beta J_1) + e^{\beta J_2}\sqrt{\sinh^2(\beta J_1) + e^{-4\beta J_2}} \right]^N .$$

In the limit $N \to \infty$, the correlation function $\langle S_i S_{i+1} \rangle$ should be the same at all positions i along the chain, and so

$$\langle S_i S_{i+1} \rangle = \langle b_i \rangle = \frac{1}{N}\sum_i \langle b_i \rangle = \frac{1}{N}\frac{\partial(\ln Z)}{\partial(\beta J_1)} = \frac{\sinh(\beta J_1)}{\sqrt{\sinh^2(\beta J_1) + e^{-4\beta J_2}}}.$$

• **Problem 5.15**

(a) The partition function can be obtained straightforwardly for all N, by summing first over S_0:

$$\begin{aligned} Z(T,N) &= \sum_{S_1 \cdots S_N = \pm 1} \sum_{S_0 = \pm 1} \exp\left(\beta h S_0 + \beta(h + J S_0) \sum_{i=1}^{N} S_i \right) \\ &= e^{\beta h} \prod_{i=1}^{N} \left(\sum_{S_i = \pm 1} e^{\beta(h+J)S_i} \right) + e^{-\beta h} \prod_{i=1}^{N} \left(\sum_{S_i = \pm 1} e^{\beta(h-J)S_i} \right) \\ &= e^{\beta h}\,[2\cosh(\beta h + \beta J)]^N + e^{-\beta h}\,[2\cosh(\beta h - \beta J)]^N . \end{aligned}$$

An expression of this sort can often be simplified when N is very large because, if $p > q$, then $p^N + q^N = p^N[1 + (q/p)^N] \simeq p^N$. In the present case this does not work, because either of the two terms in $Z(T,N)$ may be the larger, depending on whether h and J have the same, or the opposite sign. In order to simplify some of the equations that follow, we define

$$A = e^{\beta h}\,[2\cosh(\beta h + \beta J)]^N \qquad B = e^{-\beta h}\,[2\cosh(\beta h - \beta J)]^N .$$

Two quantities which can be computed immediately are

$$m \equiv \sum_{i=0}^{N} \langle S_i \rangle = \frac{\partial(\ln Z)}{\partial(\beta h)} = \frac{A - B}{A + B} + N\frac{A\tanh(\beta h + \beta J) + B\tanh(\beta h - \beta J)}{A + B}$$

$$p \equiv \sum_{i=1}^{N} \langle S_0 S_i \rangle = \frac{\partial(\ln Z)}{\partial(\beta J)} = N\frac{A\tanh(\beta h + \beta J) - B\tanh(\beta h - \beta J)}{A + B}$$

and, in terms of these, the average energy is

$$\langle E\rangle = -\frac{\partial(\ln Z)}{\partial\beta} = -hm - Jp.$$

We can also calculate directly

$$\langle S_0\rangle = Z^{-1}\sum_{S_0\cdots S_N=\pm 1} S_0 \exp\left(\beta h S_0 + \beta(h+JS_1)\sum_{i=1}^{N} S_i\right) = \frac{A-B}{A+B}.$$

It is obvious that $\mathcal{H}$ is unchanged if we permute the labels of the S_i for $i \neq 0$, so all of these S_i have the same expectation value,

$$\langle S_i\rangle = \frac{1}{N}(m - \langle S_0\rangle) = \frac{A\tanh(\beta h + \beta J) + B\tanh(\beta h - \beta J)}{A+B}.$$

Similarly, again for $i \neq 0$, we have

$$\langle S_0 S_i\rangle = \frac{p}{N} = \frac{A\tanh(\beta h + \beta J) - B\tanh(\beta h - \beta J)}{A+B}.$$

The limiting cases of these average values are

$$\begin{aligned}
&\lim_{h\to 0}\langle E\rangle = -NJ\tanh(\beta J) && \lim_{J\to 0}\langle E\rangle = -(N+1)h\tanh(\beta h)\\
&\lim_{h\to 0}\langle S_i\rangle = 0 && \lim_{J\to 0}\langle S_i\rangle = \tanh(\beta h)\\
&\lim_{h\to 0}\langle S_0 S_i\rangle = \tanh(\beta J) && \lim_{J\to 0}\langle S_0 S_i\rangle = \tanh^2(\beta h).
\end{aligned}$$

When $J = 0$, of course, we simply have $N+1$ identical and independent spins. In particular, we see that $\langle S_0 S_i\rangle = \langle S_0\rangle\langle S_i\rangle$, which means that the two spins are uncorrelated.

(b) When $h = 0$, the correlation function is given by

$$\begin{aligned}
\langle S_i S_j\rangle &= Z^{-1}\sum_{S_1\cdots S_N=\pm 1} S_i S_j \sum_{S_0=\pm 1}\exp\left(\beta J\sum_{k=1}^{N} S_k S_0\right)\\
&= Z^{-1}\sum_{S_1\cdots S_N=\pm 1} S_i S_j \prod_{k=1}^{N} \mathrm{e}^{\beta J S_k} + Z^{-1}\sum_{S_1\cdots S_N=\pm 1} S_i S_j \prod_{k=1}^{N} \mathrm{e}^{-\beta J S_k}.
\end{aligned}$$

The first contribution here is a product of $N-2$ terms (for which k is not equal to i or j), each of which sums to $2\cosh(\beta J)$, and two terms (for which $k = i$ or $k = j$), each of which sums to $2\sinh(\beta J)$. The second contribution is the same, except that the terms for which $k = i$ or $k = j$ each sum to $-2\sinh(\beta J)$. Also, when $h = 0$, we have $Z = 2[2\cosh(\beta J)]^N$, and so

$$\langle S_i S_j\rangle = \tanh^2(\beta J) = \langle S_0 S_i\rangle\langle S_0 S_j\rangle.$$

Since $S_0^2 = 1$, we can write this result as

$$\langle (S_0 S_i)(S_0 S_j) \rangle = \langle S_0 S_i \rangle \langle S_0 S_j \rangle$$

which means that $S_0 S_i$ and $S_0 S_j$ are independent random variables. Thus, the correlation between S_i and S_0 is independent of that between S_j and S_0. Since $\langle S_i S_j \rangle \neq \langle S_i \rangle \langle S_j \rangle$, we see that S_i and S_j are correlated, but that this is entirely due to the correlation of each of them with S_0. These results are not surprising, since each spin interacts only with S_0.

• **Problem 5.16**

(a) We tackle this problem with the help of the identity

$$\mathrm{e}^{\beta J_i S_i S_{i+1}} \equiv \cosh(\beta J_i) + S_i S_{i+1} \sinh(\beta J_i)$$

which is easily verified by substituting the two possible values $S_i S_{i+1} = \pm 1$. Let us first use this identity to cast the partition function in the form

$$Z = \sum_{S_1 \cdots S_N = \pm 1} \prod_{i=1}^{N} [\cosh(\beta J_i) + S_i S_{i+1} \sinh(\beta J_i)].$$

Now imagine multiplying out all the terms in the product and computing the sum for each one. Since each S_i takes the values ± 1, the only terms for which the sum is non-zero are those which contain an even power of each spin. In fact, there are only two such terms: the product of all the hyperbolic cosines is not multiplied by any spins, and the product of all the hyperbolic sines is multiplied by the factor

$$S_1 S_2 S_2 S_3 S_3 S_4 \cdots S_{N-1} S_N S_N S_{N+1}.$$

Since $S_{N+1} = S_1$, we see that each of the S_i occurs exactly twice, so this factor is actually equal to 1. Thus we find that

$$\begin{aligned} Z &= \sum_{S_1 \cdots S_N = \pm 1} \left(\prod_{i=1}^{N} \cosh(\beta J_i) + \prod_{i=1}^{N} \sinh(\beta J_i) \right) \\ &= 2^N \left(\prod_{i=1}^{N} \cosh(\beta J_i) + \prod_{i=1}^{N} \sinh(\beta J_i) \right). \end{aligned}$$

This result can also be obtained by the transfer matrix method (see problem 5.11). In this case, the transfer matrix is site dependent and so are its eigenvalues $\lambda_+^{(i)} = 2\cosh(\beta J_i)$ and $\lambda_-^{(i)} = 2\sinh(\beta J_i)$. However, its eigenvectors, and therefore also the matrix $\mathbf{U}$ which diagonalizes it, are site independent, so one gets $Z = \mathrm{Tr}(\mathbf{T}^N) = \prod_i \lambda_+^{(i)} + \prod_i \lambda_-^{(i)}$, which reproduces the above result. Readers may check, however, that in the presence of a magnetic field the eigenvectors become site dependent, and the transfer matrix is of little use (and the technique used here also becomes very cumbersome!).

Calculating $\langle S_i \rangle$ by the same method, we have

$$\langle S_i \rangle = Z^{-1} \sum_{S_1 \ldots S_N = \pm 1} S_i \prod_{k=1}^{N} [\cosh(\beta J_k) + S_k S_{k+1} \sinh(\beta J_k)]$$

and this vanishes, because every term contains an odd number of spins.

To work out the average $\langle S_i S_j \rangle$, we suppose that $j > i$ (if it is not, we just interchange the labels). We have

$$\langle S_i S_j \rangle = Z^{-1} \sum_{S_1 \ldots S_N = \pm 1} S_i S_j \prod_{k=1}^{N} [\cosh(\beta J_k) + S_k S_{k+1} \sinh(\beta J_k)]$$

and we need to construct products of spins which do not sum to zero, by using S_i, S_j and $S_k S_{k+1}$ for some sequence of values of k. A little thought will show that this can be done in just two ways, namely

$$S_i (S_i S_{i+1} S_{i+1} S_{i+2} \cdots S_{j-1} S_j) S_j$$

and

$$(S_1 S_2 S_2 S_3 \cdots S_{i-1} S_i) S_i \times S_j (S_j S_{j+1} \cdots S_N S_{N+1}).$$

Corresponding to these, $\langle S_i S_j \rangle$ consists of two terms. The first is

$$Z^{-1} 2^N \prod_{k=1}^{i-1} \cosh(\beta J_k) \prod_{k=i}^{j-1} \sinh(\beta J_k) \prod_{k=j}^{N} \cosh(\beta J_k)$$

and the second has the same structure, but with $\sinh(\beta J_k)$ and $\cosh(\beta J_k)$ interchanged. Using our previous result for the partition function, the correlation function can be written as

$$\langle S_i S_{i+r} \rangle = \frac{P_1 + P_2 P_3}{1 + P_1 P_2 P_3}$$

with

$$P_1 = \prod_{k=i}^{i+r-1} \tanh(\beta J_k) \qquad P_2 = \prod_{k=1}^{i-1} \tanh(\beta J_k) \qquad P_3 = \prod_{k=i+r}^{N} \tanh(\beta J_k).$$

In the limit $N \to \infty$ with r fixed, the product $P_2 P_3$ becomes negligibly small, because $|\tanh x| < 1$ for any finite x and $P_2 P_3$ contains $N - r$ factors of this kind, and the correlation function reduces to

$$\langle S_i S_{i+r} \rangle = \prod_{k=i}^{i+r-1} \tanh(\beta J_k).$$

Note that this conclusion does not depend upon i being small: the sites i and $j = i + r$ can be anywhere on the chain, provided that $N \gg r$, and $\langle S_i S_{i+r} \rangle$ depends only on the links joining them.

(b) The probability $\mathcal{P}_i(S)$ for the spin at site i to have the value S must satisfy the two relations

$$\mathcal{P}_i(1) + \mathcal{P}_i(-1) = 1 \qquad 1 \times \mathcal{P}_i(1) + (-1) \times \mathcal{P}_i(-1) = \langle S_i \rangle$$

and from these we deduce that

$$\mathcal{P}_i(1) = \frac{1 + \langle S_i \rangle}{2} = \tfrac{1}{2}.$$

The probability of finding the spins at sites i and j with the same value is

$$\mathcal{P}(S_i = S_j) = \mathcal{P}_{ij}(1, 1) + \mathcal{P}_{ij}(-1, -1)$$

where $\mathcal{P}_{ij}(S, S')$ is the probability for these two spins to have the values S and S' respectively. This probability distribution must satisfy

$$\mathcal{P}_{ij}(1, 1) + \mathcal{P}_{ij}(1, -1) + \mathcal{P}_{ij}(-1, 1) + \mathcal{P}_{ij}(-1, -1) = 1$$
$$\mathcal{P}_{ij}(1, 1) - \mathcal{P}_{ij}(1, -1) - \mathcal{P}_{ij}(-1, 1) + \mathcal{P}_{ij}(-1, -1) = \langle S_i S_j \rangle$$

from which we find that

$$\mathcal{P}(S_i = S_j) = \frac{1 + \langle S_i S_j \rangle}{2}.$$

(c) When the coupling is independent of position along the chain, our previous result for the correlation function in the limit of an infinite chain becomes

$$\langle S_i S_{i+r} \rangle = [\tanh(\beta J)]^r.$$

In the case of ferromagnetic coupling, $J > 0$, this can be written as

$$\langle S_i S_{i+r} \rangle = \exp[-r/\xi(T)]$$

where the correlation length is $\xi(T) = -1/\ln[\tanh(\beta J)]$. For antiferromagnetic coupling, $J < 0$, we have instead

$$\langle S_i S_{i+r} \rangle = (-1)^r \exp\left(\frac{-r}{\xi(T)}\right)$$

with $\xi(T) = -1/\ln[\tanh(\beta|J|)]$. In the low-temperature limit, the correlation length diverges as $\xi(T) \simeq \frac{1}{2} e^{2\beta|J|}$. For ferromagnetic coupling, this means that, as $T \to 0$, the probability of finding any two spins in the same state is $\mathcal{P}(S_i = S_{i+r}) = 1$, so the zero-temperature state is one in which all the spins are parallel. For antiferromagnetic coupling, the limit is

$$\mathcal{P}(S_i = S_{i+r}) = \begin{cases} 1 & r \text{ even} \\ 0 & r \text{ odd} \end{cases}$$

and the zero-temperature state is one in which neighbouring spins are antiparallel.

a { + + + + + + + . . .
    - - - - - - - . . .

b { + - + - + - + . . .
    - + - + - + - . . .

c { - + + - - + + . . .
    + - - + + - - . . .

Figure 5.5. Ground-state configurations for (a) ferromagnetic interaction ($J_i = J > 0$), (b) antiferromagnetic interaction ($J_i = J < 0$) and (c) $J_i = (-1)^i J$.

(d) When $J_i = (-1)^i J$, with $J > 0$, the correlation function becomes

$$\langle S_i S_{i+r} \rangle = (-1)^\sigma [\tanh(\beta J)]^r = (-1)^\sigma \exp\left(\frac{-r}{\xi(T)}\right)$$

where the index σ is given by

$$\sigma = \sum_{k=i}^{i+r-1} k = (i-1)r + \tfrac{1}{2}r(r+1).$$

At $T = 0$, the probability that S_{i+r} has the same value as S_i is

$$\mathcal{P}(S_i = S_{i+r}) = \tfrac{1}{2}[1 + (-1)^\sigma].$$

That is, it is 1 if σ is even and zero if σ is odd. The possible ground-state configurations can be deduced from this result, but the following equivalent argument may be more intuitive. The lowest energy is obtained when the energy of every link has its lowest value $(-1)^{i+1} J S_i S_{i+1} = -J$. This means that $S_{i+1} = (-1)^i S_i$. Suppose that $S_1 = 1$. Then setting $i = 1, 2, 3, \ldots$ in succession we deduce $S_2 = -S_1 = -1$, $S_3 = +S_2 = -1$, $S_4 = -S_3 = +1$, and so on, producing the third pattern in figure 5.5. In all three cases, of course, we get two ground states, obtained from each other by reversing all the spins.

In principle, we should enquire whether the ground-state configurations that we have deduced are compatible with the boundary condition that $S_{N+1} \equiv S_1$.

A moment's thought will show that the ferromagnetic state is indeed compatible for any N, that the antiferromagnetic state is compatible only if N is even and that the last state is compatible only if N is a multiple of 4, because the total length must comprise a whole number of repeated units of the pattern. If this condition is not met, then there is no configuration in which every link has its minimum energy, and the system is said to be 'frustrated'. In a one-dimensional chain, at most one link need have the wrong energy, however the couplings are distributed along the chain, and this is negligible for a long chain. In lattices of higher dimensions, however, frustration may be much more important, as is easily seen by considering very short closed paths on the lattice. This is characteristic of glassy systems in which couplings vary at random throughout the lattice, and the ground state of such a system may be extremely difficult to ascertain.

• **Problem 5.17** Allowing for spins with magnetic moment μ to interact with an external magnetic field H, the Hamiltonian of this system is

$$\mathcal{H} = -h\sum_i S_i - \frac{g}{N}\sum_{ij} S_i S_j$$

with $S_i = \pm 1$ and $h = \mu H$. Note that, with identical interactions between every pair of sites, this Hamiltonian is independent of the structure of the lattice, and of the number of dimensions. The partition function has the form

$$Z = \sum_{\{S_i=\pm 1\}} \exp\left[\beta h \sum_i S_i + \frac{\beta g}{2N}\left(\sum_i S_i\right)^2\right]$$

in which we see that the exponent is a quadratic expression in the quantity $\Sigma \equiv \sum_i S_i$. A problem of this kind can be treated using an integral transformation (known as the Hubbard–Stratonovich transformation). Writing $a = \sqrt{\beta g/2N}\,\Sigma$, we re-express the factor e^{a^2} by means of the identity

$$\frac{1}{\sqrt{\pi}}\int_{-\infty}^{\infty} \mathrm{e}^{-x^2-2ax}\,\mathrm{d}x = \frac{\mathrm{e}^{a^2}}{\sqrt{\pi}}\int_{-\infty}^{\infty} \mathrm{e}^{-(x+a)^2}\,\mathrm{d}x = \frac{\mathrm{e}^{a^2}}{\sqrt{\pi}}\int_{-\infty}^{\infty} \mathrm{e}^{-y^2}\,\mathrm{d}y = \mathrm{e}^{a^2}$$

and then the partition function can be calculated as

$$\begin{aligned} Z &= \frac{1}{\sqrt{\pi}}\int_{-\infty}^{\infty} \mathrm{d}x\,\mathrm{e}^{-x^2} \sum_{\{S_i=\pm 1\}} \exp\left(\beta h\sum_i S_i - x\sqrt{\frac{2\beta g}{N}}\sum_i S_i\right) \\ &= \frac{1}{\sqrt{\pi}}\int_{-\infty}^{\infty} \mathrm{d}x\,\mathrm{e}^{-x^2}\left[2\cosh\left(\beta h - x\sqrt{\frac{2\beta g}{N}}\right)\right]^N. \end{aligned}$$

In the remaining integral, we make a further change in integration variable, $x\sqrt{2\beta g/N} - \beta h = w$, to get

$$Z = 2^N\sqrt{\frac{N}{2\pi\beta g}}\int_{-\infty}^{\infty} \mathrm{d}w \exp\left[-N\left(\frac{(w+\beta h)^2}{(2\beta g)} - \ln\cosh w\right)\right].$$

In the thermodynamic limit $N \to \infty$, we can use the saddle-point approximation, expanding the integrand about its maximum value. In detail, writing the integrand as $\exp[-Nf(w)]$, we first locate the minimum, w_0, of $f(w)$ and then make a Taylor series expansion, using $w = w_0 + u/\sqrt{N}$, to obtain

$$\begin{aligned}\int_{-\infty}^{\infty} \mathrm{d}w\, \mathrm{e}^{-Nf(w)} &= \frac{1}{\sqrt{N}} \int_{-\infty}^{\infty} \mathrm{d}u \exp\Big(- Nf(w_0) - \tfrac{1}{2} f''(w_0)u^2 \\ &\quad - \frac{1}{6\sqrt{N}} f'''(w_0)u^3 + \cdots \Big) \\ &= \mathrm{e}^{-Nf(w_0)} \sqrt{\frac{2\pi}{Nf''(w_0)}} [1 + \mathrm{O}(N^{-1/2})].\end{aligned}$$

Inserting this result into our previous expression for Z, we find for the Helmholtz free energy

$$F = -kT \ln Z = NkT[f(w_0) - \ln 2 + \mathrm{O}(N^{-1})].$$

So in the thermodynamic limit F is determined completely by $f(w_0)$. The magnetization per spin is now given by

$$m = -\frac{\mu}{N} \frac{\partial F}{\partial h} = -\mu \frac{\partial f(w_0)}{\partial(\beta h)} = -\mu \frac{w_0 + h}{\beta g}.$$

In calculating the derivative, we have kept w_0 fixed because, although w_0 depends on h, the derivative $\mathrm{d}f(w)/\mathrm{d}w$ vanishes at $w = w_0$, by definition. At $h = 0$, there is a spontaneous magnetization provided that $w_0 \neq 0$. But $f(w)$ is minimized when w_0 is a solution of the equation

$$w_0 = \beta g \tanh w_0.$$

By plotting a graph of $\tanh w_0$, is is easily seen that there is both a positive and a negative solution provided that $\beta g > 1$, but that the only solution for $\beta g < 1$ is $w_0 = 0$. Consequently, the critical temperature is $T_\mathrm{c} = g/k$.

The equation for w_0 is essentially the same as that occurring in the Weiss mean-field theory of ferromagnetism. Indeed, mean-field approximations to a variety of models can be obtained by using the saddle-point approximation. In this model, the saddle-point method actually gives the exact free energy in the thermodynamic limit, and so mean-field theory is also exact. This is due to the fact that each spin interacts equally with all the others. It is found that mean-field theory is also exact for a variety of other models with sufficiently long-range interactions.

• **Problem 5.18** Assuming that the chain has periodic boundary conditions, the partition function for an interaction of this type can be found using the transfer matrix method discussed in problem 5.11. It is given by

$$Z = \mathrm{Tr}\left(\prod_{i=1}^{N} \langle S_i|\mathbf{T}|S_{i+1}\rangle\right) = \mathrm{Tr}\mathbf{T}^N = \lambda_1^N + \lambda_2^N + \lambda_3^N$$

where $\mathbf{T}$ is a 3×3 transfer matrix and λ_1, λ_2 and λ_3 its eigenvalues. This matrix is given by

$$\mathbf{T} = \begin{pmatrix} e^{\beta J} & 1 & e^{-\beta J} \\ 1 & 1 & 1 \\ e^{-\beta J} & 1 & e^{\beta J} \end{pmatrix}$$

with $J > 0$, and the eigenvalues, which are solutions of the cubic equation $\det(\mathbf{T} - \lambda\mathbf{I}) = 0$, are

$$\lambda_1 = 2\sinh(\beta J)$$

$$\lambda_{\frac{2}{3}} = \tfrac{1}{2}[1 + 2\cosh(\beta J) \pm \sqrt{4\cosh^2(\beta J) - 4\cosh(\beta J) + 9}].$$

By plotting graphs of these eigenvalues, it may be found that $\lambda_2 > \lambda_1 > \lambda_3$. Therefore, for large values of N, we keep only the second eigenvalue. At low temperatures, $\beta J \to \infty$, we have $\lambda_2 \simeq e^{\beta J}$ and $Z \simeq e^{N\beta J}$. Then we obtain for the internal energy

$$U = -\left(\frac{\partial(\ln Z)}{\partial\beta}\right)_N \to -NJ$$

for $\beta J \to \infty$. This result is expected for ferromagnetic interactions $J > 0$, since the lowest-energy states are those in which either all the S_i are equal to 1 or they are all equal to -1.

• **Problem 5.19** Let us first write each unit vector in the lattice in terms of angular variables (θ, ϕ) as

$$\boldsymbol{S}_i = \sin\theta_i \cos\phi_i\, \hat{\boldsymbol{x}} + \sin\theta_i \sin\phi_i\, \hat{\boldsymbol{y}} + \cos\theta_i\, \hat{\boldsymbol{z}}.$$

The partition function is then computed as the integral of $e^{-\beta\mathcal{H}}$ over all these variables:

$$\begin{aligned} Z &= \sum_{\{S_i\}} e^{-\beta\mathcal{H}} \\ &= \int d\Omega_1 \cdots \int d\Omega_N \exp\left(\beta J \sum_{\langle ij\rangle} \boldsymbol{S}_i(\Omega_i)\cdot\boldsymbol{S}_j(\Omega_j)\right) \\ &= \int d\Omega_1 \cdots \int d\Omega_N \prod_{i=1}^{N} \exp[\beta J \boldsymbol{S}_i(\Omega_i)\cdot\boldsymbol{S}_{i+1}(\Omega_{i+1})] \end{aligned}$$

where $d\Omega_i = \sin\theta_i\, d\theta_i\, d\phi_i$ is the element of solid angle for the direction of the ith vector, and $\boldsymbol{S}(\Omega)$ stands for $\boldsymbol{S}(\theta, \phi)$. Since the interaction energy is simply

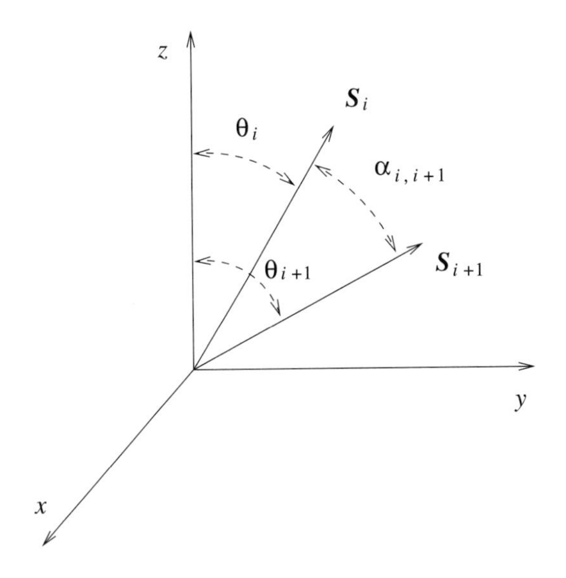

Figure 5.6. Relative angle between two nearest-neighbour vector spins.

proportional to the scalar product of neighbouring spin vectors, the integrand depends only on relative angles between these vectors (figure 5.6). That is,

$$\exp[\beta J \boldsymbol{S}_i(\Omega_i) \cdot \boldsymbol{S}_{i+1}(\Omega_{i+1})] = \exp(\beta J \cos \alpha_{i,i+1}).$$

It will prove convenient to express this as an expansion in Legendre polynomials, $P_l(x) = (2^l l!)^{-1}(\mathrm{d}^l/\mathrm{d}x^l)(x^2 - 1)^l$. Thus, we can write

$$\mathrm{e}^{\beta J \cos \alpha} = \sum_{l=0}^{\infty} A_l P_l(\cos \alpha) \qquad A_l = \frac{2l+1}{2} \int_{-1}^{1} \mathrm{e}^{\beta J x} P_l(x)\, \mathrm{d}x.$$

Then the partition function becomes

$$Z = \int \mathrm{d}\Omega_1 \cdots \int \mathrm{d}\Omega_N \prod_{i=1}^{N} \sum_{l_i=0}^{\infty} A_{l_i} P_{l_i}(\cos \alpha_{i,i+1}).$$

We now make use of the addition theorem for spherical harmonics, namely

$$\sum_{m=-l}^{l} Y_{l,m}(\Omega_i)\, Y^*_{l,m}(\Omega_j) = \frac{2l+1}{4\pi} P_l(\cos \alpha_{i,j})$$

to write

$$Z = \int \mathrm{d}\Omega_1 \cdots \int \mathrm{d}\Omega_N \prod_{i=1}^{N} \sum_{l_i=0}^{\infty} \sum_{m_i=-l_i}^{l_i} \frac{4\pi A_{l_i}}{2l_i+1} Y_{l_i,m_i}(\Omega_i) Y^*_{l_i,m_i}(\Omega_{i+1}).$$

Of all the factors in the product, those involving a particular pair of neighbours, say, $i = 1$ and $i = 2$, are

$$\sum_{l_1,m_1} \frac{4\pi A_{l_1}}{2l_1+1} Y_{l_1,m_1}(\Omega_1) Y^*_{l_1,m_1}(\Omega_2) \sum_{l_2,m_2} \frac{4\pi A_{l_2}}{2l_2+1} Y_{l_2,m_2}(\Omega_2) Y^*_{l_2,m_2}(\Omega_3)$$

and we see that the integration variable Ω_2 appears just in the product $Y^*_{l_1,m_1}(\Omega_2)Y_{l_2,m_2}(\Omega_2)$ and nowhere else. Carrying out this integral, we obtain simply

$$\int \mathrm{d}\Omega_2\, Y^*_{l_1,m_1}(\Omega_2) Y_{l_2,m_2}(\Omega_2) = \delta_{l_1,l_2}\delta_{m_1,m_2}.$$

This procedure can be repeated for all the sites, yielding the partition function in the form

$$Z = \sum_{l_1}\sum_{l_2}\cdots\sum_{l_N}\sum_{m_1}\sum_{m_2}\cdots\sum_{m_N} \delta_{l_1,l_2}\delta_{l_2,l_3}\cdots\delta_{l_N,l_1}$$
$$\times\, \delta_{m_1,m_2}\delta_{m_2,m_3}\cdots\delta_{m_N,m_1} \frac{4\pi A_{l_1}}{2l_1+1}\frac{4\pi A_{l_2}}{2l_2+1}\cdots\frac{4\pi A_{l_N}}{2l_N+1}.$$

Because of the periodic boundary conditions, the product of delta symbols forms a closed chain, whose net effect is to set all the l_i equal to the same value, say l, and all the m_i equal to, say, m. There remains just one sum over l and one over m:

$$Z = \sum_{l=0}^{\infty}\sum_{m=-l}^{l}\left(\frac{4\pi A_\ell}{2l+1}\right)^N = (2\pi)^N \sum_{l=0}^{\infty}(2l+1)\left(\int_{-1}^{1} \mathrm{e}^{\beta J x}\, P_l(x)\, \mathrm{d}x\right)^N.$$

More generally, we might consider a model in which the coupling strength J is different for each link in the chain, say, J_i for the link between sites i and $i+1$. In that case, the same computation would yield the result

$$Z(T,N) = (2\pi)^N \sum_{l=0}^{\infty}(2l+1)\prod_{i=1}^{N}\int_{-1}^{1} \mathrm{e}^{\beta J_i x}\, P_l(x)\, \mathrm{d}x.$$

• **Problem 5.20** In general, the isothermal magnetic susceptibility is a tensor, defined as

$$(\chi_T)_{\mu,\nu} = \left(\frac{\partial M_\mu}{\partial H_\nu}\right)_{T,N} = -\left(\frac{\partial^2 F}{\partial H_\mu\, \partial H_\nu}\right)_{T,N}$$

where M_μ is the μth component of the magnetization, H_μ is the μth component of the external magnetic field and $F(T,H,N)$ is the Helmholtz free energy. To connect this thermodynamic definition with statistical mechanics, we naturally use the identification $F(T, H, N) = -kT \ln[Z(T, H, N)]$, where $Z(T, H, N)$ is the canonical partition function, to obtain the important expression

$$(\chi_T)_{\mu,\nu} = kT\left[\frac{1}{Z}\left(\frac{\partial^2 Z}{\partial H_\mu \partial H_\nu}\right)_{T,N} - \frac{1}{Z^2}\left(\frac{\partial Z}{\partial H_\mu}\right)_{T,N}\left(\frac{\partial Z}{\partial H_\nu}\right)_{T,N}\right].$$

The Ising system models a highly anisotropic magnet, whose spins are effectively restricted to one spatial direction, and we consider only the component of the magnetic field in that direction. With this understanding, the magnetic susceptibility is a single function, given by

$$\chi_T = kT\left[\frac{1}{Z}\left(\frac{\partial^2 Z}{\partial H^2}\right)_{T,N} - \frac{1}{Z^2}\left(\frac{\partial Z}{\partial H}\right)^2_{T,N}\right].$$

If each spin has an associated magnetic moment μ, then the Ising Hamiltonian has a magnetic interaction term $-\mu H \sum_i S_i$, and we readily evaluate the partial derivatives as

$$\frac{1}{Z}\left(\frac{\partial Z}{\partial H}\right)_{T,N} = \beta\mu\sum_i \langle S_i\rangle \qquad \frac{1}{Z}\left(\frac{\partial^2 Z}{\partial H^2}\right)_{T,N} = \beta^2\mu^2\sum_{i,j}\langle S_i S_j\rangle$$

where $\langle\cdots\rangle$ denotes the ensemble average. Therefore

$$\chi_T = \beta\mu^2\sum_{i,j}[\langle S_i S_j\rangle - \langle S_i\rangle\langle S_j\rangle] = \beta\mu^2\sum_{i,j} G_{ij}(T, H, N).$$

For a very large system, say, of N spins in d dimensions, separated by a nearest-neighbour distance a, it is reasonable to approximate the lattice as a continuum. We call $\boldsymbol{x}$ the position of site i and $\boldsymbol{y}$ the position of site j and replace the sum $\sum_i$ with $a^{-d}\int \mathrm{d}^d x$. In the thermodynamic limit ($N \to \infty$ and $V = Na^d \to \infty$ with N/V fixed), the system becomes translationally invariant, so we have

$$G_{ij} = G(\boldsymbol{x} - \boldsymbol{y}).$$

Therefore

$$\chi_T = \beta m^2 \int \mathrm{d}^d x\, \mathrm{d}^d y\, G(\boldsymbol{x} - \boldsymbol{y}) = \beta m^2 V \int \mathrm{d}^d r\, [\langle S(\boldsymbol{r})S(\boldsymbol{0})\rangle - \langle S(\boldsymbol{0})\rangle^2]$$

where $m = \mu a^{-d}$ is the magnetic moment per unit volume. In obtaining the second expression, we have again made use of translational invariance to write $\langle S(\boldsymbol{x})S(\boldsymbol{y})\rangle = \langle S(\boldsymbol{r})S(\boldsymbol{0})\rangle$, with $\boldsymbol{r} = \boldsymbol{x} - \boldsymbol{y}$, which means that the correlation between two spins depends on their separation, but not on their actual positions, and $\langle S(\boldsymbol{x})\rangle = \langle S(\boldsymbol{y})\rangle = \langle S(\boldsymbol{0})\rangle$, since the mean value of each spin is the same. We expect that spins separated by a large distance should become uncorrelated, so that $\langle S(\boldsymbol{r})S(\boldsymbol{0})\rangle \to \langle S(\boldsymbol{0})\rangle^2$ as $\boldsymbol{r} \to \infty$. Then, provided that the correlation decays sufficiently rapidly, the integral over $\boldsymbol{r}$ is finite, and the susceptibility is proportional to the volume.

Finally, defining the Fourier transform as $\tilde{G}(\boldsymbol{k}) = \int \mathrm{d}^3 r\, G(\boldsymbol{r})\, \mathrm{e}^{-\mathrm{i}\boldsymbol{k}\cdot\boldsymbol{r}}$, we conclude that the magnetic susceptibility is proportional to the zero Fourier mode of the correlation function. More specifically, in the thermodynamic limit, we obtain the susceptibility per unit volume as

$$\lim_{V\to\infty}\left(\frac{\chi_T}{V}\right) = \beta m^2 \tilde{G}(\mathbf{0}).$$

It is proportional to m^2 because both the magnetization and the strength of its interaction with the field are proportional to m. The magnetic susceptibility is a response function: it indicates how the magnetization varies as one changes the external magnetic field. In thermal equilibrium, we have just seen that it is determined by the correlation function, which may be said to provide a measure of correlated fluctuations of spins about their mean values. In particular, the susceptibility is associated with long-wavelength fluctuations (those with wavevector $\boldsymbol{k} = \mathbf{0}$). Near the Curie temperature, these fluctuations become very large, and the susceptibility diverges.

• **Problem 5.21** We evaluate the two terms in the correlation function $G(j, j+1) = \langle S_j S_{j+1}\rangle - \langle S_j\rangle\langle S_{j+1}\rangle$ separately. For convenience, we define $b = \beta\mu H$ and $K = \beta J$, and then the first term is

$$\langle S_j S_{j+1}\rangle = \frac{1}{Z_N(b, K)} \sum_{\{S_i=\pm 1\}} S_j S_{j+1} \exp\left(\sum_{i=1}^{N}(bS_i + K S_i S_{i+1})\right)$$

with $S_{N+1} \equiv S_1$, and where $Z_N(b, K)$ is the partition function. With periodic boundary conditions, the system has a translation symmetry which implies that this expectation value is the same for any pair of neighbouring spins regardless of their position on the chain, so summing it over all values of j produces N times the value we want. Therefore we can write

$$\begin{aligned}\langle S_j S_{j+1}\rangle &= \frac{1}{N Z_N(b, K)} \sum_{\{S_i=\pm 1\}} \left(\sum_{k=1}^{N} S_k S_{k+1}\right) \exp\left(\sum_{i=1}^{N}(bS_i + K S_i S_{i+1})\right)\\ &= \frac{1}{N}\frac{\partial\{\ln[Z_N(b, K)]\}}{\partial K}.\end{aligned}$$

The partition function can be obtained using the transfer matrix method (see problem 5.11). In this case, the transfer matrix is

$$\mathbf{T} = \begin{pmatrix} \mathrm{e}^{K+b} & \mathrm{e}^{-K+b} \\ \mathrm{e}^{-K-b} & \mathrm{e}^{K-b} \end{pmatrix}$$

and its eigenvalues are $\lambda_\pm = \mathrm{e}^K\left(\cosh b \pm \sqrt{\sinh^2 b + \mathrm{e}^{-4K}}\right)$. In the thermodynamic limit ($N \to \infty$), the partition function is given by $Z_N(b, K) \simeq$

λ_+^N, the largest eigenvalue being λ_+. In this way, we find that

$$\langle S_j S_{j+1}\rangle = \frac{\partial(\ln\lambda_+)}{\partial K} = 1 - \frac{2\,\mathrm{e}^{-4K}\left(\sqrt{\sinh^2 b + \mathrm{e}^{-4K}}\right)^{-1}}{\cosh b + \sqrt{\sinh^2 b + \mathrm{e}^{-4K}}}.$$

For the second term in the correlation function, translation symmetry implies that $\langle S_j\rangle = \langle S_{j+1}\rangle = N^{-1}\sum_j \langle S_j\rangle = N^{-1}\partial\{\ln[Z_N(b,K)]\}/\partial b$, and our result for the partition function yields

$$\langle S_j\rangle = \frac{\sinh b}{\sqrt{\sinh^2 b + \mathrm{e}^{-4K}}}.$$

Putting these results together, we obtain the correlation function in the form

$$G(j,j+1) = 1 - \frac{2\,\mathrm{e}^{-4K}\left(\sqrt{\sinh^2 b + \mathrm{e}^{-4K}}\right)^{-1}}{\cosh b + \sqrt{\sinh^2 b + \mathrm{e}^{-4K}}} - \frac{\sinh^2 b}{\sinh^2 b + \mathrm{e}^{-4K}}.$$

For non-interacting spins ($K = 0$), we can use the identity $\cosh^2 b = 1 + \sinh^2 b$ to find that the first two terms exactly cancel the third term, which means that the correlation function vanishes identically. This is the result that we should have expected: if the spins do not interact, then they are independent variables, so $\langle S_j S_{j+1}\rangle = \langle S_j\rangle\langle S_{j+1}\rangle$ and the correlation function vanishes. (The definition of the correlation function is chosen so that it vanishes when the spins are uncorrelated!) When there is no external magnetic field ($b = 0$), we recover the well-known result that $\langle S_j\rangle = 0$, so the one-dimensional Ising model has no spontaneous magnetization at any finite temperature. The first term in the correlation function does not vanish, however, and we get $G(j,j+1) = \langle S_j S_{j+1}\rangle = \tanh K$. This means that, although each spin has equal probabilities of being up or down, and therefore a mean value of zero, the probability that neighbouring spins are in the same state is different from the probability that they are in opposite states. If the interactions are ferromagnetic ($K > 0$), then neighbouring spins prefer to be in the same state and the correlation is positive. With antiferromagnetic interactions ($K < 0$), they prefer to be in opposite states and the correlation is negative.

• Problem 5.22

(a) The Hamiltonian for this system is

$$\mathcal{H} = -\mu H\Big(\sum_{i\in \mathrm{A}} S_i^{\mathrm{A}} + \sum_{i\in \mathrm{B}} S_i^{\mathrm{B}}\Big) - J_1\Big(\sum_{\langle i,j\rangle} S_i^{\mathrm{A}} S_j^{\mathrm{A}} + \sum_{\langle i,j\rangle} S_i^{\mathrm{B}} S_j^{\mathrm{B}}\Big) + J_2 \sum_{\langle i,j\rangle} S_i^{\mathrm{A}} S_j^{\mathrm{B}}$$

where the first two terms give the interaction between the spins and the applied magnetic field H, the next two terms give the (ferromagnetic) interaction between nearest neighbours on each of the two sublattices, while the last term is the (antiferromagnetic) interaction between neighbouring spins on different sublattices. Note that the nearest neighbours of each type-A site are of type B and vice versa, while the next-nearest neighbours of any site are sites of the same type.

(b) In the mean-field approximation, one considers the influence on any one spin of all the others as being equivalent to the effect of an internal magnetic field, which is added to the true applied external magnetic field. Thus, the interaction energy of the spin S_i^{A} with its neighbours is

$$-\left(J_1 \sum_j S_j^{\mathrm{A}} - J_2 \sum_k S_k^{\mathrm{B}}\right) S_i^{\mathrm{A}} \equiv -\mu H_{\mathrm{int}}^{\mathrm{A}}(i) S_i^{\mathrm{A}}$$

where j runs over the nearest type-A sites to site i, while k runs over the nearest type-B sites. This expression is exact and, of course, $H_{\mathrm{int}}^{\mathrm{A}}$ depends on the site i. The approximation consists in replacing the spins in $H_{\mathrm{int}}^{\mathrm{A}}$ with their average values which, on account of translational invariance, are site independent. Of course, we do the same for spins on the B sublattice. In this way, we obtain an approximate 'mean-field' Hamiltonian of the form

$$\mathcal{H}_{\mathrm{MF}} = -\mu\left(H_{\mathrm{eff}}^{\mathrm{A}} \sum_{i\in\mathrm{A}} S_i^{\mathrm{A}} + H_{\mathrm{eff}}^{\mathrm{B}} \sum_{i\in\mathrm{B}} S_i^{\mathrm{B}}\right).$$

Writing $\sigma_{\mathrm{A}} = \langle S_i^{\mathrm{A}}\rangle$ and $\sigma_{\mathrm{B}} = \langle S_i^{\mathrm{B}}\rangle$, the effective fields are

$$\begin{aligned} H_{\mathrm{eff}}^{\mathrm{A}} &= H + \frac{J_1}{\mu} z_{\mathrm{AA}}\sigma_{\mathrm{A}} - \frac{J_2}{\mu} z_{\mathrm{AB}}\sigma_{\mathrm{B}} \\ H_{\mathrm{eff}}^{\mathrm{B}} &= H + \frac{J_1}{\mu} z_{\mathrm{BB}}\sigma_{\mathrm{B}} - \frac{J_2}{\mu} z_{\mathrm{BA}}\sigma_{\mathrm{A}} \end{aligned}$$

where the z_{ab} are the coordination numbers (all equal to 4 in the lattice considered here), which give the appropriate numbers of nearest neighbours.

With this approximation, the partition function is easy to compute, since $\mathcal{H}_{\mathrm{MF}}$ is a sum of terms, each linear in just one of the spins. Supposing that there are N sites of each type (and we shall consider the limit $N \to \infty$ in which the translational invariance we assumed above becomes valid), it is given by

$$Z = \sum_{\{S_i=\pm 1\}} \mathrm{e}^{-\beta\mathcal{H}_{\mathrm{MF}}} = [2\cosh(\beta\mu H_{\mathrm{eff}}^{\mathrm{A}})]^N [2\cosh(\beta\mu H_{\mathrm{eff}}^{\mathrm{B}})]^N$$

and we shall need the Helmholtz free energy per spin given by

$$\begin{aligned} f(T, H) &= -\lim_{N\to\infty} [(2N)^{-1} kT \ln Z] \\ &= -\tfrac{1}{2}kT[\ln\cosh(b + K_1\sigma_{\mathrm{A}} - K_2\sigma_{\mathrm{B}}) \\ &\quad + \ln\cosh(b + K_1\sigma_{\mathrm{B}} - K_2\sigma_{\mathrm{A}}) + 2\ln 2] \end{aligned}$$

where $b = \beta\mu H$, $K_1 = 4\beta J_1$ and $K_2 = 4\beta J_2$. The mean values of the spins are also easily found to be

$$\sigma_A = Z^{-1} \sum_{\{S_i\}} S_i^A \, e^{-\beta\mathcal{H}_{MF}} = \tanh(b + K_1\sigma_A - K_2\sigma_B)$$

$$\sigma_B = Z^{-1} \sum_{\{S_i\}} S_i^B \, e^{-\beta\mathcal{H}_{MF}} = \tanh(b + K_1\sigma_B - K_2\sigma_A)$$

and these equations can be solved self-consistently (although not in closed form) to determine the actual values of σ_A and σ_B.

(c) When $H = 0$, the system has two symmetries: the energy of a configuration is unchanged either when all the spins are reversed or when the A and B sublattices are interchanged. These symmetries imply (as a little thought should show) that the equations for σ_A and σ_B should have three kinds of solution: (i) a paramagnetic (p) state with $\sigma_A = \sigma_B = 0$; (ii) a ferromagnetic (f) state with $\sigma_A = \sigma_B \neq 0$; (iii) an antiferromagnetic (a) state with $\sigma_A = -\sigma_B \neq 0$. Setting $b = 0$ in these equations, it is obvious that $\sigma_A = \sigma_B = 0$ is a solution. Setting $\sigma_A = \sigma_B = \sigma_f$, both equations are satisfied if

$$\sigma_f = \tanh[(K_1 - K_2)\sigma_f]$$

which is possible if $T < T_f = 4(J_1 - J_2)/k$. On the other hand, setting $\sigma_A = -\sigma_B = \sigma_a$, we find that both equations are again satisfied if

$$\sigma_a = \tanh[(K_1 + K_2)\sigma_a]$$

which is possible if $T < T_a = 4(J_1 + J_2)/k$. Since $0 < J_2 < J_1$, we have $0 < T_f < T_a$.

As far as these equations are concerned, it seems that there is a temperature range $T > T_a$ in which only the paramagnetic state is possible, a second range $T_f < T < T_a$ in which the paramagnetic and antiferromagnetic states are possible, and a third range $0 < T < T_f$ in which all three states are possible. Intuitively, however, it is clear that the antiferromagnetic state has the lowest energy, and might be expected to be the most stable state, when it exists. Thermodynamically, the stable state is the state with the lowest free energy, so we calculate $f(T, H = 0)$ for all three states:

$$f_p = -kT \ln 2$$

$$f_f = -kT\{\ln\cosh[(K_1 - K_2)\sigma_f] + \ln 2\}$$

$$f_a = -kT\{\ln\cosh[(K_1 + K_2)\sigma_a] + \ln 2\}.$$

Now, although we do not have exact solutions for σ_f and σ_a, these are solutions of equations of the form $\sigma = \tanh(K\sigma)$, and it can be seen (for example, by plotting a graph) that σ is an increasing function of K. For σ_f, we have $K = K_1 - K_2$ and for σ_a we have $K = K_1 + K_2$, so, when σ_a and σ_f both exist, we get $\sigma_a > \sigma_f$. From this it follows that, when the antiferromagnetic state exists, it has the lowest free energy. Therefore, the state above T_a is the paramagnetic state, while at all temperatures below T_a (including those below T_f) the stable state is the antiferromagnetic state. The ferromagnetic state is never stable.

(d) The magnetization per spin of our system is $M = \frac{1}{2}\mu[\langle S_i^A\rangle + \langle S_i^B\rangle] = \frac{1}{2}\mu(\sigma_A + \sigma_B)$, so the magnetic susceptibility at zero field can be expressed as

$$\chi = \lim_{H\to 0}\left[\frac{1}{2}\left(\frac{\partial M}{\partial H}\right)_T\right] = \tfrac{1}{2}\beta\mu^2(\chi_A + \chi_B)$$

where

$$\chi_A = \lim_{b\to 0}\left[\left(\frac{\partial\sigma_A}{\partial b}\right)_T\right] \qquad \chi_B = \lim_{b\to 0}\left[\left(\frac{\partial\sigma_B}{\partial b}\right)_T\right].$$

At temperatures above T_a, we have seen that $\sigma_A = \sigma_B = 0$ when $b = H = 0$, so in a small field, we can assume that $\sigma_A \simeq \chi_A b$ and $\sigma_B \simeq \chi_B b$. Substituting these into our equations for σ_A and σ_B, and using the expansion $\tanh x = x + O(x^3)$ (where, in each case, $x \propto b$ is small), we get the pair of equations

$$\chi_A = 1 + K_1\chi_A - K_2\chi_B \qquad \chi_B = 1 + K_1\chi_B - K_2\chi_A$$

whose solution is $\chi_A = \chi_B = (1 - K_1 + K_2)^{-1} = T/(T - T_f)$. The total susceptibility is then

$$\chi = \tfrac{1}{2}\beta\mu^2(\chi_A + \chi_B) = \frac{\mu^2}{k(T - T_f)}.$$

This is, of course, the Curie–Weiss law which applies in mean-field theory to a simple ferromagnet with the Curie temperature T_f. At first sight, this may seem puzzling, since we found above that the ferromagnetic state is unstable, even below T_f. The divergence of χ at $T = T_f$ ought not, therefore, to be a real effect, since there is no transition to a ferromagnetic state at T_f. To understand this better, we first note that our result for χ is valid only for $T > T_a > T_f$. Next, let us find the susceptibility in the antiferromagnetic state, which is stable for $T < T_a$. In zero field, we have $\sigma_A = -\sigma_B = \sigma_a$, so in a small field we assume that $\sigma_A = \sigma_a + \chi_A b + O(b^2)$ and $\sigma_B = -\sigma_a + \chi_B b + O(b^2)$ and repeat our previous calculation. The equation for σ_A becomes

$$\begin{aligned}\sigma_a + \chi_A b + O(b^2) &= \tanh[F\sigma_a + Gb + O(b^2)]\\ &= \tanh[F\sigma_a] + \operatorname{sech}^2[F\sigma_a]\,Gb + O(b^2)\end{aligned}$$

where $F = K_1 + K_2$ and $G = 1 + K_1\chi_A - K_2\chi_B$. From previous results, we know that $\tanh(F\sigma_a) = \sigma_a$ and hence that $\operatorname{sech}^2(F\sigma_a) = 1 - \tanh^2(F\sigma_a) = 1 - \sigma_a^2$. Then, at $b = 0$, our equation becomes

$$\chi_A = (1 - \sigma_a^2)(1 + K_1\chi_B - K_2\chi_A)$$

and the equation for σ_B takes the same form, with χ_A and χ_B interchanged. The solution of this pair of equations is

$$\chi_A = \chi_B = \frac{1-\sigma_a^2}{1-(1-\sigma_a^2)(K_1-K_2)} = \frac{T(1-\sigma_a^2)}{T-(1-\sigma_a^2)T_f}.$$

Together with our first result, this yields the susceptibility as

$$\chi(T) = \begin{cases} \dfrac{\mu^2(1-\sigma_a^2)}{k[T-(1-\sigma_a^2)T_f]} & \text{for } T < T_a \\ \dfrac{\mu^2}{k(T-T_f)} & \text{for } T > T_a. \end{cases}$$

At $T = T_f < T_a$, this has the finite value $\chi(T_f) = \mu^2(1-\sigma_a^2)/\sigma_a^2 kT_f$. At $T = T_a$, where $\sigma_a = 0$, both expressions have the same value. Thus, in the mean-field approximation, the susceptibility is finite and continuous at the antiferromagnetic transition temperature T_a (although a more sophisticated approximation would reveal that it actually has a weak singularity there). This is in contrast with a ferromagnet, for which the susceptibility diverges at the transition (in the mean-field approximation, according to the Curie–Weiss law). The difference arises from the fact that H is an 'ordering field' for a ferromagnet. That is to say, application of H tends to make all the spins point in the same direction, enhancing the effect of ferromagnetic interactions. On the other hand, antiferromagnetic interactions favour states in which neighbouring spins point in opposite directions, and this tendency is opposed by H. Mathematically, one can introduce an ordering field $H^\dagger$ for an antiferromagnet. In the present case, it is a 'staggered' magnetic field, which has, say, the value $+H^\dagger$ on sites of type A and $-H^\dagger$ on sites of type B. Correspondingly, we think of a staggered magnetization $M^\dagger = \mu(\sigma_A - \sigma_B)$ and a staggered susceptibility $\chi^\dagger = \partial M^\dagger/\partial H^\dagger$. It is then found that $\chi^\dagger$ follows the modified Curie–Weiss law $\chi^\dagger \sim 1/(T - T_a)$ and does indeed diverge at the antiferromagnetic transition at $T = T_a$. Interested readers may like to derive this result in detail. Experimentally, a staggered field is not normally available in the laboratory, although there are rare (and complicated) instances in which a similar effect can be created by indirect methods.

• Problem 5.23

(a) Suppose that the external field is in the $\hat{z}$ direction, so that $\boldsymbol{h} = h\hat{z}$ with $h > 0$. In the mean-field approximation (see problem 5.22) we focus on a spin at site i, say, and replace the neighbours with which it interacts with their ensemble averages $\langle \boldsymbol{S}_j \rangle = \sigma\hat{z}$. This gives us an approximate Hamiltonian

$$\mathcal{H}_{MF} = -J\sum_i \boldsymbol{S}_i \cdot \sum_j \sigma\hat{z} - h\sum_i \boldsymbol{S}_i \cdot \hat{z} = -(h + cJ\sigma)\sum_i S_i^z.$$

In the first expression, j runs over the c nearest neighbours of site i, giving c equal contributions in the second expression. For a hypercubic lattice in

d dimensions, the coordination number is $c = 2d$. This approximation is equivalent to a set of spins interacting independently with an effective magnetic field $H_{\text{eff}} = H + cJ\sigma/\mu$, and the value of σ is to be found self-consistently by calculating $\langle \boldsymbol{S}_i \rangle$ in the approximate ensemble. The partition function for a lattice of N sites is just the product of N identical partition functions for a single spin, namely

$$Z_1 = 2\pi \int_0^\pi \sin\theta \, \mathrm{d}\theta \, \mathrm{e}^{\beta(h+cJ\sigma)\cos\theta} = \frac{4\pi \sinh[\beta(h+cJ\sigma)]}{\beta(h+cJ\sigma)}$$

$$Z = Z_1^N = \left\{ \frac{4\pi \sinh[\beta(h+cJ\sigma)]}{\beta(h+cJ\sigma)} \right\}^N$$

and the mean value σ of each spin is

$$\sigma = \frac{\partial(\ln Z_1)}{\partial(\beta h)} = \coth[\beta(h+cJ\sigma)] - [\beta(h+cJ\sigma)]^{-1}.$$

To find the critical temperature, we look for the conditions under which there is a non-zero spontaneous magnetization in the absence of an external field. When $h = 0$, the equation to be solved for σ is

$$\sigma = \coth(a\sigma) - \frac{1}{a\sigma}$$

where $a = \beta cJ$. Plotting the two sides of this equation we see that it has non-zero solutions when

$$\frac{\mathrm{d}}{\mathrm{d}\sigma}\left[\coth(a\sigma) - \frac{1}{a\sigma}\right]_{\sigma=0} = \frac{a}{3} > 1$$

or when $T < T_{\mathrm{c}} = cJ/3k$.

Thus, when $h = 0$, we have $\sigma = 0$ for $T > T_{\mathrm{c}}$ and $\sigma \neq 0$ for $T < T_{\mathrm{c}}$. Note that the mean-field theory predicts a phase transition for $d = 1$ which is entirely wrong, as can be shown by solving the model exactly, and for $d = 2$, which is also wrong, as shown by a celebrated general theorem due to N D Mermin and H Wagner.

(b) In order to study how the magnetization and susceptibility behave near the critical point, we make a Taylor expansion of the equation for σ:

$$\sigma = \coth(a\sigma + \beta h) - \frac{1}{a\sigma + \beta h} = \tfrac{1}{3}(a\sigma + \beta h) - \tfrac{1}{45}(a\sigma + \beta h)^3 + \cdots$$

and invert this series to find βh as a power series expansion in σ:

$$\beta h = (3 - a)\sigma + \tfrac{3}{5}\sigma^3 + \mathrm{O}(\sigma^5).$$

At $T = T_c$, we have $a = 3$. So, on approaching the critical point where $\sigma \to 0$ we get $\beta h \simeq \frac{3}{5}\sigma^3$ and hence $\delta = 3$. (The magnetization per spin is $m = \mu\sigma$, where μ is the magnetic moment of each spin.)

For $h = 0$ and $T < T_c$ we know that $\sigma \neq 0$, so we can divide by a factor of σ to get $a - 3 = \frac{3}{5}\sigma^2 + O(\sigma^4)$. From (a) we know that $a - 3 = 3(T_c - T)/T$, which gives us $\sigma \sim (T_c - T)^{1/2}$ as $T \to T_c$, and therefore $\beta = \frac{1}{2}$.

Finally, to study the susceptibility, we differentiate the equation for βh:

$$\frac{1}{\chi} = \frac{1}{\chi_0}\left(\frac{\partial \beta h}{\partial \sigma}\right)_T = \frac{1}{\chi_0}[(3 - a) + \tfrac{9}{5}\sigma^2 + O(\sigma^4)]$$

where χ_0 is a constant whose precise value does not matter to us. For $T > T_c$ and $h = 0$, we have $\sigma = 0$ and so

$$\chi = \chi_0(3 - a)^{-1} \propto (T - T_c)^{-1}$$

giving $\gamma = 1$. For $T < T_c$ and $h = 0$, we have $\sigma^2 \simeq \frac{5}{3}(a - 3)$ and, when $T \to T_c$,

$$\chi \simeq \tfrac{1}{2}\chi_0(a - 3)^{-1} \propto (T_c - T)^{-1}$$

giving $\gamma' = 1 = \gamma$.

Suppose that we define amplitudes A_+ and A_- by $\chi \simeq A_+(T - T_c)^{-\gamma}$ for $T > T_c$ and $\chi \simeq A_-(T_c - T)^{-\gamma'}$ for $T < T_c$. Then we see from the above calculations that $A_+/A_- = 2$. The critical exponents and this amplitude ratio have values which are universal, in the sense that they do not depend on parameters such as J which vary from one system to another. Both experimental measurements and more sophisticated approximations show that all these quantities do indeed have universal values, although these values are somewhat different from those obtained in the mean-field approximation.

• Problem 5.24

(a) Using the mean-field method described in problems 5.22 and 5.23, we replace the Ising Hamiltonian with the approximate version $\mathcal{H}_{\mathrm{MF}} = -(H\mu + zJ\sigma)\sum_i S_i$, where $\sigma = \langle S_i \rangle$ is to be determined self-consistently by calculating $\langle S_i \rangle$ in the mean-field ensemble. The approximate partition function is a product of single-spin partition functions, from which we readily obtain the mean value of each spin

$$\sigma = Z_1^{-1} \sum_{S=\pm 1} S\,\mathrm{e}^{\beta(H\mu + zJ\sigma)S} = \tanh[\beta(H\mu + zJ\sigma)]$$

and the internal energy per spin

$$u = -\left\langle \left(H\mu + \frac{z}{2}J\sigma\right) S \right\rangle = -\left(H\mu\sigma + \frac{z}{2}J\sigma^2\right).$$

A tricky issue is involved in this expression for u. In obtaining the total mean energy Nu, for a system of N spins, we have to take into account that the interaction energy arises from bonds, each connecting a pair of nearest-neighbour spins, and the total number of bonds is $Nz/2$. On the other hand, the probability of finding a particular spin with the value S_i is determined by the interaction of that spin with all of its z nearest neighbours, and this is the interaction that we use in $\mathcal{H}_{\rm MF}$ to construct the partition function.

For $H = 0$, σ is determined, as usual, as a solution of the transcendental equation $\sigma = \tanh(\beta z J \sigma)$ and there are non-zero solutions only when the slope of $\tanh(\beta z J\sigma)$ at $\sigma = 0$ is greater than unity. This happens when T is smaller than the value $T_{\rm c} = zJ/k$, which we identify as the Curie temperature.

To determine the behaviour of the spontaneous magnetization per spin, $m = \mu\sigma$, just below $T_{\rm c}$, it is convenient to introduce the reduced temperature $t = (T_{\rm c} - T)/T$. When σ and t are both small, we can use the expansion

$$\sigma = \tanh[(1+t)\sigma] = (1+t)\sigma - \tfrac{1}{3}(1+t)^3\sigma^3 + \cdots \simeq (1+t)\sigma - \tfrac{1}{3}\sigma^3$$

to find $\sigma = \sqrt{3t}[1 + {\rm O}(t)]$, which means that the magnetization exponent is $\beta = \frac{1}{2}$.

When H is non-zero, it is again convenient to introduce a reduced magnetic field $h = \beta\mu H$, in terms of which the magnetic susceptibility per spin is

$$\chi = \left(\frac{\partial m}{\partial H}\right)_T = \frac{\mu^2}{kT}\bar{\chi} \qquad \bar{\chi} = \left(\frac{\partial \sigma}{\partial h}\right)_T .$$

The equation to be solved for σ is now

$$\sigma = \tanh\left(h + \frac{T_{\rm c}}{T}\sigma\right).$$

On differentiating this equation with respect to h and setting $h = 0$, we get

$$\bar{\chi} = \mathrm{sech}^2\left(\frac{T_{\rm c}}{T}\sigma\right)\left(1 + \frac{T_{\rm c}}{T}\bar{\chi}\right) = (1-\sigma^2)\left(1 + \frac{T_{\rm c}}{T}\bar{\chi}\right)$$

which can be solved to express $\bar{\chi}$ as

$$\bar{\chi} = \frac{1}{(1-\sigma^2)^{-1} - T_{\rm c}/T}.$$

For $T > T_{\rm c}$, we have $\sigma = 0$ and $\bar{\chi} = T/(T - T_{\rm c})$, from which we identify the susceptibility exponent as $\gamma = 1$. For $T < T_{\rm c}$, we use $T_{\rm c}/T = 1 + t$ and $(1-\sigma^2)^{-1} = 1 + 3t + {\rm O}(t^2)$ to obtain $\bar{\chi} = (2t)^{-1}[1 + {\rm O}(t)]$, which shows that the same exponent $\gamma = 1$ is valid above and below the transition. Comparing these results with those of problem 5.23, we see that both the exponents and the amplitude ratio A_+/A_- have the same values for the Ising and Heisenberg models. This is true only within the mean-field approximation: more accurate methods give values which differ slightly both from the mean-field results and from each other.

(b) When $H = 0$, the internal energy per spin is $u = -(z/2)J\sigma^2 = -(kT_c/2)\sigma^2$. For $T > T_c$, we clearly have $u = 0$, so the specific heat per spin is $c_H = 0$. For $T < T_c$, we evaluate the specific heat as

$$c_H = \frac{du}{dT} = -\frac{T_c}{T^2}\frac{du}{dt} = \frac{3k}{2}[1 + O(t)].$$

Evidently, c_H has a discontinuity at the transition, given by

$$\Delta c_H = \lim_{\delta T \to 0}[c_H(T_c + \delta T) - c_H(T_c - \delta T)] = -3k/2.$$

Again, this result holds only in the mean-field theory. More accurate approximations show that c_H has a singularity of the form $c_H \sim |T - T_c|^{-\alpha}$, where α is a small exponent (approximately equal to 0.12 for the three-dimensional Ising model).

• Problem 5.25

(a) We treat the interaction between any two given contiguous spins explicitly and consider their interaction with the rest of the spins within the mean-field approximation, as indicated in figure 5.7, case a. The Hamiltonian is that of the standard Ising model, so the effective magnetic field seen by each of the chosen spins is

$$H_{\text{eff}} = H + \frac{3J}{\mu}\sigma$$

where H is the external field, J is the coupling constant, μ is the magnetic moment associated with each spin and $\sigma = \langle S_i \rangle$ is the mean value of each spin. Each of the chosen spins has three neighbouring spins which are replaced by their mean values, its fourth neighbour being treated explicitly, so the interaction energy included in the effective field is $-3J\sigma S_i$. We now obtain the partition function for the system of two spins. It is

$$Z_2 = \sum_{\{S_1, S_2 = \pm 1\}} e^{\beta(S_1+S_2)\mu H_{\text{eff}} + \beta J S_1 S_2} = 2\,e^{\beta J}\cosh(2\beta\mu H_{\text{eff}}) + 2\,e^{-\beta J}.$$

In the spirit of the mean-field approximation, we use this partition function to find the mean value of one of the spins:

$$\sigma = Z_2^{-1} \sum_{\{S_1, S_2 = \pm 1\}} e^{\beta(S_1+S_2)\mu H_{\text{eff}} + \beta J S_1 S_2}\, S_1 = \frac{\sinh(2\beta\mu H_{\text{eff}})}{\cosh(2\beta\mu H_{\text{eff}}) + e^{-2\beta J}}.$$

At zero external field, σ is therefore the solution to the transcendental equation

$$\sigma = \frac{\sinh(6\beta J\sigma)}{\cosh(6\beta J\sigma) + e^{-2\beta J}}.$$

The function on the right-hand side is qualitatively similar to the function $\tanh(4\beta J\sigma)$ found in the standard mean-field approximation, so we may draw

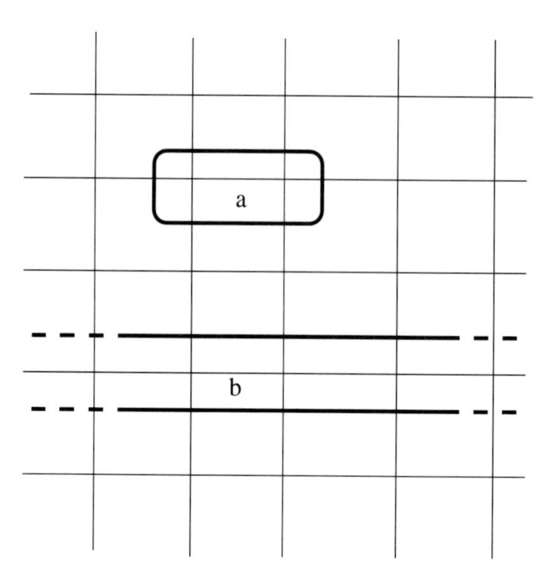

Figure 5.7. Two sites treated explicitly (case a), and a whole line treated explicitly (case b) for a two-dimensional Ising model in the mean-field approximation.

qualitatively the same conclusion. Namely, above a critical temperature T_c, the only solution to the equation is $\sigma = 0$, which means that there is no net magnetization. Below T_c there are non-zero solutions corresponding to a spontaneous magnetization. The borderline between these two cases occurs when the slope of the function on the right at $\sigma = 0$ is equal to 1, so we have

$$1 = \frac{6\beta_c J}{1 + e^{-2\beta_c J}}.$$

A solution for $\beta_c J$ can be found by plotting a graph or by solving the equation numerically. From a numerical solution, we find that $\beta_c J \approx 0.264\,80$, which is a little larger than the value $\beta_c J = 0.25$ found in the standard mean-field approximation.

(b) In principle, we proceed as in (a), but now all spins in a given infinite row are treated explicitly (figure 5.7, case b). Each spin in the chosen row has two neighbours (one above and one below) which are replaced by their mean values. Accounting for the interaction energy with these neighbours by an effective field, we see that the effective field seen by each spin in the chosen row is $H_{\text{eff}} = H + 2J\sigma/\mu$. The approximate partition function for the row of N spins (as usual, we shall take the limit $N \to \infty$) is given by

$$Z_{\text{row}} = \sum_{\{S_i=\pm 1\}} \exp\left(\beta\mu H_{\text{eff}} \sum_{i=1}^{N} S_i + \beta J \sum_{\langle i,j\rangle} S_i S_j\right)$$

and this, of course, is just the partition function for the one-dimensional Ising model, with H_{eff} replacing the external field. We can therefore use the given expression for the mean value of the spin in the one-dimensional model to write down an equation to be solved for σ in our new approximation to the two-dimensional model, namely

$$\sigma = \sinh(\beta\mu H_{\text{eff}})\,[\sinh^2(\beta\mu H_{\text{eff}}) + \mathrm{e}^{-4\beta J}]^{-1/2}.$$

Setting the external field H to zero, we obtain the transcendental equation

$$\sigma = \sinh(2\beta J\sigma)\,[\sinh^2(2\beta J\sigma) + \mathrm{e}^{-4\beta J}]^{-1/2}.$$

Again, the critical temperature is that for which the slope of the function on the right-hand side is equal to one when $\sigma = 0$, that is

$$1 = 2\beta_{\text{c}} J\,\mathrm{e}^{2\beta_{\text{c}} J}.$$

A numerical solution of this equation gives $\beta_{\text{c}} J \approx 0.283\,57$, which is a little larger than the value obtained in (a).

(c) Onsager's exact solution to the two-dimensional Ising model shows that the system exhibits a spontaneous magnetization below a critical temperature given by $\beta_{\text{c}} J = 0.440\,69\ldots$. Evidently, the sequence of mean-field-like approximations considered in this problem shows the trend that we might have expected: by treating more spins at a time in an exact manner, we get a result for $\beta_{\text{c}} J$ which is somewhat closer to the exact value.

• Problem 5.26

(a) According to the strategy of the mean-field approximation described in several earlier problems, we construct an approximate Hamiltonian

$$\mathcal{H}_{\text{MF}} = -(h + zJ\sigma)\sum_{i=1}^{N} S_i$$

where z is the coordination number, $\sigma = \langle S_i\rangle$ and S_i takes the values $-s, \ldots, s$. The partition function factorizes into a product of single-spin partition functions, each given by

$$Z_1 = \sum_{S=-s}^{s} \mathrm{e}^{xS} = \frac{\sinh[(s+\frac{1}{2})x]}{\sinh[(\frac{1}{2}x)]}$$

where $x = \beta(h + zJ\sigma)$. This result, valid for both integer and half-integer values of s, is readily obtained as the sum of a geometric series.

As usual, we obtain an implicit equation for σ by computing the ensemble average $\sigma = \langle S\rangle = \sum_{S=-s}^{s} S\,\mathrm{e}^{xS} = \mathrm{d}(\ln Z_1)/\mathrm{d}x$. In particular, at $h = 0$, we find that

$$\sigma = (s + \tfrac{1}{2})\coth[(s + \tfrac{1}{2})\beta zJ\sigma] - \tfrac{1}{2}\coth(\tfrac{1}{2}\beta zJ\sigma).$$

The function on the right-hand side is, essentially, the Brillouin function which describes the paramagnetic behaviour of non-interacting spin-s atoms, except that here the magnetic field acting on these atoms is the effective internal field. In the usual way, we find that there are non-zero solutions for σ when the slope of this function (considered as a function of σ) at $\sigma = 0$ is greater than 1, that is, when

$$\frac{s(s+1)}{3}\beta zJ > 1 \quad \text{or} \quad T < T_{\mathrm{c}} \equiv \frac{s(s+1)}{3}\frac{zJ}{k}.$$

(b) The standard Ising model corresponds to the case $s = \frac{1}{2}$, when S_i can assume two possible values. The usual formulation can be recovered by defining $\bar{S}_i = 2S_i$, so that $\bar{S}_i = \pm 1$ and $\bar{\sigma} = \langle\bar{S}_i\rangle = 2\sigma$. The interaction energy will then have the standard form when written in terms of the coupling constant $\bar{J} = J/4$. With these definitions, the equation for σ assumes the form $\bar{\sigma} = \tanh(\beta z\bar{J}\bar{\sigma})$ and the critical temperature becomes $T_{\mathrm{c}} = z\bar{J}/k$, in agreement with the results of problem 5.24.

The Heisenberg model describes classical three-dimensional unit vectors. Loosely speaking, the correspondence principle suggests that we might recover this model in the limit $s \to \infty$. In this limit, however, the quantum-mechanical spins have a magnitude $\sqrt{s(s+1)}$, so we define $\bar{S}_i = S_i/\sqrt{s(s+1)}$, $\bar{\sigma} = \sigma/\sqrt{s(s+1)}$ and, to get the correct interaction energy, $\bar{J} = s(s+1)J$. In terms of these variables, and in the limit $s \to \infty$, the equation for σ becomes

$$\bar{\sigma} = \coth(\beta z\bar{J}\bar{\sigma}) - \frac{1}{\beta z\bar{J}\bar{\sigma}}$$

and the critical temperature is $T_{\mathrm{c}} = z\bar{J}/3k$, reproducing the results of problem 5.23. Alert readers will realize that the limit described here actually reproduces only one part of the interaction energy of the Heisenberg model, namely $-\bar{J}\sum_{\langle i,j\rangle}\bar{S}_i^z\bar{S}_j^z$, say, the x and y spin components being absent. In general, therefore, the limit $s \to \infty$ of the present model does not correctly reproduce the Heisenberg model. As can be seen from problem 5.23, however, only one spin component (that in the direction of an applied magnetic field) is actually relevant within the mean-field approximation, so we do recover the correct mean-field results.

• **Problem 5.27** The partition function Z for the interacting system is

$$Z = \sum_{n_1\cdots n_N=0}^{\infty} \exp\left(-\beta\sum_{k=1}^{N}\varepsilon_k n_k - \frac{\beta\lambda}{2}\sum_{i=1}^{N}\sum_{j=1}^{N}V_{ij}n_in_j\right)$$

and, on expanding this to first order in λ, we obtain

$$Z = \sum_{n_1 \dots n_N = 0}^{\infty} \left(1 - \frac{\beta\lambda}{2} \sum_{ij} V_{ij} n_i n_j + \mathrm{O}(\lambda^2)\right) \exp\left(-\beta \sum_{k=1}^{N} \varepsilon_k n_k\right).$$

Since $V_{ii} = 0$, we can use the results

$$\sum_{n=0}^{\infty} \mathrm{e}^{-\beta\varepsilon n} = \frac{1}{1 - \mathrm{e}^{-\beta\varepsilon}} \qquad \sum_{n=0}^{\infty} n\,\mathrm{e}^{-\beta\varepsilon n} = \frac{\mathrm{e}^{-\beta\varepsilon}}{(1 - \mathrm{e}^{-\beta\varepsilon})^2}$$

to find

$$Z = \left(1 - \beta\lambda \sum_{i=1}^{N} \varepsilon_i^{(1)}(T) \frac{\mathrm{e}^{-\beta\varepsilon_i}}{1 - \mathrm{e}^{-\beta\varepsilon_i}} + \mathrm{O}(\lambda^2)\right) \prod_{k=1}^{N} \frac{1}{1 - \mathrm{e}^{-\beta\varepsilon_k}}$$

where

$$\varepsilon_i^{(1)}(T) = \tfrac{1}{2} \sum_{j=1}^{N} V_{ij} \frac{\mathrm{e}^{-\beta\varepsilon_j}}{1 - \mathrm{e}^{-\beta\varepsilon_j}}.$$

On the other hand, the partition function of the approximate non-interacting system is

$$\begin{aligned} Z_0 &= \prod_{i=1}^{N} \frac{1}{1 - \mathrm{e}^{-\beta(\varepsilon_i + \lambda\varepsilon_i^{(1)})}} \\ &= \prod_{i=1}^{N} \left(1 - \beta\lambda\varepsilon_i^{(1)} \frac{\mathrm{e}^{-\beta\varepsilon_i}}{1 - \mathrm{e}^{-\beta\varepsilon_i}} + \mathrm{O}(\lambda^2)\right) \frac{1}{1 - \mathrm{e}^{-\beta\varepsilon_i}}. \end{aligned}$$

Noting that $\prod_i (A_i B_i) = (\prod_i A_i)(\prod_k B_k)$ and that

$$\prod_{i=1}^{N} \left(1 - \beta\lambda\varepsilon_i^{(1)} \frac{\mathrm{e}^{-\beta\varepsilon_i}}{1 - \mathrm{e}^{-\beta\varepsilon_i}} + \mathrm{O}(\lambda^2)\right) = 1 - \beta\lambda \sum_{i=1}^{N} \varepsilon_i^{(1)} \frac{\mathrm{e}^{-\beta\varepsilon_i}}{1 - \mathrm{e}^{-\beta\varepsilon_i}} + \mathrm{O}(\lambda^2)$$

we can write Z_0 in the same form as Z, namely as

$$Z_0 = \left(1 - \beta\lambda \sum_{i=1}^{N} \varepsilon_i^{(1)}(T) \frac{\mathrm{e}^{-\beta\varepsilon_i}}{1 - \mathrm{e}^{-\beta\varepsilon_i}} + \mathrm{O}(\lambda^2)\right) \prod_{k=1}^{N} \frac{1}{1 - \mathrm{e}^{-\beta\varepsilon_k}}$$

provided that we identify $\varepsilon_i^{(1)}(T)$ as above. At the first order of perturbation theory, therefore, the net effect of the interactions is just a temperature-dependent correction to the energy levels of the non-interacting system.

• Problem 5.28

(a) Let us first relabel the spin variables, so as to distinguish those on odd-numbered sites from those on even-numbered sites:

$$S_{2j} = \bar{S}_j \qquad S_{2j+1} = \sigma_j .$$

Then the reduced Hamiltonian can be written as

$$\begin{aligned}\tilde{\mathcal{H}} &= -b\sum_j(\bar{S}_j + \sigma_j) - K\sum_j(\bar{S}_j\sigma_j + \sigma_j\bar{S}_{j+1}) \\ &= -\sum_j\{\tfrac{1}{2}b(\bar{S}_j + \bar{S}_{j+1}) + [b + K(\bar{S}_j + \bar{S}_{j+1})]\sigma_j\}\end{aligned}$$

and the partition function as

$$Z(K,b) = \sum_{\{\bar{S}_i,\sigma_i\}} \mathrm{e}^{-\tilde{\mathcal{H}}} = \sum_{\{\bar{S}_i\}}\prod_j z(K,b,\bar{S}_j,\bar{S}_{j+1})$$

where

$$\begin{aligned}z(K,b,\bar{S}_j,\bar{S}_{j+1}) &= \mathrm{e}^{b(\bar{S}_j+\bar{S}_{j+1})/2}\sum_{\sigma_j=\pm1}\exp\{[b + K(\bar{S}_j + \bar{S}_{j+1})]\sigma_j\} \\ &= \mathrm{e}^{b}\exp\{(K + b/2)(\bar{S}_j + \bar{S}_{j+1})\} \\ &\quad + \mathrm{e}^{-b}\exp\{-(K - b/2)(\bar{S}_j + \bar{S}_{j+1})\}.\end{aligned}$$

Suppose that $z(K,b,\bar{S}_j,\bar{S}_{j+1})$ can be expressed in the form

$$z(K,b,\bar{S}_j,\bar{S}_{j+1}) = \mathrm{e}^{-f}\exp[\tfrac{1}{2}\bar{b}(\bar{S}_j + \bar{S}_{j+1}) + \bar{K}\bar{S}_j\bar{S}_{j+1}].$$

Then the partition function becomes

$$Z(K,b) = Z_0(K,b)Z(\bar{K},\bar{b})$$

with $Z_0 = \mathrm{e}^{-Nf}$, where N is one half of the original number of spins. On an infinite lattice, N is of course infinite, but f is a finite contribution to the free energy per spin. In fact, it is easy to see that the four possible values $z(K,b,\pm1,\pm1)$ are correctly reproduced by the desired expression, provided that the three relations

$$\begin{aligned}\mathrm{e}^{-f+\bar{K}+\bar{b}} &= \mathrm{e}^{2(K+b)} + \mathrm{e}^{-2K} \\ \mathrm{e}^{-f+\bar{K}-\bar{b}} &= \mathrm{e}^{2(K-b)} + \mathrm{e}^{-2K} \\ \mathrm{e}^{-f-\bar{K}} &= \mathrm{e}^{b} + \mathrm{e}^{-b}\end{aligned}$$

hold, which they will if f, $\bar{K}$ and $\bar{b}$ are appropriately chosen functions of K and b. In terms of the variables $\tau = \mathrm{e}^{-4K}$ and $\eta = \mathrm{e}^{-2b}$, the solution to these three equations is

$$\bar{\tau}(\tau,\eta) = \frac{\tau(2+\eta+\eta^{-1})}{1+(\eta+\eta^{-1})\tau+\tau^2}$$
$$\bar{\eta}(\tau,\eta) = \frac{\eta+\tau}{\eta^{-1}+\tau}$$
$$f(\tau,\eta) = \tfrac{1}{4}\ln[\bar{\tau}(\tau,\eta)] - \ln(\eta^{1/2}+\eta^{-1/2}).$$

(b) It is easy to verify that $\bar{\tau}(0,1) = 0$ and $\bar{\eta}(0,1) = 1$, so that $(\tau^*, \eta^*) = (0, 1)$ is a fixed point of the transformation. Since $\eta = 1$ corresponds to $b = 0$, it is convenient to examine small perturbations about the fixed point by taking $\delta\tau = \tau$ and $\delta\eta = -2b$. By expanding in powers of τ and b, we obtain linearized versions of the transformation equations:

$$\bar{\tau} \simeq 4\tau \qquad \bar{b} \simeq 2b.$$

If τ and b are small enough for these linearized equations to remain valid when the transformation is repeated n times, we clearly get the sequence of renormalized parameters $\tau_n \simeq 2^{2n}\tau$ and $b_n \simeq 2^n b$. As n increases, both τ_n and b_n grow, so the fixed point is unstable.

(c) Let us call a the distance between neighbouring sites of the original lattice. On that lattice, correlations between spins a distance r apart decay, when r is large, as

$$\langle S_i S_j\rangle - \langle S_i\rangle\langle S_j\rangle \sim \mathrm{e}^{-r/\xi(\tau,b)a}$$

where ξ is the correlation length measured in units of a. In particular, consider the correlation of two spins which have still not been summed over after n iterations of the renormalization-group transformation. The transformation does not change the state of the system—it just allows us to describe the same state in terms of different parameters and so these two spins are still correlated in exactly the same way. According to the new description, the spacing between the remaining lattice points is $2^n a$. The effective Hamiltonian that describes the remaining spins has exactly the same form as the original Hamiltonian; so, if we use it to calculate the correlation length, we must get $\xi(\tau_n, b_n)$, with the same functional form $\xi(\cdot,\cdot)$ as before. Because the actual correlation length is unchanged, we must have

$$\xi(\tau_n, b_n)2^n a = \xi(\tau, b)a.$$

When τ and b are small enough for the linearized equations to be valid, this implies that

$$\xi(2^{2n}\tau, 2^n b) \simeq 2^{-n}\xi(\tau, b)$$

and this relation can be satisfied only if the function $\xi(\tau, b)$ is of the form

$$\xi(\tau, b) = \tau^{-\nu}X(b\tau^{-\Delta})$$

where the exponents are $\nu = \frac{1}{2}$ and $\Delta = \frac{1}{2}$. From the renormalization group alone, we cannot find out what the scaling function $X(x)$ is. When the magnetic field vanishes, however, we have $b = 0$, and $X(0)$ must have some finite value. Indeed, the explicit calculation of problem 5.16 shows that, at low temperatures, $\xi \simeq \frac{1}{2}\mathrm{e}^{2\beta J} = \frac{1}{2}\tau^{-1/2}$, which agrees with our exponent $\nu = \frac{1}{2}$ and tells us that in fact $X(0) = \frac{1}{2}$.

The magnetization per spin is $m = \mu\langle S_i\rangle$, where μ is the magnetic moment of each spin. Because our system is translationally invariant, it does not matter which spin we take for S_i, so let us consider S_0, which is one of the spins that remain after any number of renormalization-group transformations. Its expectation value is the same, whether we calculate it with the original Hamiltonian or with the renormalized Hamiltonian; so we must have

$$m(\tau_n, b_n) = m(\tau, b).$$

The same argument that we used for the correlation length shows that m must have the scaling form $m(\tau, b) \simeq \mathcal{M}(b\tau^{-\Delta})$. Again, we cannot immediately deduce what the scaling function is. However, the exact calculation of problem 5.14 shows us that in fact

$$m = \mu \frac{\sinh b}{\sqrt{\sinh^2 b + \tau}}$$

and, when b is small, this reduces to

$$m \simeq \frac{\mu}{\sqrt{1 + (b\tau^{-1/2})^{-2}}}$$

which confirms the scaling form that we have deduced and exhibits the scaling function $\mathcal{M}$.

It should be apparent from our discussion of the correlation length that a fixed point of the renormalization group corresponds to a state in which either $\xi = 0$ or $\xi = \infty$. In this system, the correlation length diverges only at $T = 0$ (where $\tau = \mathrm{e}^{-4J/kT} = 0$). In more complicated systems (for example, the three-dimensional Ising model), ξ diverges at a critical point, and the renormalization group provides a means of estimating critical exponents and deducing the scaling properties introduced in problems 5.9, 5.23 and 5.24 when exact calculations are not available. When the critical point occurs at a non-zero temperature, perturbations from the fixed point will typically involve $T - T_c$. The appearance in our calculation of the parameter τ, which vanishes exponentially with temperature, is an atypical feature, arising only when the fixed point is at $T = 0$. The basic renormalization-group strategy illustrated in this problem can be implemented in many different ways, and for most systems of practical interest the transformation must be carried out in some approximate manner.

6

NON-EQUILIBRIUM METHODS AND TRANSPORT PROCESSES

The problems in this chapter concern physical systems which are not in thermal equilibrium. Non-equilibrium situations can be extremely difficult to deal with in any rigorous manner, and we are able to explore only some of the simpler phenomenological approximations that are available. The first two problems, 6.1 and 6.2, study particle diffusion modelled as a random walk and, in problems 6.3 and 6.4, we have taken the opportunity to illustrate how similar ideas can be applied in the context of financial analysis. Problems 6.5–6.9 examine Markov processes (for which the probabilities of transitions between microstates at a given instant of time are assumed to be independent of the system's prior history), the associated master equations for probability distributions and the approach to thermal equilibrium. In problems 6.10 and 6.11, the behaviour of classical gases is studied in near-equilibrium situations where scattering is relatively unimportant. The opposite situation, exemplified by Brownian motion, where the dynamics of a system of interest are principally controlled by friction and random forces owing to its environment, can be investigated through the Langevin equation. This is illustrated by problems 6.12 and 6.13, where the diffusion equation is seen to re-emerge. Two further analogues of particle diffusion are dealt with in problems 6.14 and 6.15, which concern time-dependent temperature distributions in solids, and in problem 6.16, which introduces the time-dependent Ginzburg–Landau equation. Finally, problems 6.17–6.24 provide examples of transport processes in gases and metals, based on the relaxation time approximation to the Boltzmann transport equation.

6.1 Questions

• **Problem 6.1** A man walks along a straight line, taking a step of length a every τ seconds. Each step is either forwards, with probability p, or backwards, with probability $q = 1 - p$.

(a) For a large number of steps, find the Gaussian probability distribution $n(x, t)$, such that $n(x, t)\,\mathrm{d}x$ is the probability of finding him at a point between x and $x + \mathrm{d}x$ at time t, assuming that he started from $x = 0$ at time $t = 0$.

(b) Verify that this probability distribution satisfies a differential equation of the form

$$\frac{\partial n}{\partial t} + v\frac{\partial n}{\partial x} - D\frac{\partial^2 n}{\partial x^2} = 0$$

and find the constants v and D.

(c) Write down the equation found in (b) in the form of a continuity equation, identifying the associated probability current.

• **Problem 6.2** A drunken man walks at random along a street, taking steps (each of length a) forwards and backwards with equal probability, every τ seconds. His random walk starts at an initial time $t = 0$ from a position $x(t = 0) = x_0 > 0$. On the street, say, at $x = 0$, there is a hole into which the man may eventually fall and never get out (a typical Buenos Aires street!). Assuming that $a \ll x_0$, compute the probability $\mathcal{P}(t)$ of finding the man in the hole at time t.

• **Problem 6.3** A simple binomial model for the stochastic behaviour of a stock price over time assumes that the *spot price* S changes at regular intervals of length τ, starting with the value S_0 at time $t = 0$. At each time step, there is a probability p for S to go up to uS, with $u = 1+\alpha$ and a probability $q = 1-p$ for S to go down to dS, with $d = 1 - \beta$. We must have $\beta < 1$, so that S is always non-negative.

(a) Find the spot price $S_N(\nu)$ after N steps, if exactly ν of the price changes have been upwards. What is the probability $\mathcal{P}_N(\nu)$ that the price has increased exactly ν times?

(b) In the limit that N is very large, find the probability distribution $n(S, t)$, such that $n(S, t)\,\mathrm{d}S$ gives the probability for the price at time t to be between S and $S + \mathrm{d}S$. Show that, at time t, the mean value of $\ln S$ and its root mean square deviation can be expressed as $\langle \ln S\rangle = \ln S_0 + vt$ and $\Delta(\ln S) = \sigma\sqrt{t}$, and find v and σ in terms of the parameters of the model given above.

(c) Show that $n(S, t)$ obeys a variant of the diffusion equation.

• **Problem 6.4** In the context of financial analysis, the value of a *derivative security* is determined as a function $f(S, t)$ of time and of the price $S \geq 0$ of an underlying stochastic asset by solving the Black–Scholes differential equation

$$\frac{\partial f}{\partial t} + rS\frac{\partial f}{\partial S} + \tfrac{1}{2}\sigma^2 S^2\frac{\partial^2 f}{\partial S^2} = rf.$$

Here, the parameter σ is a measure of the volatility of the market, and r is a risk-free interest rate. The value $f(S, T)$ at a future time T at which the investment matures is a specified function of S, and the solution $f(S, t)$ subject to this boundary condition yields a pricing formula for valuation of the security at some earlier time $t \leq T$.

(a) Compare this equation with that for particle diffusion in one dimension and establish a correspondence between the solutions of these two equations.

(b) A *European call option* is a particular derivative security defined by the key boundary condition $f(S, T) = \max(S - X, 0)$, where X is a constant, called the *strike price*. Obtain the pricing formula for this option.

• **Problem 6.5** A set of N lights is controlled by a computer in the following way: once every second, one of the lights is selected at random and its state is reversed (that is if the light is on it is turned off, and if it is off it is turned on). Consider the evolution of the set of probabilities $\{P_N(n,t)\}$, where $P_N(n,t)$ is the probability that n lights are on at time t. Verify that this evolution is a Markov chain. Find the transition matrix, and show that there is a unique stationary state, given by a binomial distribution.

• **Problem 6.6** Two Ising spins S_1 and S_2 interact with a magnetic field so that (with a suitable choice of units) the energy of each spin is $-\varepsilon$ when the spin is parallel to the field ($S = 1$) and ε when it is antiparallel to the field ($S = -1$). The spins are acted on independently by a random external force, which can be modelled by a Markov chain, as follows: at each time step, there is a probability $a < 1$ for the transition $S = 1 \to S = -1$ and a probability $b < 1$ for the transition $S = -1 \to S = 1$.

(a) Obtain the transition matrix for the stochastic variable $E(t)$ which represents the total energy at time t.

(b) Verify that the transition matrix has the following eigenvalues λ_i and left eigenvectors $\boldsymbol{p}_i$:

$$\lambda_1 = 1 \qquad \boldsymbol{p}_1 = \frac{1}{(a+b)^2}(b^2, 2ab, a^2)$$

$$\lambda_2 = 1 - a - b \qquad \boldsymbol{p}_2 = \frac{1}{(a+b)}(b, (a-b), -a)$$

$$\lambda_3 = (1 - a - b)^2 \qquad \boldsymbol{p}_3 = (1, -2, 1).$$

(The order of the elements in these eigenvectors depends, of course, on the ordering of rows and columns in the transition matrix.)

(c) Show that there is a unique stationary probability distribution for E and that this distribution is approached after many steps from any initial distribution. Interpret your results for the special cases $a = 0$ and $a = b$.

(d) Show that the stationary distribution can be interpreted as a state of thermal equilibrium and find the temperature of this state.

(e) Find the relaxation rate which, at long times, characterizes the approach of the mean energy towards its equilibrium value.

• **Problem 6.7** A particle lives on the sites of a one-dimensional lattice. At any moment, there is a probability per unit time α that it will hop to the site on its right and a probability per unit time β that it will hop to the site on its left. Write down the master equation governing the evolution of the set of probabilities $p_n(t)$ of finding this particle at the nth site ($n = -\infty, \ldots, \infty$) at time t. By solving this equation, assuming that the particle is at the site $n = 0$ at $t = 0$, find its mean position $\langle n \rangle_t$ after a time t and its root mean square deviation $\Delta n(t) = \sqrt{\langle n^2 \rangle_t - \langle n \rangle_t^2}$ from this mean position.

• **Problem 6.8** A large number of radioactive samples of an originally inactive substance is produced by irradiation, which ceases at time $t = 0$. Each of the active nuclei produced has a decay rate α.

(a) Find the master equation for the probabilities $p_n(t)$ that a randomly selected sample has n undecayed nuclei at time $t > 0$.

(b) Show that the generating functional $F(z, t) = \sum_n z^n p_n(t)$ depends only on the single variable $u = (z - 1)\,\mathrm{e}^{-\alpha t}$.

(c) Show that the mean number $\bar{n}(t) = \langle n \rangle_t$ of undecayed nuclei at time t obeys $\bar{n}(t) = \bar{n}(0)\,\mathrm{e}^{-\alpha t}$, regardless of the initial distribution.

(d) Find the mean number of undecayed nuclei and the variance of the distribution at time t for a sample which initially has exactly n_0 active nuclei. What is $p_n(t)$ in this case?

(e) If the initial numbers of active nuclei in the samples are given by a Poisson distribution whose mean is n_0, show that the corresponding distribution at time t is a Poisson distribution whose mean is $n_0\,\mathrm{e}^{-\alpha t}$.

• **Problem 6.9** A system of N idealized atoms, each having two energy levels $E = \pm\varepsilon$, is brought into contact with a heat bath at temperature T. The atoms do not interact with each other, but each atom interacts with the heat bath so as to have a probability $w_+(T)$ per unit time for a transition to the upper level if it is in the lower level, and a probability $w_-(T)$ per unit time for the reverse transition. Find the relaxation time τ (in terms of w_+ and w_-) that characterizes the approach of this system to equilibrium.

• **Problem 6.10** A container is divided into two two compartments by a plane wall. One of these compartments contains a classical ideal gas in thermal equilibrium at a temperature T, with a number density n_0. At a time $t = 0$, the wall is removed. For the idealized case that each compartment can be considered infinitely large, and that the wall disappears instantaneously, find the density $n(\boldsymbol{r}, t)$ at all subsequent times, in terms of a suitable integral. Obtain approximate explicit expressions for the density of particles at a distance x from the original position of the wall and for the mean energy of these particles, when $t \gg |x|\sqrt{m/kT}$, where m is the mass of each particle.

• **Problem 6.11** A box of volume V contains a classical ideal gas whose particles have mass m, and whose number density and temperature are initially n_0 and T_0 respectively. A very small hole of area A is made in one of the walls, allowing particles to escape into a surrounding vacuum.

(a) If the box is thermally isolated, find the temperature $T(t)$ and the number density $n(t)$ inside the box at subsequent times, assuming that collisions between molecules are frequent enough to maintain thermal equilibrium, but can otherwise be neglected.

(b) If the box is in contact with a heat reservoir so as to keep the temperature fixed at the value T_0, find the rate of heat flow $\dot{Q}(t)$ into the box.

• **Problem 6.12** Drops of oil, each of mass m, are allowed to fall to the floor in still air at temperature T. Each drop is released at rest from the same point, at a height h above the floor. Estimate the radius $r(h)$ of the patch of oil formed on the floor as a function of the height h, assuming (i) that this radius is determined principally by Brownian motion of the falling drops and (ii) that the time that a drop takes to reach its terminal velocity is much less than the time that it takes to reach the floor. Do you think that the result of this calculation could be tested experimentally?

• **Problem 6.13** Consider a fluid at a temperature T such that a small fraction of its molecules are ionized, in an electric field $\boldsymbol{E}(\boldsymbol{x})$. We study the motion of an ion of charge q, assuming that it collides frequently with neutral molecules (which are uniformly distributed), but very rarely with other ions, and that the time intervals of interest are much longer than the characteristic relaxation time.

(a) Starting from a suitable Langevin equation, show that the number density $n(\boldsymbol{x}, t)$ of ions of charge q evolves with time as a diffusion process, governed by the equations

$$\frac{\partial n}{\partial t} + \boldsymbol{\nabla} \cdot \boldsymbol{j} = 0 \quad \text{with} \quad \boldsymbol{j} = -D\boldsymbol{\nabla} n + \frac{q}{kT} Dn\boldsymbol{E}$$

where D is the diffusion constant.

(b) Verify that these equations have a steady-state solution corresponding to the canonical ensemble of equilibrium statistical mechanics.

(c) If there are ions of charge q with diffusion constant D_+ and ions of charge $-q$ with diffusion constant D_-, find the conductivity σ of this fluid. Compare this result with the Drude expression $\sigma = ne^2\tau_c/m$ for the conductivity of a metal, in which n is the number density of electrons and τ_c is the mean time between collisions.

• **Problem 6.14** A long copper rod is attached at one end to a copper cup, which makes good thermal contact with the rod, and the whole apparatus is well insulated. The cup is initially filled with boiling water. After 10 min, the boiling water is replaced with freezing water. After a further 10 min, the freezing water is replaced with boiling water, and the cycle is repeated for several hours. During the course of this experiment, the temperature of the rod at a point 50 cm from the cup is monitored and, to a good approximation, is found to vary sinusoidally. Why is this? Estimate the amplitude of the sinusoidal variation.

The specific heat of copper is $C = 385$ J K^{-1} kg^{-1}, its density is $\rho = 8930$ kg m^{-3} and its thermal conductivity is $\kappa = 390$ W K^{-1} m^{-1}.

• **Problem 6.15**

(a) A long well-insulated rod of thermal diffusivity D is initially at a uniform temperature T_0. At a time $t = 0$, the temperature of one end of the rod is raised by an amount ΔT and then held fixed. Find the temperature distribution in the rod at subsequent times, in terms of a suitable integral. (Hint: first find an expression for $\partial T(x,t)/\partial x$, which obeys the same equation as $T(x,t)$, and then integrate this expression.)

(b) The ends of a well-insulated rod of length ℓ are held at fixed temperatures T_0 and T_ℓ. If the temperature gradient at intermediate points is not uniform, find the time dependence of the Fourier components of the temperature distribution.

(c) A well-insulated pipe, 1 m long, made from a plastic material of negligible thermal conductivity, is full of ice at a temperature of $-5\,^\circ$C. One end is maintained at this temperature while, at $t = 0$, the other end is immersed in water, which is maintained at its boiling point. Clearly, the ice will begin to melt. Find a differential equation for the position of the water–ice interface at time t, assuming that the temperature gradients in the water and in the ice are always approximately uniform. What is the maximum length of ice that can be melted in this way? How long does it take for the first 10 cm of ice to melt?

The thermal conductivities of water and ice are $\kappa_w = 0.6$ W m^{-1} K^{-1} and $\kappa_i = 2$ W m^{-1} K^{-1} respectively; their specific heats are $C_w = 4190$ J K^{-1} kg^{-1} and $C_i = 2100$ J K^{-1} kg^{-1}; their densities are $\rho_w = 998$ kg m^{-3} and $\rho_i = 920$ kg m^{-3}, and the latent heat of fusion is $L = 3 \times 10^5$ J kg^{-1}. Neglect any complications due to the contraction of the ice as it melts.

(d) Use the results of (a) and (b) to assess whether the assumption of uniform temperature gradients is reasonable.

• **Problem 6.16** Within the Landau theory of phase transitions, the equilibrium free energy of a uniaxial ferromagnet is obtained by minimizing a functional of the form

$$\Phi(\phi) = \int \mathrm{d}x \left[\frac{1}{2} \left(\frac{\mathrm{d}\phi}{\mathrm{d}x} \right)^2 + V(\phi) \right]$$

(for simplicity, we consider only one spatial dimension), the function $M(x) = \phi_{\min}(x)$ for which the minimum is achieved being the magnetization. The potential $V(\phi)$ has the form

$$V(\phi) = -\frac{a}{2}\phi^2 + \frac{b}{4}\phi^4 - h\phi$$

where a is proportional to $T_c - T$, h is proportional to the applied magnetic field and b is a positive constant. To study ferromagnetic states, we take a to be positive also. A phenomenological treatment of small departures from thermal equilibrium takes the magnetization $M(x, t)$ to be a solution of the *time-dependent Ginzburg–Landau equation*

$$\frac{\partial M}{\partial t} = D \left(\frac{\partial^2 M}{\partial x^2} - \frac{\partial V(M)}{\partial M} \right)$$

where D is a suitable (positive) transport coefficient, which is a generalization of the diffusion equation.

(a) Show that small but possibly non-uniform perturbations away from an equilibrium state of uniform magnetization always relax back to the correct equilibrium state.

(b) For $h = 0$, show that there are equilibrium configurations of the form

$$M(x) = M_0 \tanh[q(x - x_0)]$$

representing two domains of opposite magnetization separated by a domain wall, and find the parameters M_0 and q.

(c) Show that, in the presence of a small magnetic field h, there are solutions of the form

$$M(x, t) = M_1 \tanh[q(x - x_0 - vt)] + m$$

and find the parameters M_1, m, q and v to leading order of an expansion in powers of h.

(d) Show that small perturbations of this moving domain wall always relax towards the given steady-state solution. (Note: this proof requires some simple ideas from the theory of solitons.)

• **Problem 6.17** The relaxation time approximation to the Boltzmann equation replaces the collision integral with an expression of the form $-(f - f_0)/\tau$. For a monatomic classical gas, with no external forces, take f_0 to be

$$f_0(\boldsymbol{r}, \boldsymbol{p}, t) = \frac{n(\boldsymbol{r}, t)}{[2\pi mkT(\boldsymbol{r}, t)]^{3/2}} \exp\left(- \frac{|\boldsymbol{p} - m\boldsymbol{v}(\boldsymbol{r}, t)|^2}{2mkT(\boldsymbol{r}, t)} \right)$$

and investigate the transport properties of the gas in the following way.

(a) Use the conservation of the number of particles, the conservation of momentum and the conservation of energy to show that the number density, the particle current and the energy density in the state described by f are the same as those in the state described by f_0. You should find that the particle current is $\boldsymbol{j}(\boldsymbol{r}, t) = n(\boldsymbol{r}, t)\boldsymbol{v}(\boldsymbol{r}, t)$ and the energy density is $\varepsilon(\boldsymbol{r}, t) = \frac{3}{2}n(\boldsymbol{r}, t)kT(\boldsymbol{r}, t) + \frac{1}{2}n(\boldsymbol{r}, t)mv^2(\boldsymbol{r}, t)$.

(b) Obtain an approximate solution for $f(\boldsymbol{r}, \boldsymbol{p}, t)$ of the form $f(\boldsymbol{r}, \boldsymbol{p}, t) = f_0(\boldsymbol{r}, \boldsymbol{p}, t) + \tau f_1(\boldsymbol{r}, \boldsymbol{p}, t) + \mathrm{O}(\tau^2)$. In this approximation, show that

$$\frac{\partial(nv_i)}{\partial t} + \frac{1}{m}\frac{\partial(nkT)}{\partial x_i} + \sum_j \frac{\partial(nv_i v_j)}{\partial x_j} = 0.$$

(c) For the case $\boldsymbol{v}(\boldsymbol{r}, t) = \boldsymbol{0}$, evaluate the heat current

$$\boldsymbol{q}(\boldsymbol{r}, t) = \int \mathrm{d}^3 p \, \frac{|\boldsymbol{p}|^2}{2m}\frac{\boldsymbol{p}}{m} f(\boldsymbol{r}, \boldsymbol{p}, t)$$

to first order in τ and hence determine the thermal conductivity κ.

(d) Evaluate the stress tensor

$$\theta_{ij}(\boldsymbol{r}, t) = \frac{1}{m}\int \mathrm{d}^3 p \, p_i p_j f(\boldsymbol{r}, \boldsymbol{p}, t)$$

to first order in τ and hence determine the shear viscosity η. Verify that $\kappa/\eta C_V = \frac{5}{3}$, where C_V is the specific heat per unit mass at constant volume.

• **Problem 6.18** Consider the relaxation time approximation to the Boltzmann equation for a monatomic gas with an external force $\boldsymbol{F}(\boldsymbol{r}) = -\boldsymbol{\nabla}\phi(\boldsymbol{r})$ that acts independently on each molecule. Use the expression given in problem 6.17 for the function $f_0(\boldsymbol{r}, \boldsymbol{p}, t)$.

(a) Generalize the results of problem 6.17(a) by showing that the number density, the particle current and the energy density in the state described by the distribution function $f(\boldsymbol{r}, \boldsymbol{p}, t)$ are the same as those in the state described by $f_0(\boldsymbol{r}, \boldsymbol{p}, t)$.

(b) Find the generalization of the relation given in problem 6.17(b).

(c) For a stationary gas ($\boldsymbol{v}(\boldsymbol{r}, t) = 0$) show that, to first order of the expansion in powers of the relaxation time τ, the heat current is given by

$$\boldsymbol{q}(\boldsymbol{r}, t) = -\frac{5}{2}\frac{n(\boldsymbol{r}, t)k^2 T(\boldsymbol{r}, t)\tau}{m}\boldsymbol{\nabla} T(\boldsymbol{r}, t).$$

• **Problem 6.19** An idealized model of the Earth's atmosphere is based on the following assumptions:

(i) the Earth's surface temperature has a uniform value of $T_0 = 288$ K;
(ii) the atmosphere is a monatomic gas of molecular weight 29;
(iii) the system is spherically symmetric and the temperature and pressure vary only in the radial direction;
(iv) the atmosphere is in a steady state and is stationary—there is no convection.

The thermal conductivity at sea level is $\kappa_0 = 2.5 \times 10^{-2}$ W m^{-1} K^{-1} and the pressure at sea level is $P_0 = 10^5$ Pa. The radius of the Earth is $r_0 = 6.4 \times 10^3$ km. Use the results of problem 6.18 to investigate the following.

(a) Suppose that the height of the atmosphere is sufficiently small that one can consider just a vertical column of air with a constant value of g, the acceleration due to gravity. Find the variation in pressure with altitude z for an isothermal atmosphere (in which the temperature is constant).

(b) Interpreting the relaxation time τ as the mean collision time, and assuming that the scattering cross-section σ is constant, how does τ depend on pressure and temperature?

(c) Near the Earth's surface, the atmosphere actually has a temperature gradient $\mathrm{d}T/\mathrm{d}z = -\alpha_0$, with $\alpha_0 \simeq 6 \times 10^{-3}$ K m^{-1}. Modify the analysis of (a) to find the variations in pressure and temperature with altitude, using this boundary condition. You should find that there is a maximum altitude z_{max} beyond which this model does not work. What is the value of z_{max}?

(d) Why does the analysis in (c) go wrong? Estimate the altitude up to which the analysis might be reasonable.

(e) Find the variation in pressure with altitude in an isothermal atmosphere, taking into account that the Earth is (very nearly) spherical, and the variation in the gravitational force with distance from the centre of the Earth. You should find that a steady-state solution is possible only if the Earth is immersed in an infinite gaseous medium. What are the pressure and number density of this medium a long way from the Earth?

• **Problem 6.20** A monatomic classical gas is contained between two concentric spherical walls. The inner wall, of radius a, is maintained at a temperature T_1, while the outer wall, of radius b, is maintained at a temperature T_2. For the idealized case when there is no convection, find the temperature profile in the steady state,

(a) assuming that the thermal conductivity κ is constant and

(b) assuming a suitable temperature dependence for κ.

Verify that the two profiles are approximately the same when the temperature difference $T_2 - T_1$ is small enough.

• **Problem 6.21** For a quantum-mechanical system, the Wigner distribution function $f_W(\boldsymbol{r}, \boldsymbol{p}, t)$ provides an analogue of the classical distribution function, which can be taken approximately to obey a Boltzmann-like equation. The classical Boltzmann equation contains a collision integral of the form

$$\left.\frac{\partial f}{\partial t}\right|_{\mathrm{coll}} = -\int \mathrm{d}^3 p_2\, \mathrm{d}^3 p_1'\, \mathrm{d}^3 p_2'\, W_{(\boldsymbol{p},\boldsymbol{p}_2)\to(\boldsymbol{p}_1',\boldsymbol{p}_2')}(f f_2 - f_{1'} f_{2'})$$

where W is the transition probability per unit time for a scattering process in which particles with momenta $\boldsymbol{p}$ and $\boldsymbol{p}_2$ collide, producing outgoing particles with momenta $\boldsymbol{p}_1'$ and $\boldsymbol{p}_2'$. Assuming that energy and momentum are conserved in each collision, this scattering integral vanishes in the equilibrium state for which $f(\boldsymbol{r}, \boldsymbol{p}, t)$ is the Maxwell–Boltzmann distribution. Suggest a suitable generalization of this scattering integral for a system of spin-$\frac{1}{2}$ fermions, and show that it vanishes when $f_W(\boldsymbol{r}, \boldsymbol{p}, t)$ is the Fermi–Dirac distribution.

• **Problem 6.22** Use the relaxation time approximation to the Boltzmann equation to obtain expressions for the electrical conductivity σ, the Hall coefficient R_H, the thermoelectric power S and the thermal conductivity κ of a metal, considering the electrons as a highly degenerate gas of particles that move freely except for collisions with lattice impurities and phonons. Assume that the chemical potential in the background distribution function $f_W^{(0)}$ is constant. (Under conditions where no electric current flows, a temperature gradient causes an electrical potential to appear across the sample, and the thermoelectric power S can be defined as $\boldsymbol{E} = S\nabla T$, where $\boldsymbol{E}$ is the electric field.)

• **Problem 6.23** For the simple model of a metal described in problem 6.22, find the frequency-dependent electrical conductivity $\sigma(\omega)$ governing the response to an applied electric field that oscillates with angular frequency ω.

• **Problem 6.24** Generalize the analysis of problem 6.22 to obtain expressions for the electrical conductivity, the thermopower and the thermal conductivity when the relaxation time τ depends on energy. For a highly degenerate electron gas, with $T \ll T_F$, you should find that these quantities depend only on $\tau(\varepsilon_F)$ and $\tau'(\varepsilon_F) = \mathrm{d}\tau/\mathrm{d}\varepsilon|_{\varepsilon=\varepsilon_F}$.

6.2 Answers

• Problem 6.1

(a) Suppose that the man has taken N steps. We first find the probability $\mathcal{P}_N(\nu)$ that he has taken ν steps forwards and $N - \nu$ steps backwards (which means that he has moved forward by a net distance $x = (2\nu - N)a$). The probability of taking these steps in a given sequence is $p^\nu q^{N-\nu}$, and the number of sequences in which they could be taken is $N!/\nu!(N-\nu)!$, so we get

$$\mathcal{P}_N(\nu) = p^\nu q^{N-\nu} \frac{N!}{\nu!(N-\nu)!}.$$

This probability is correctly normalized, because $\sum_{\nu=0}^{N} \mathcal{P}_N(\nu) = (p+q)^N = 1$. We expect to be able to obtain a continuum approximation to this probability when N, ν and $N - \nu$ are all large. In these circumstances, we can use Stirling's approximation but, to do everything systematically, it is necessary to use a more accurate version than we adopted in earlier chapters, namely $\ln N! \simeq N \ln N - N + \frac{1}{2}\ln(2\pi N)$. After a little algebra, we find that

$$\ln[\mathcal{P}_N(\nu)] \simeq \nu \ln\left(\frac{pN}{\nu}\right) + (N-\nu)\ln\left(\frac{qN}{N-\nu}\right) + \tfrac{1}{2}\ln\left(\frac{N}{2\pi\nu(N-\nu)}\right).$$

By differentiating with respect to ν, we find that the most probable value of ν is $\nu_{\max} = Np[1 + \mathrm{O}(N^{-1})]$, and the expected Gaussian distribution is obtained by expanding about this value. Specifically, we define the deviation u by $\nu = Np + \sqrt{N}u$, and expand $\ln[\mathcal{P}_N(\nu)]$ in powers of N:

$$\ln[\mathcal{P}_N(\nu)] = -\tfrac{1}{2}\ln(2\pi Npq) - \frac{u^2}{2pq} + \mathrm{O}(N^{-1/2}).$$

The factor of $\sqrt{N}$ in the definition of u was chosen, of course, so that the term proportional to u^2 in this last expression is independent of N. To obtain the continuum approximation, we suppose that $\mathcal{P}_N(\nu)$ is approximately constant over a range of values from ν to $\nu + \mathrm{d}\nu$. The total probability that ν has one of the values in this range is $\mathcal{P}_N(\nu)\,\mathrm{d}\nu$, so we identify

$$n(x,t)\,\mathrm{d}x = \mathcal{P}_N(\nu)\,\mathrm{d}\nu = \mathcal{P}_N(\nu)\frac{\mathrm{d}\nu}{\mathrm{d}x}\,\mathrm{d}x = \frac{\mathcal{P}_N(\nu)}{2a}\,\mathrm{d}x.$$

Then, on expressing u in terms of x, and using $t = N\tau$, we find that

$$n(x,t) = \frac{\exp[-(x-vt)^2/4Dt]}{\sqrt{4\pi Dt}}$$

where $v = (2p-1)a/\tau = (p-q)a/\tau$ and $D = 2pqa^2/\tau$. It is easy to check that $\int_{-\infty}^{\infty} n(x,t)\,\mathrm{d}x = 1$, so this probability distribution is correctly normalized. Evidently, to maintain the correct normalization, it was essential to keep all the terms in $\ln[\mathcal{P}_N(\nu)]$ that do not vanish when $N \to \infty$, including the correction term in Stirling's approximation. The probability distribution that we have obtained clearly has its maximum at $x = vt$, and this is intuitively reasonable since, on average, the man moves a distance $(p-q)a$ in each time interval τ, corresponding to the drift velocity v. If we measure a deviation $y = x - vt$ from the position of this maximum, then the root mean square deviation $y_{\mathrm{rms}} = \sqrt{\langle y^2 \rangle} = \sqrt{2Dt}$ increases as $\sqrt{t}$, which is characteristic of a random walk or diffusion process, with diffusion coefficient D. In the limit when $p \to 1$ and $q \to 0$ (or the other way round), we have $D \to 0$ and $n(x,t) \to \delta(x - vt)$. In this limit, the man always walks forwards (or backwards) and is at the position $x = vt$ with probability 1.

(b) By substituting the distribution $n(x,t)$ found in (a) into the given differential equation, we easily verify that it is satisfied, with the values of v and D given above. It is tempting to try to derive this equation from the following argument. Regarding $n(x,t)$ as equivalent to the exact probability from which we started, the probability of being at x at time t is the sum of the probability for being at $x-a$ at time $t-\tau$ and moving forwards at the next step, and that for being at $x+a$ at time $t-\tau$ and moving backwards at the next step, that is

$$n(x,t) = p\,n(x-a, t-\tau) + q\,n(x+a, t-\tau).$$

Now, treating x and t as continuous variables, and taking a and τ to be small, we write the first few terms of the Taylor series on the right-hand side:

$$n(x,t) = n(x,t) - (p-q)an'(x,t) + \tfrac{1}{2}a^2 n''(x,t) - \tau\dot{n}(x,t) + \cdots$$

where $\dot{n} = \partial n/\partial t$ and $n' = \partial n/\partial x$. If we neglect all the remaining terms, this is equivalent to

$$\frac{\partial n}{\partial t} + (p-q)\frac{a}{\tau}\frac{\partial n}{\partial x} - \frac{1}{2}\frac{a^2}{\tau}\frac{\partial^2 n}{\partial x^2} = 0$$

which is similar to our previous equation but with the *wrong* diffusion coefficient, except in the symmetrical case $p = q = \frac{1}{2}$. It is instructive to understand why this argument does not work in general, which can be seen as follows. The correct probability density was obtained in (a) by considering large values of N. One might suppose that this is equivalent to taking a continuum limit in which $a \to 0$ and $\tau \to 0$, with some appropriate ratio of these quantities held fixed. However, this cannot be done in general. If we keep the ratio a/τ fixed, then the drift velocity v remains fixed (for definite values of p and q), but the diffusion coefficient D vanishes as $D \sim a \to 0$. On the other hand, if we keep a^2/τ fixed, then D remains fixed, but v diverges as $v \sim a^{-1} \to \infty$. Now consider the above Taylor series. In order to justify neglecting the higher-order terms, we divide the equation by τ, and hope to take a and τ to zero, in such a way that the terms that we kept remain finite and non-zero, while the remaining terms vanish. We see, however, that this is not possible in general. It *is* possible if $p = q = \frac{1}{2}$, in which case $v = 0$ and we can take D to have the fixed value $a^2/2\tau$. It is also possible if p and q are only slightly different from $\frac{1}{2}$, say if $p = \frac{1}{2} + \lambda a$ and $q = \frac{1}{2} - \lambda a$, where λ is a constant. In that case, we can take a continuum limit with a^2/τ fixed, so that $v = 2\lambda a^2/\tau$ remains finite. In this case too, the diffusion coefficient has the limit $D \to a^2/2\tau$. In general, however, if p and q differ from $\frac{1}{2}$ by some finite amount, we cannot take a continuum limit in this way. Although it is legitimate, after many steps, to treat x and t as continuous, the microscopic parameters a and τ must still be taken to have definite non-zero values if both the macroscopic quantities v and D are to be finite and non-zero.

(c) The equation found in (b) can be written in the form

$$\frac{\partial n}{\partial t} + \frac{\partial j}{\partial x} = 0$$

where the probability current is $j(x,t) = vn(x,t) - D\,\partial n(x,t)/\partial x$. This is the equation of continuity, which expresses the conservation of probability. Alternatively, we might interpret $n(x,t)$ as the (linear) density of a collection of diffusing particles, in which case it should be normalized so that $\int_{-\infty}^{\infty} n(x,t)\,\mathrm{d}x = N$. Then the equation of continuity expresses the conservation of the total number N of particles because

$$\frac{\mathrm{d}}{\mathrm{d}t}\left(\int_{-\infty}^{\infty} n(x,t)\,\mathrm{d}x\right) = -\int_{-\infty}^{\infty} \frac{\partial j(x,t)}{\partial x}\,\mathrm{d}x = j(-\infty,t) - j(+\infty,t) = 0$$

since particles do not diffuse to $x = \pm\infty$ in any finite time. We see that the first term in the current of particles, $vn(x,t)$, describes the motion of free particles, all moving with the drift velocity v, while the second term, $-D\,\partial n(x,t)/\partial x$, describes the diffusion of particles in the opposite direction to the concentration gradient, namely from regions of high concentration towards regions of low concentration.

• **Problem 6.2** With $a \ll x_0$, very many steps are needed to reach the hole at $x = 0$, so it is permissible to treat the position x and time t as continuous. As shown in problem 6.1 (with $p = q = \frac{1}{2}$), the probability density $n(x,t)$ for finding the man near the point x at time t obeys the diffusion equation

$$\frac{\partial n}{\partial t} - D\frac{\partial^2 n}{\partial x^2} = 0$$

with the diffusion constant $D = a^2/2\tau$ and the initial condition $n(x,0) = \delta(x - x_0)$, provided that $x > 0$. However, the man can never reach negative values of x, because he will be 'absorbed' by the hole at $x = 0$, so we have $n(x,t) = 0$ for $x \leq 0$.

Without the hole, the solution would be $n(x,t) = n_0(x - x_0, t)$, where (as found in problem 6.1)

$$n_0(x,t) = \frac{e^{-x^2/4Dt}}{\sqrt{4\pi Dt}}.$$

An elegant way of taking account of the hole at $x = 0$ is to adapt the method of images, which is well known in the treatment of analogous problems in electrostatics. Because the diffusion equation is linear, any superposition of solutions is also a solution. To get the correct probability density for $x > 0$, we need to subtract the contributions from walks that pass into the region $x < 0$ and then return, and it is reasonable to guess that these are the same as walks which start from $x = -x_0$ at $t = 0$ and eventually get to positive values of x. So the particular superposition that we consider is

$$n(x,t) = n_0(x - x_0, t) - n_0(x + x_0, t) \qquad \text{for } x \geq 0.$$

Because $n_0(x,t) = n_0(-x,t)$, it is clear that $n(x,t)$ vanishes at $x = 0$ for all times t, which is what we need, and, since $n_0(x + x_0, 0) = \delta(x + x_0) = 0$ for $x > 0$, this superposition also satisfies our initial condition for positive x.

At time t, the probability that the man has fallen into the hole is

$$\mathcal{P}(t) = 1 - \int_0^\infty n(x,t)\,\mathrm{d}x$$

the integral being the total probability of finding him at some positive value of x. After a little rearrangement, this probability can be expressed as

$$\mathcal{P}(t) = \frac{2}{\sqrt{\pi}} \int_{y_0}^\infty e^{-y^2}\,\mathrm{d}y = \mathrm{erfc}(y_0)$$

where $y_0 = x_0/\sqrt{4Dt}$ and erfc($\cdot$) is the complementary error function. This probability is sketched in figure 6.1, and we see that it approaches unity when $t \gg (x_0^2/a^2)\tau$.

• Problem 6.3

(a) If the spot price has increased ν times and decreased $N - \nu$ times, then its final value is $S_N(\nu) = S_0 u^\nu d^{N-\nu}$, regardless of the order in which the increases and decreases occurred. The probability for exactly ν increases and $N - \nu$ decreases is given by the binomial distribution

$$\mathcal{P}_N(\nu) = p^\nu q^{N-\nu} \frac{N!}{\nu!(N-\nu)!}.$$

(b) By defining the variable $x = \ln(S/S_0)$, we obtain a very close analogy with the random walk studied in problem 6.1. The value of x after ν increases and

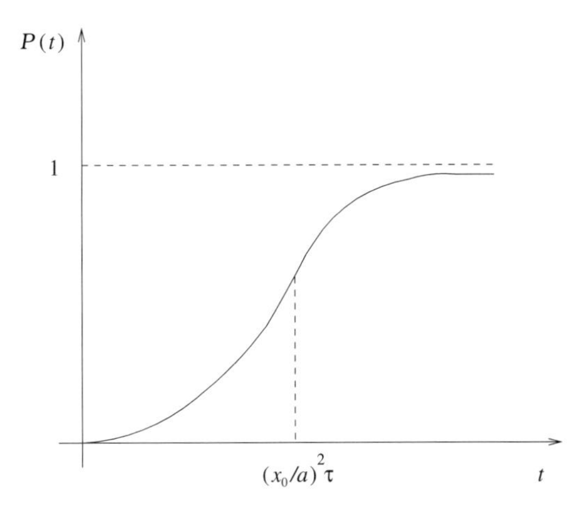

Figure 6.1. Probability of falling into the hole as a function of time.

$N - \nu$ decreases in price is $x(N, \nu) = N \ln d + \nu \ln(u/d) = \nu \ln u + (N - \nu) \ln d$. The only difference between the two problems is that here forward and backward steps have different lengths, $\ln u$ and $-\ln d$ respectively (since d is smaller than 1, the positive length of a backward step is $-\ln d$). The analysis of problem 6.1 can be repeated exactly. For large N, the probability $\mathcal{P}_N(\nu)$ has, at a given time $t = N\tau$, a maximum value at $\nu = Np + \mathrm{O}(N^{-1})$, corresponding to a most probable value of x, which is $x_0(t) = N(p \ln u + q \ln d) \equiv vt$, where the drift velocity is $v = (p \ln u + q \ln d)/\tau$. Expanding about this maximum probability, we have

$$\ln[\mathcal{P}_N(\nu)] = -\tfrac{1}{2} \ln(2\pi Npq) - \frac{\mu^2}{2pq} + \mathrm{O}(N^{-1/2})$$

where $\mu = N^{-1/2}(\nu - Np)$, and this can be converted into a probability density $\tilde{n}(x, t)$. Requiring that $\tilde{n}(x, t)\,\mathrm{d}x = \mathcal{P}_N(\nu)\,\mathrm{d}\nu$, and using $\mathrm{d}x/\mathrm{d}\nu = \ln(u/d)$, we get

$$\tilde{n}(x, t) = \frac{\exp[-(x - vt)^2/4Dt]}{\sqrt{4\pi Dt}}$$

where the diffusion coefficient is $D = pq[\ln(u/d)]^2/2\tau$. It may be observed that both v and D have the correct dimensions, $[\text{time}]^{-1}$. Finally, we can convert this into a probability density for S, requiring that $n(S, t)\,\mathrm{d}S = \tilde{n}(x, t)\,\mathrm{d}x$. Since $\mathrm{d}x/\mathrm{d}S = S^{-1}$, we find that

$$n(S, t) = \frac{\exp\{-[\ln(S/S_0) - vt]^2/4Dt\}}{S\sqrt{4\pi Dt}}.$$

The mean value and root mean square deviation of $\ln S$ are obtained as standard integrals from the Gaussian distribution for x. The results are

$$\langle \ln S \rangle = \ln S_0 + vt \qquad \Delta(\ln S) = \sqrt{\langle [\ln S - \langle \ln S \rangle]^2 \rangle} = \sigma \sqrt{t}$$

with $\sigma = \sqrt{2D} = 2\sqrt{pq/\tau}\ln(u/d)$. In financial analysis, this continuum limit is known as *geometric Brownian motion*. It is a particular case of an Ito process, for which the stochastic differential equation is

$$\mathrm{d}S = vS\,\mathrm{d}t + \sigma S\,\mathrm{d}z$$

the stochastic variable $z(t)$ having a mean of zero and a standard deviation $\sqrt{t}$. The parameter v is the expected rate of return per unit time from the stock and the parameter σ is the volatility of the stock price.

(c) From problem 6.1, we know that $\tilde{n}(x, t)$ obeys the diffusion equation

$$\frac{\partial \tilde{n}}{\partial t} + v\frac{\partial \tilde{n}}{\partial x} - D\frac{\partial^2 \tilde{n}}{\partial x^2} = 0.$$

It is straightforward to convert this into an equation for $n(S, t)$, using $\tilde{n} = Sn$ and $\partial/\partial x = S\,\partial/\partial S$, with the result

$$\frac{\partial n}{\partial t} - DS^2\frac{\partial^2 n}{\partial S^2} + (v - 3D)S\frac{\partial n}{\partial S} + (v - D)n = 0.$$

• Problem 6.4

(a) We would like to show that the Black–Scholes equation can be written in the form of a generalized diffusion equation

$$\frac{\partial}{\partial t'}[n(x, t')] = \frac{\partial}{\partial x}\left(D(x)\frac{\partial n(x, t')}{\partial x} - v(x)n(x, t')\right) + \mathcal{S}(n, x, t')$$

with a position-dependent diffusion coefficient $D(x)$ and drift velocity $v(x)$, where $\mathcal{S}$ is a source for 'particles'. Comparing the relative signs in the two equations, we see that it will be advantageous to use the time variable $t' = T - t$, corresponding to the fact that the equation is to be solved in reversed time from the date of maturity. After a little rearrangement, we then find that the equation can be written as

$$\frac{\partial f}{\partial t'} = \frac{\partial}{\partial S}\left(D(S)\frac{\partial f}{\partial S} - v(S)f\right) + (\sigma^2 - 2r)f$$

with $D(S) = \sigma^2 S^2/2$ and $v(S) = (\sigma^2 - r)S$. Recalling from problem 6.1 that the diffusion coefficient D for a random walk with step length a is proportional to a^2, we see that stock market prices are taken to follow a *geometric Brownian motion* with a step length proportional to the current position.

For the purpose of solving the equation, it would be useful to re-express it as a diffusion equation with a constant D, and this can be done. Indeed, the appearance of the derivative $S\partial/\partial S$ suggests the change of variable $S = e^y$, which ensures that the stock price S is positive for any real y. In terms of this variable, the equation becomes

$$\frac{\partial f}{\partial t'} = \tfrac{1}{2}\sigma^2 \frac{\partial^2 f}{\partial y^2} - (\tfrac{1}{2}\sigma^2 - r)\frac{\partial f}{\partial y} - rf.$$

The residual drift velocity $v = \sigma^2/2 - r$ can be removed by using a moving frame of reference, with the coordinate $x = y - vt'$, and finally the unwanted term $-rf$ can be eliminated by defining $f(S,t) = e^{-rt'} n(x,t')$. Then $n(x,t')$ obeys the diffusion equation

$$\frac{\partial n(x,t')}{\partial t'} = D\frac{\partial^2 n(x,t')}{\partial x^2}$$

with $D = \sigma^2/2$. A solution to this equation yields a corresponding solution to the Black–Scholes equation given by

$$f(S,t) = e^{-r(T-t)}\, n(x(S,t), T-t) \qquad \text{with } x(S,t) = \ln S + v(t-T).$$

(b) The diffusion equation has a solution

$$n_0(x,t') = \frac{e^{-x^2/4Dt'}}{\sqrt{4\pi Dt'}}$$

for which $n_0(x,0) = \delta(x)$. The solution to the Black–Scholes equation for which, say, $f(S,T) = \mathcal{C}(S)$ corresponds to a solution of the diffusion equation for which $n(x,0) = \mathcal{C}(e^x)$ and can be expressed as

$$f(S,t) = e^{-rt'}\int_{-\infty}^{\infty} dx'\, \mathcal{C}(e^{x'}) n_0(x-x',t')$$

where, as above, the parameters of the diffusion problem are related to the original ones by $D = \sigma^2/2$, $x = \ln S - v(T-t)$, $v = \sigma^2/2 - r$ and $t' = T - t$. For the European call option, we have $\mathcal{C}(S) = (S-X)\theta(S-X)$, and thus

$$f(S,t) = e^{-rt'}\int_{\ln X}^{\infty} dx'\, (e^{x'} - X)\frac{\exp[-(x'-x)^2/4Dt']}{\sqrt{4\pi Dt'}}.$$

The integrals needed can be expressed in terms of the probability integral

$$N(u) = \frac{1}{\sqrt{2\pi}}\int_{-\infty}^{u} ds\, e^{-s^2/2}$$

to yield the final result

$$f(S,t) = SN(u_+) - X\, e^{-r(T-t)}\, N(u_-)$$

where

$$u_{\pm} = \frac{\ln(S/X) + (r \pm \sigma^2/2)(T-t)}{\sigma\sqrt{T-t}}.$$

This solution for $f(S, t)$ is an exact formula for the price of a European call option.

• **Problem 6.5** Let us call $Z(t)$ the number of lights that are on at time t. The variable Z is a stochastic variable, which means that its value at time t cannot be predicted with certainty from a knowledge of its values at any set of earlier times: it can be established only as the outcome of an experiment performed at time t. Suppose that we know that there are n lights on at time t, so $Z(t) = n$. Then the probability that one of these lights will be turned off is n/N, while the probability that one of the remaining $N - n$ lights will be turned on is $(N - n)/N$. The conditional probability of finding n' lights on at time $t + 1$, given that there were n lights on at time t (or the probability that $Z(t + 1) = n'$, given that $Z(t) = n$) is therefore

$$\mathcal{P}(n, t \mid n', t + 1) = \frac{n}{N}\delta_{n',n-1} + \frac{N-n}{N}\delta_{n',n+1}$$

and this is called the *transition probability*. Clearly, this transition probability is unaffected by any information that we might have about earlier values of Z, that is the probability $\mathcal{P}(n_1, t_1; n_2, t_2; \ldots; n, t \mid n', t + 1)$ that $Z(t + 1) = n'$ given that $Z(t) = n$ *and* that $Z(t_1) = n_1$, etc, for some set of times $\{t_1, t_2, \ldots\}$ earlier than t, is the same[1] as $\mathcal{P}(n, t \mid n', t+1)$. This, by definition, is a Markov process. In fact, it is a Markov *chain*, because the evolution is in discrete time steps.

The net probability $P_N(n', t+1)$ of finding n' of the N lights on at time $t+1$ is obviously related to $P_N(n, t)$ by

$$P_N(n', t + 1) = \sum_{n=0}^{N} P_N(n, t)\mathcal{P}(n, t \mid n', t + 1).$$

Expressing things in a slightly different way, we can define a probability distribution vector $\boldsymbol{p}(t) = (p_0(t), p_1(t), \ldots, p_N(t))$, whose elements $p_n(t) = P_N(n, t)$ give, at each time t, the probabilities for all the possible values of Z. Then the time evolution of this distribution is

$$\boldsymbol{p}(t + 1) = \boldsymbol{p}(t) \cdot \mathbf{T} \quad \text{or} \quad p_{n'}(t + 1) = \sum_{n=0}^{N} p_n(t)T_{nn'}$$

[1] There is a possible source of confusion here. Since only one light changes at each time step, the possible values of $Z(t + 1)$ must lie between $n_1 - (t+1-t_1)$ and $n_1 + (t+1-t_1)$ if $Z(t_1) = n_1$, so the probability distribution for $Z(t+1)$ is not completely independent of n_1 and t_1. What this means for the transition probability is that the given conditions $(n_1, t_1; n_2, t_2; \ldots; n, t)$ must be consistent. If we specify impossible conditions, for example $n > n_1 + (t - t_1)$, then the conditional probability will be *meaningless*. However, if we specify a meaningful set of conditions, then only the latest one matters.

where the transition matrix $\mathbf{T}$ is the matrix of all the transition probabilities:

$$T_{nn'} = \mathcal{P}(n,t \mid n',t+1) = \frac{n}{N}\delta_{n',n-1} + \frac{N-n}{N}\delta_{n',n+1}.$$

A stationary state is one for which $\boldsymbol{p}(t+1) = \boldsymbol{p}(t)$ (so $\boldsymbol{p}$ is actually independent of t). For such a state, therefore, we have

$$p_{n'} = \sum_{n=0}^{N} p_n T_{nn'} = p_{n'+1}\frac{n'+1}{N} + p_{n'-1}\frac{N-n'+1}{N}.$$

This recursion relation holds for all the values $n' = 0, 1, \ldots, N$, so long as we adopt the convention that $p_{-1} = p_{N+1} = 0$. Using the first few values of n', we deduce that $p_1 = Np_0$, $p_2 = N(N-1)p_0/2$ and so on. We can then verify by direct substitution that

$$p_{n'} = \frac{N!}{n'!(N-n')!}p_0.$$

The one unknown parameter p_0 is fixed by the requirement that $\sum_{n=0}^{N} p_n = 1$, so this stationary distribution is unique. In fact, comparing our result with the binomial expansion

$$\sum_{n=0}^{N} \frac{N!}{n!(N-n)!}q^n(1-q)^{N-n} = [q+(1-q)]^N = 1$$

valid for any q, we see that p_n is the nth term of this series, provided that we choose $q = \frac{1}{2}$ and $p_0 = 2^{-N}$. In this stationary state, therefore, the most probable number of lights shining at any moment is $N/2$.

• Problem 6.6

(a) The system obviously has four possible states, $(S_1, S_2) = (1,1)$, $(1,-1)$, $(-1,1)$ and $(-1,-1)$, and the energy has three possible values, $E_{1,1} = -2\varepsilon$, $E_{1,-1} = E_{-1,1} = 0$ and $E_{-1,-1} = 2\varepsilon$. The transition matrix is

$$\mathbf{T} = \begin{pmatrix} (1-a)^2 & 2a(1-a) & a^2 \\ b(1-a) & (1-a)(1-b)+ab & a(1-b) \\ b^2 & 2b(1-b) & (1-b)^2 \end{pmatrix}$$

where the rows and columns correspond to energy values in the order $(-2\varepsilon, 0, 2\varepsilon)$ and can be obtained as follows. Suppose first that the energy at time t is $E = -2\varepsilon$, so that the state must be $(1,1)$. The probability of remaining in the same state is $(1-a)^2$; there are equal probabilities $a(1-a)$ of transitions to the states $(1,-1)$ and $(-1,1)$, giving a total probability $2a(1-a)$ of a transition to a state with $E = 0$, and the probability of a transition to the state $(-1,-1)$ with energy $E = 2\varepsilon$ is a^2. These probabilities give the top row of $\mathbf{T}$.

To find the second row, suppose that the energy at time t is $E(t) = 0$, then the state is either $(1, -1)$ or $(-1, 1)$. Suppose that it is $(1, -1)$. Then the probability of a transition to the state $(1, 1)$ with energy $E = -2\varepsilon$ is $b(1-a)$. For the energy to remain equal to zero, the state must either remain unchanged or change to $(-1, 1)$ and the total probability for these two possibilities is $(1-a)(1-b) + ab$. The probability of a transition to the state $(-1, -1)$ with $E = 2\varepsilon$ is $a(1-b)$. If the state at time t is $(-1, 1)$, then we obtain exactly the same set of probabilities. If this were not so (and it would not be if, for example, the transition probabilities for S_1 and S_2 were different), then it would not be possible to define a transition matrix for E.

The third row of $\mathbf{T}$ can be found in the same way as the top row.

(b) It is a matter of straightforward, if slightly tedious, matrix algebra to verify that the quoted eigenvalues and eigenvectors satisfy $\boldsymbol{p}_i \cdot \mathbf{T} = \lambda_i \boldsymbol{p}_i$. Note that $\boldsymbol{p}_1$ is normalized so that the sum of its elements is 1, but that the elements of $\boldsymbol{p}_2$ and $\boldsymbol{p}_3$ sum to zero.

(c) A stationary probability distribution $\boldsymbol{\pi}$ must satisfy $\boldsymbol{\pi} \cdot \mathbf{T} = \boldsymbol{\pi}$, and its elements must sum to 1. Clearly, there is just the one possibility $\boldsymbol{\pi} = \boldsymbol{p}_1$. Now, the three eigenvectors are linearly independent, so the probability distribution at $t = 0$ can be expressed as $\boldsymbol{p}(0) = \boldsymbol{p}_1 + \alpha \boldsymbol{p}_2 + \beta \boldsymbol{p}_3$, where α and β are constants. From the properties noted in (b), the elements of $\boldsymbol{p}(0)$ automatically sum to 1, but α and β must be such that these elements are all non-negative. For simplicity, let us use units of time such that the interval between transitions is $\Delta t = 1$. Then, at time t, the probability distribution is

$$\boldsymbol{p}(t) = \boldsymbol{p}(0)\mathbf{T}^t = \boldsymbol{p}_1 + \alpha \lambda_2^t \boldsymbol{p}_2 + \beta \lambda_3^t \boldsymbol{p}_3$$

since $\lambda_1^t = 1$. Because $a < 1$ and $b < 1$, we have $\lambda_3 = \lambda_2^2 < |\lambda_2| < 1$, so $\lim_{t\to\infty} \lambda_2^t = \lim_{t\to\infty} \lambda_3^t = 0$ and $\lim_{t\to\infty} \boldsymbol{p}(t) = \boldsymbol{p}_1 = \boldsymbol{\pi}$. Thus, the probability distribution approaches the stationary state $\boldsymbol{\pi}$ from any initial state.

In the case $a = 0$, only transitions from $S = -1$ to $S = 1$ are allowed, so we might expect both spins eventually to approach the value of one and stay there. This is indeed so, because the distribution $\boldsymbol{\pi}$ indicates that $E = -2\varepsilon$ with a probability of one. If $a = b$, then transitions in either direction are equally likely, so we might expect all four states to become equally populated, and this is again true.

(d) In a state of thermal equilibrium, the probability of finding the value E for the energy is $p(E) = g(E)\,\mathrm{e}^{-E/kT}/Z$, where $g(E)$ is the degeneracy of the energy level and Z is the partition function. In the present case, we have $g(-2\varepsilon) = g(2\varepsilon) = 1$ and $g(0) = 2$, so the ratio of probabilities for the three energy values must be

$$\frac{p(2\varepsilon)}{p(0)} = \frac{\mathrm{e}^{-2\varepsilon/kT}}{2} \quad \text{and} \quad \frac{p(0)}{p(-2\varepsilon)} = 2\,\mathrm{e}^{-2\varepsilon/kT}.$$

We can match one of these ratios to the probabilities in π by choosing a suitable value for the temperature T but, if both are to be correct, then the three elements of π must obey the temperature-independent condition

$$\frac{p(0)^2}{p(2\varepsilon)p(-2\varepsilon)} = 4.$$

In terms of the known elements of π, we have $p(-2\varepsilon) = b^2/(a+b)^2$, $p(0) = 2ab/(a+b)^2$ and $p(2\varepsilon) = a^2/(a+b)^2$, so this condition is indeed satisfied and we can identify the temperature as $T = 2\varepsilon/[k\ln(b/a)]$.

(e) From the probability vector $\boldsymbol{p}(t)$ found in (c), we can calculate the mean energy at time t:

$$\langle E(t)\rangle = (-2\varepsilon)p(-2\varepsilon,t) + 2\varepsilon p(2\varepsilon,t) = -2\varepsilon\left(\frac{b-a}{a+b} + \alpha\lambda_2^t\right).$$

It happens that the contribution from $\boldsymbol{p}_3$ vanishes identically. Even if it did not, it would become negligible after a long enough time, on account of the factor λ_3^t, λ_3 being the smallest eigenvalue. Suppose that $a+b < 1$, so that $\lambda_2 = 1-a-b$ is positive. Then we can write

$$\langle E(t)\rangle = -2\varepsilon\left[\tanh\left(\frac{\varepsilon}{kT}\right) + \alpha\,\mathrm{e}^{-t/\tau}\right]$$

where $\tau = -1/\ln(1-a-b)$ is the characteristic relaxation time (or $1/\tau$ is the relaxation rate). We see that $\langle E(t)\rangle$ approaches its equilibrium value $-2\tanh(\varepsilon/kT)$ in a smooth exponential manner. If $1 < a+b < 2$, then λ_2 is negative. In that case, the non-equilibrium term still has an amplitude which decays exponentially, with $\tau = -1/\ln|1-a-b|$, but it oscillates, changing sign at each time step. In the special case when $a+b = 1$, we have $\tau = 0$. In fact, equilibrium is then reached after only one time step, and readers may like to examine in detail the mechanism by which this happens.

• **Problem 6.7** In the time interval between t and $t+\delta t$, there is a probability $p_{n-1}(t)\alpha\,\delta t$ that the particle hops from site $n-1$ to site n and a probability $p_{n+1}(t)\beta\,\delta t$ that it hops from site $n+1$ to site n. These are gain terms, which increase the probability p_n. There is also a loss term $p_n(t)(\alpha+\beta)\,\delta t$ which is the probability that the particle hops from site n to one of the neighbouring sites. Obviously, the net rate of change in p_n is given by the master equation

$$\dot{p}_n(t) = \alpha p_{n-1}(t) + \beta p_{n+1}(t) - (\alpha+\beta)p_n(t).$$

A convenient method of solving this equation is to introduce the generating function

$$F(z,t) = \sum_{n=-\infty}^{\infty} p_n(t)z^n.$$

This function must always obey $F(1,t) = 1$, since this gives the total probability of finding the particle somewhere. Also, if the particle is at $n = 0$ when $t = 0$, then $p_n(0) = \delta_{n,0}$ and so $F(z,0) = 1$. On multiplying the master equation by z^n and summing over all values of n, we obtain the equation

$$\frac{\partial F}{\partial t} = [\alpha z + \beta z^{-1} - (\alpha + \beta)]F$$

which is easily solved to give

$$F(z,t) = \mathrm{e}^{-(\alpha+\beta)t}\, \mathrm{e}^{\alpha z t}\, \mathrm{e}^{\beta z^{-1} t}\, F_0(z).$$

The amplitude $F_0(z)$ is determined by the initial condition $F(z,0) = 1$ to be $F_0(z) = 1$. We can find the probabilities $p_n(t)$ by expanding the exponentials in powers of z and z^{-1}, which gives

$$F(z,t) = \mathrm{e}^{-(\alpha+\beta)t} \sum_{k=0}^{\infty} \sum_{l=0}^{\infty} \frac{\alpha^k \beta^l t^{k+l}}{k!\,l!} z^{k-l}.$$

The probability $p_n(t)$ is the coefficient of z^n. If $n \geq 0$, we can set $k = n + l$ to get

$$p_n(t) = \mathrm{e}^{-(\alpha+\beta)t}\, (\alpha t)^n \sum_{l=0}^{\infty} \frac{(\alpha\beta t^2)^l}{l!(l+n)!}$$

while, if $n \leq 0$, we set $l = k - n$ to get

$$p_n(t) = \mathrm{e}^{-(\alpha+\beta)t}\, (\beta t)^{-n} \sum_{k=0}^{\infty} \frac{(\alpha\beta t^2)^k}{k!(k-n)!}.$$

The simplest way to find $\langle n \rangle_t$ and $\langle n^2 \rangle_t$ is to use the generating function directly:

$$\langle n \rangle_t = \sum_{n=-\infty}^{\infty} n p_n(t) = \left. \frac{\partial F}{\partial z} \right|_{z=1} = (\alpha - \beta)t$$

$$\langle n^2 \rangle_t = \sum_{n=-\infty}^{\infty} n^2 p_n(t) = \left. \frac{\partial}{\partial z} \left(z \frac{\partial F}{\partial z} \right) \right|_{z=1} = (\alpha + \beta)t + (\alpha - \beta)^2 t^2.$$

Thus, the mean position of the particle is $(\alpha - \beta)t$ and its root mean square deviation from this position is $\Delta n = \sqrt{(\alpha + \beta)t}$.

This problem is clearly related to the random walk studied in problem 6.1. Indeed, if we identify the probability density $n(x,t) = a^{-1} p_n(t)$, where a is the distance between neighbouring sites of the lattice and $x = na$, then we can obtain a continuum approximation by writing $p_{n\pm1} = an(x \pm a, t) \simeq a[n(x,t) \pm an'(x,t) + \frac{1}{2}a^2 n''(x,t)]$, and the master equation reproduces the diffusion equation

$$\dot{n}(x,t) + vn'(x,t) - Dn''(x,t) = 0$$

with the drift velocity $v = (\alpha-\beta)a$ and the diffusion coefficient $D = \frac{1}{2}(\alpha+\beta)a^2$. However, it is not possible to relate the probabilities p and q in problem 6.1 unambiguously to the α and β used here. The reason is that, under the conditions of problem 6.1, the 'particle' was forced to move after each time interval τ, whereas here there is a non-zero probability $\exp[-(\alpha+\beta)t]$ that it has not moved, after any finite time t.

• Problem 6.8

(a) In the time interval between t and $t+\delta t$, the probability p_n increases by the amount $(n+1)\,p_{n+1}(t)\alpha\,\delta t$ (which is the joint probability that there were $(n+1)$ active nuclei at time t and that one of these decayed) and decreases by $n\,p_n(t)\alpha\,\delta t$ (which is the probability that there were n active nuclei, one of which decayed). The master equation which gives the rate of change of $p_n(t)$ is therefore

$$\dot{p}_n(t) = \alpha(n+1)p_{n+1}(t) - \alpha n p_n(t).$$

(b) To obtain an equation for the generating functional, we multiply the master equation by z^n and sum over all values of n:

$$\begin{aligned}\frac{\partial F(z,t)}{\partial t} &= \alpha\sum_{n=0}^{\infty}(n+1)z^n\,p_{n+1}(t) - \alpha\,z\sum_{n=0}^{\infty}nz^{n-1}\,p_n(t)\\ &= \alpha(1-z)\frac{\partial F(z,t)}{\partial z}.\end{aligned}$$

In terms of the variable $y = \ln(z-1)$, this equation is

$$\frac{\partial F}{\partial t} = -\alpha\frac{\partial F}{\partial y}$$

and this is just the statement that F depends on the single variable $y-\alpha t$. Equivalently, we can write

$$F(z,t) = \phi(u) \qquad u = \mathrm{e}^{y-\alpha t} = (z-1)\,\mathrm{e}^{-\alpha t}$$

where $\phi(\cdot)$ is a function determined by the initial probability distribution.

(c) The mean number of undecayed nuclei is

$$\bar{n}(t) = \sum_{n=0}^{\infty}np_n(t) = \left.\frac{\partial F}{\partial z}\right|_{z=1} = \mathrm{e}^{-\alpha t}\left.\frac{\mathrm{d}\phi(u)}{\mathrm{d}u}\right|_{u=0}$$

and this is obviously equal to $\mathrm{e}^{-\alpha t}\,\bar{n}(0)$, where $\bar{n}(0) = \phi'(0)$ is determined by the initial distribution.

(d) If the sample initially has exactly n_0 active nuclei, then we have $p_n(0) = \delta_{n,n_0}$ and

$$\phi(z-1) = F(z,0) = \sum_{n=0}^{\infty}z^n\delta_{n,n_0} = z^{n_0}$$

from which we discover that $\phi(u) = (1+u)^{n_0}$. The result in (c) then becomes $\bar{n}(t) = n_0\,\mathrm{e}^{-\alpha t}$. To find the variance $(\Delta n_t)^2 = \langle n^2\rangle_t - \bar{n}^2(t)$, we note that

$$\mathrm{e}^{-2\alpha t}\left.\frac{\mathrm{d}^2\phi(u)}{\mathrm{d}u^2}\right|_{u=0} = \left.\frac{\partial^2 F(z,t)}{\partial z^2}\right|_{z=1} = \sum_{n=0}^{\infty} n(n-1)p_n(t) = \langle n^2\rangle_t - \bar{n}(t)$$

and so

$$(\Delta n_t)^2 = \mathrm{e}^{-2\alpha t}\left.\frac{\mathrm{d}^2\phi(u)}{\mathrm{d}u^2}\right|_{u=0} + \bar{n}(t) - \bar{n}^2(t) = n_0\,\mathrm{e}^{-\alpha t}(1-\mathrm{e}^{-\alpha t}).$$

The probabilities themselves are given by

$$p_n(t) = \left.\frac{1}{n!}\frac{\partial^n F}{\partial z^n}\right|_{z=0} = \frac{1}{n!}(\mathrm{e}^{-\alpha t})^n \left.\frac{\mathrm{d}^n\phi}{\mathrm{d}u^n}\right|_{u=-\mathrm{e}^{-\alpha t}}$$
$$= \frac{n_0!}{n!(n_0-n)!}(\mathrm{e}^{-\alpha t})^n(1-\mathrm{e}^{-\alpha t})^{n_0-n}.$$

This is clearly the right answer, namely the joint probability that $n_0 - n$ nuclei (which can be chosen in $\binom{n_0}{n_0-n}$ ways) have decayed while the remaining n nuclei have not decayed.

(e) If the initial distribution is a Poisson distribution with mean n_0,

$$p_n(0) = \frac{(n_0)^n\,\mathrm{e}^{-n_0}}{n!}$$

then we have

$$F(z,0) = \sum_{n=0}^{\infty} z^n p_n(0) = \sum_{n=0}^{\infty}\frac{(zn_0)^n}{n!}\,\mathrm{e}^{-n_0} = \mathrm{e}^{(z-1)n_0}.$$

To get $F(z,t)$, we have to replace $z-1$ by $(z-1)\,\mathrm{e}^{-\alpha t}$ and this gives us just the initial probability distribution with n_0 replaced by $n_0\,\mathrm{e}^{-\alpha t}$, that is a Poisson distribution with mean $\bar{n}(t) = n_0\,\mathrm{e}^{-\alpha t}$. If the irradiation process is such that n_0 is large, but still only a small fraction of the total number of nuclei in each sample, then the initial numbers of active nuclei in the different samples should indeed be a Poisson distribution.

• **Problem 6.9** Let $n_+(t)$ be the number of atoms with energy $+\varepsilon$, and $n_-(t)$ the number with energy $-\varepsilon$. The rates of change in these numbers are clearly given by the master equations

$$\frac{\mathrm{d}n_+}{\mathrm{d}t} = -n_+w_- + n_-w_+ \qquad \frac{\mathrm{d}n_-}{\mathrm{d}t} = n_+w_- - n_-w_+.$$

At equilibrium, the right-hand sides of these equations vanish, so we can deduce the ratio of the transition probabilities:

$$\frac{w_+}{w_-} = \left.\frac{n_+}{n_-}\right|_{\text{eq}} = \mathrm{e}^{-2\beta\varepsilon}$$

with $\beta = 1/kT$ as usual. Since the total number of atoms $n_+ + n_- = N$ is fixed, it is convenient to describe the non-equilibrium state in terms of the difference $n(t) = n_-(t) - n_+(t)$, for which we obtain the master equation

$$\frac{\mathrm{d}n(t)}{\mathrm{d}t} = -\frac{1}{\tau}[n(t) - n_{\text{eq}}]$$

where $\tau = 1/(w_+ + w_-)$ and

$$n_{\text{eq}} = N\frac{w_- - w_+}{w_- + w_+} = N\tanh(\beta\varepsilon).$$

The solution of this equation is

$$n(t) = n_{\text{eq}} + \mathrm{e}^{-t/\tau}[n(0) - n_{\text{eq}}].$$

So τ is indeed the relaxation time that characterizes the exponential approach of the system to equilibrium from any initial state.

• **Problem 6.10** We define a single-particle distribution function $f(\boldsymbol{r}, \boldsymbol{p}, t)$ such that $f(\boldsymbol{r}, \boldsymbol{p}, t)\,\mathrm{d}^3r\,\mathrm{d}^3p$ is the number of particles which, at time t, are located in the volume element d^3r containing the point $\boldsymbol{r}$ and have momentum in the range d^3p around the value $\boldsymbol{p}$. Since we are dealing with an ideal gas, there is no scattering; so the change in f with time arises just from the free-streaming motion of the particles. It is very easy to see how the distribution changes, because the momentum of each particle is constant. Consequently, the particles which at time t are close to the point $\boldsymbol{r}$ and have momentum $\boldsymbol{p}$ are those which, at time $t - \Delta t$, had the same momentum but were near the point $\boldsymbol{r} - (\boldsymbol{p}/m)\,\Delta t$, that is

$$f(\boldsymbol{r}, \boldsymbol{p}, t) = f\left(\boldsymbol{r} - \frac{\boldsymbol{p}}{m}\,\Delta t, \boldsymbol{p}, t - \Delta t\right).$$

So long as no forces act, this relation is valid for any time interval Δt. In particular, we can choose $\Delta t = t$, so as to get the distribution at time t in terms of that at time $t = 0$:

$$f(\boldsymbol{r}, \boldsymbol{p}, t) = f\left(\boldsymbol{r} - \frac{\boldsymbol{p}}{m}t, \boldsymbol{p}, 0\right).$$

Now, at $t = 0$, the particles had a uniform spatial density n_0 for $x < 0$ and there were no particles in the region $x > 0$ (we take the wall to be in the plane

$x = 0$). Since the gas was in thermal equilibrium, the momenta were distributed according to the Maxwell–Boltzmann function, so we have

$$f(\boldsymbol{r}, \boldsymbol{p}, 0) = n_0(2\pi mkT)^{-3/2}\,\mathrm{e}^{-|\boldsymbol{p}|^2/2mkT}\,\theta(-x)$$

and hence

$$f(\boldsymbol{r}, \boldsymbol{p}, t) = n_0(2\pi mkT)^{-3/2}\,\mathrm{e}^{-|\boldsymbol{p}|^2/2mkT}\,\theta\left(\frac{p_x t}{m} - x\right).$$

The number density is

$$\begin{aligned} n(\boldsymbol{r}, t) = \int \mathrm{d}^3p\, f(\boldsymbol{r}, \boldsymbol{p}, t) &= n_0(2\pi mkT)^{-1/2}\int_{mx/t}^{\infty} \mathrm{e}^{-p_x/2mkT}\,\mathrm{d}p_x \\ &= \frac{n_0}{\sqrt{\pi}}\int_{x/ct}^{\infty} \mathrm{e}^{-u^2}\,\mathrm{d}u \end{aligned}$$

where $c = \sqrt{2kT/m}$ is a characteristic molecular speed. To obtain the last expression, we used the substitution $p_x = \sqrt{2mkT}u$. For the mean energy of particles near the point $\boldsymbol{r}$, we have

$$\bar{E}(\boldsymbol{r}, t) = \frac{\int \mathrm{d}^3p\,[(p_x^2 + p_y^2 + p_z^2)/2m]\, f(\boldsymbol{r}, \boldsymbol{p}, t)}{\int \mathrm{d}^3p\, f(\boldsymbol{r}, \boldsymbol{p}, t)}.$$

The term involving p_y^2 is

$$\frac{\int \mathrm{d}^3p\,(p_y^2/2m)\, f(\boldsymbol{r}, \boldsymbol{p}, t)}{\int \mathrm{d}^3p\, f(\boldsymbol{r}, \boldsymbol{p}, t)} = \frac{1}{2m}\frac{\int_{-\infty}^{\infty} p_y^2\,\mathrm{e}^{-p_y^2/2mkT}\,\mathrm{d}p_y}{\int_{-\infty}^{\infty} \mathrm{e}^{-p_y^2/2mkT}\,\mathrm{d}p_y} = \frac{kT}{2}$$

since the p_x and p_z integrals cancel, and the term involving p_z^2 has the same value. In the term involving p_x^2, we make the same substitution as before to obtain

$$\bar{E}(\boldsymbol{r}, t) = kT + \frac{kT\int_{x/ct}^{\infty} u^2\,\mathrm{e}^{-u^2}\,\mathrm{d}u}{\int_{x/ct}^{\infty} \mathrm{e}^{-u^2}\,\mathrm{d}u}.$$

When $|x| \ll ct$, the remaining integrals can be evaluated approximately in the following way. When x is positive, we write

$$\begin{aligned} \int_{x/ct}^{\infty} u^n\,\mathrm{e}^{-u^2}\,\mathrm{d}u &= \int_0^{\infty} u^n\,\mathrm{e}^{-u^2}\,\mathrm{d}u - \int_0^{x/ct} u^n\,\mathrm{e}^{-u^2}\,\mathrm{d}u \\ &= \frac{\sqrt{\pi}}{c_n} - \int_0^{x/ct} u^n\,\mathrm{e}^{-u^2}\,\mathrm{d}u \end{aligned}$$

with $c_0 = 2$ and $c_2 = 4$. In the last integral, we have $u \le x/ct \ll 1$, so it is legitimate to expand the exponential and integrate term by term:

$$\int_{x/ct}^{\infty} u^n\,\mathrm{e}^{-u^2}\,\mathrm{d}u = \frac{\sqrt{\pi}}{c_n} - \left(\frac{x}{ct}\right)^{n+1}\left[\frac{1}{n+1} - \frac{1}{n+3}\left(\frac{x}{ct}\right)^2 + \cdots\right].$$

When x is negative, the result is exactly the same. Keeping just the leading power of x/ct, we find that

$$n(\boldsymbol{r},t) = n_0\left\{\frac{1}{2} - \frac{1}{\sqrt{\pi}}\frac{x}{ct} + \mathrm{O}\left[\left(\frac{x}{ct}\right)^2\right]\right\}$$

$$\bar{E}(\boldsymbol{r},t) = kT\left\{\frac{3}{2} + \frac{1}{\sqrt{\pi}}\frac{x}{ct} + \mathrm{O}\left[\left(\frac{x}{ct}\right)^2\right]\right\}.$$

In a truly infinite system, these approximations are valid for any positive or negative x after a sufficiently long time. As $t \to \infty$, one half of the particles (those for which p_x is negative) have disappeared towards $x = -\infty$ and the steady state at any finite x is a spatially uniform distribution of the remaining particles, all with $p_x > 0$. Before this steady state is reached, the region $x > 0$, which was originally vacuum, has (not surprisingly) a density a little less than $n_0/2$, while the region $x < 0$ is a little more dense. On the other hand, the mean energy is greater than the steady-state value of $3kT/2$ for $x > 0$, but smaller for $x < 0$, reflecting the fact that fast-moving particles are redistributed more quickly than slower particles. In a finite-sized container, of course, the situation becomes more complicated once a significant number of particles have been reflected from the walls.

• Problem 6.11

(a) For a beam of particles in which there are n particles per unit volume, all having the same velocity $\boldsymbol{v}$, the particle current (number of particles crossing unit normal area per unit time) is $\boldsymbol{j} = n\boldsymbol{v}$. More generally, the contribution to this current at the point $\boldsymbol{r}$ from particles with momentum in the range d^3p near $\boldsymbol{p}$ is easily seen to be $\boldsymbol{j}(\boldsymbol{r},\boldsymbol{p},t)\,\mathrm{d}^3p$, where $\boldsymbol{j}(\boldsymbol{r},\boldsymbol{p},t) = f(\boldsymbol{r},\boldsymbol{p},t)\boldsymbol{p}/m$ and f is the single-particle distribution function discussed in problem 6.10. In the case at hand, this distribution function is

$$f(\boldsymbol{r},\boldsymbol{p},t) = n(t)[2\pi mkT(t)]^{-3/2}\,\mathrm{e}^{-|\boldsymbol{p}|^2/2mkT(t)}$$

for particles inside the box. Suppose that the outward normal to the hole is the positive x direction. Then the total number of particles passing out through the hole per unit time is given (bearing in mind that only those with $p_x > 0$ get out!) by

$$\dot{N}_{\mathrm{out}}(t) = A\int_{-\infty}^{\infty}\mathrm{d}p_y\int_{-\infty}^{\infty}\mathrm{d}p_z\int_0^{\infty}\mathrm{d}p_x\, j_x(\boldsymbol{r},\boldsymbol{p},t) = An(t)\sqrt{\frac{kT(t)}{2\pi m}}$$

where $j_x = f\,p_x/m$. In principle, of course, we put in for $\boldsymbol{r}$ the position of the hole but, for a uniform gas, f is actually independent of $\boldsymbol{r}$. This result is often expressed as $An\bar{c}/4$, where $\bar{c} = \sqrt{8kT/\pi m}$ is the mean molecular speed. Under the assumption that thermal equilibrium is maintained in the box, the number density is the same throughout the box, and the total number of particles is $Vn(t)$. Since particles are lost at the rate that we have just calculated, we find that

$$\dot{n}(t) = -Dn(t)T^{1/2}(t)$$

where $D = (A/V)\sqrt{k/2\pi m}$.

Each particle passing through the hole carries with it an energy $|\boldsymbol{p}|^2/2m$, so the rate at which energy is carried out through the hole is

$$\begin{aligned}\dot{E}_{\text{out}}(t) &= A\int_{-\infty}^{\infty} \mathrm{d}p_y \int_{-\infty}^{\infty} \mathrm{d}p_z \int_{0}^{\infty} \mathrm{d}p_x\, j_x(\boldsymbol{r}, \boldsymbol{p}, t)\left(\frac{p_x^2 + p_y^2 + p_z^2}{2m}\right) \\ &= 2An(t)kT^{3/2}(t)\sqrt{\frac{k}{2\pi m}}.\end{aligned}$$

Since the average energy per particle is $3kT(t)/2$, the energy of the remaining gas is $E(t) = (3kV/2)n(t)T(t)$ and, if the box is thermally isolated, we have $\dot{E}(t) = -\dot{E}_{\text{out}}(t)$ or

$$\dot{n}(t)T(t) + n(t)\dot{T}(t) = -\tfrac{4}{3}Dn(t)T^{3/2}(t).$$

Combining this with our previous result for $\dot{n}(t)$, we obtain the pair of simultaneous equations

$$\dot{T} = -\left(\frac{D}{3}\right)T^{3/2} \qquad \dot{n} = -DnT^{1/2}$$

to be solved with the initial conditions $T(0) = T_0$ and $n(0) = n_0$. The solutions are

$$T(t) = \frac{T_0}{(1 + t/\tau)^2} \qquad n(t) = \frac{n_0}{(1 + t/\tau)^6}$$

where the characteristic time is $\tau = 6/D\sqrt{T_0} = (6V/A)\sqrt{2\pi m/kT_0}$.

(b) When the temperature is held fixed, the equation for $n(t)$ becomes $\dot{n} = -D\sqrt{T_0}n$ and its solution is

$$n(t) = n_0\,\mathrm{e}^{-6t/\tau}.$$

The rate at which energy is lost through the hole is that calculated above, namely

$$\dot{E}_{\text{out}} = 2An(t)kT_0^{3/2}\sqrt{\frac{k}{2\pi m}} = \frac{8E_0}{\tau}\,\mathrm{e}^{-6t/\tau}$$

where $E_0 = 3n_0VkT_0/2$ is the initial total energy of the gas in the box. The rate of change in the energy of the remaining gas is

$$\dot{E} = \tfrac{3}{2}\dot{n}VkT_0 = -\frac{6E_0}{\tau}\,\mathrm{e}^{-6t/\tau}$$

and, if heat is supplied at a rate $\dot{Q}(t)$, then we have $\dot{E} = \dot{Q} - \dot{E}_{\text{out}}$, giving

$$\dot{Q}(t) = \frac{2E_0}{\tau}\,\mathrm{e}^{-6t/\tau}.$$

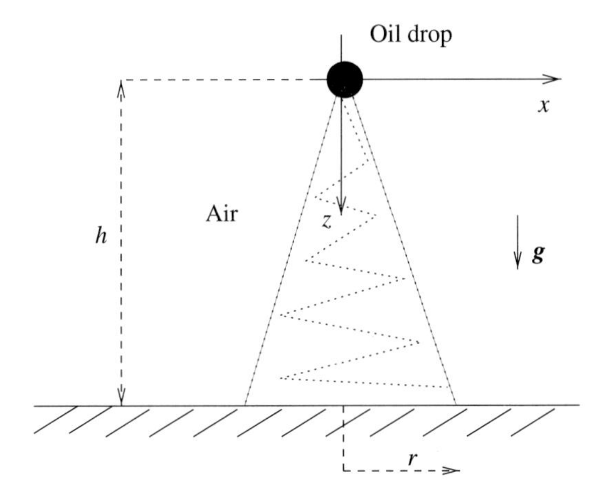

Figure 6.2. Brownian motion of a drop of oil in still air falling under the action of gravity.

• **Problem 6.12** We use the coordinates $\boldsymbol{x} = (x, y, z)$ shown in figure 6.2, so that z is measured downwards from the point at which the drops are released and x and y are horizontal.

Following the usual treatment of Brownian motion, we take the position of an oil drop to be governed by the Langevin equation

$$m\ddot{\boldsymbol{x}} + \alpha\dot{\boldsymbol{x}} = \boldsymbol{\eta}(t) + m\boldsymbol{g}$$

where α is the friction coefficient, $\boldsymbol{g} = (0, 0, g)$ is the acceleration due to gravity and $\boldsymbol{\eta}(t) = (\eta_x(t), \eta_y(t), \eta_z(t))$ is a random force due to collisions with air molecules. The mean value of this force is $\langle\boldsymbol{\eta}(t)\rangle = 0$ and its autocorrelation function is of the form

$$\langle\eta_i(t)\eta_j(t')\rangle = A\delta_{ij}\delta(t - t').$$

We shall estimate the radius r as $r = \sqrt{\langle x^2\rangle + \langle y^2\rangle} = \sqrt{2\langle x^2\rangle}$, where $\langle x^2\rangle$ is the mean square distance that an oil drop has travelled in the x direction before reaching the floor. Since the Langevin equation for this problem is linear, it is straightforward to find a formal solution. With $\boldsymbol{x}(0) = \dot{\boldsymbol{x}}(0) = 0$, the velocity is given by

$$\dot{\boldsymbol{x}}(t) = \frac{1}{m}\int_0^t \mathrm{d}t'\, \mathrm{e}^{(t'-t)/\tau}\, [\boldsymbol{\eta}(t') + m\boldsymbol{g}]$$

where the characteristic relaxation time is $\tau = m/\alpha$, and this can be integrated again to give

$$\boldsymbol{x}(t) = \frac{1}{m}\int_0^t \mathrm{d}t' \int_0^{t'} \mathrm{d}t''\, \mathrm{e}^{(t''-t')/\tau}\, [\boldsymbol{\eta}(t'') + m\boldsymbol{g}].$$

As a first application of these solutions, we deduce the value of the amplitude A of the Brownian force, by calculating

$$\begin{aligned}\langle \dot{x}^2(t)\rangle &= \frac{1}{m^2}\int_0^t \mathrm{d}t' \int_0^t \mathrm{d}t''\, \mathrm{e}^{(t'+t''-2t)/\tau}\, \langle \eta_x(t')\eta_x(t'')\rangle \\ &= \frac{A}{m^2}\int_0^t \mathrm{d}t' \int_0^t \mathrm{d}t''\, \mathrm{e}^{(t'+t''-2t)/\tau}\, \delta(t'-t'') \\ &= \frac{A\tau}{2m^2}(1-\mathrm{e}^{-2t/\tau}).\end{aligned}$$

After a long time, this approaches the constant value $A\tau/2m^2$. The motion of the oil drop should then be in thermal equilibrium with the air, so the equipartition theorem tells us that $m\langle \dot{x}^2\rangle = kT$ and we deduce that

$$A = \frac{2mkT}{\tau} = 2\alpha kT.$$

This is the simplest example of a *fluctuation–dissipation relation*. It relates the amplitude of the fluctuating force to the coefficient α which describes the dissipation of the oil drop's energy by friction with the air. Note that this relation is not an intrinsic property of the Langevin equation from which we started. Rather, it is a consistency condition that we impose on the phenomenological parameters α and A in order to recover the mean energy suggested by equilibrium statistical mechanics.

Next, we calculate the mean vertical distance through which an oil drop has fallen after a time t. It is

$$\langle z(t)\rangle = g\int_0^t \mathrm{d}t' \int_0^{t'} \mathrm{d}t''\, \mathrm{e}^{(t''-t')/\tau} = g\tau[t-\tau(1-\mathrm{e}^{-t/\tau})]$$

since $\langle \eta_z(t)\rangle = 0$. For short times $t \ll \tau$, we have $\langle z\rangle \simeq gt^2/2$, as for a freely falling particle while, after a long time $t \gg \tau$, the drop is falling with its terminal speed $g\tau$ and $\langle z\rangle \simeq g\tau t$.

The final step is to calculate $\langle x^2(t)\rangle$. This is given by

$$\begin{aligned}\langle x^2(t)\rangle &= \frac{1}{m^2}\int_0^t \mathrm{d}t' \int_0^{t'} \mathrm{d}t''\, \mathrm{e}^{(t''-t')/\tau} \int_0^t \mathrm{d}s' \int_0^{s'} \mathrm{d}s''\, \mathrm{e}^{(s''-s')/\tau}\, \langle \eta_x(t'')\eta_x(s'')\rangle \\ &= \frac{A}{m^2}\int_0^t \mathrm{d}t' \int_0^{t'} \mathrm{d}t'' \int_0^t \mathrm{d}s' \int_0^{s'} \mathrm{d}s''\, \mathrm{e}^{(t''+s''-t'-s')/\tau}\, \delta(t''-s'') \\ &= \frac{A}{m^2}\int_0^t \mathrm{d}t' \int_0^{t'} \mathrm{d}t'' \int_{t''}^t \mathrm{d}s'\, \mathrm{e}^{(2t''-t'-s')/\tau} \\ &= \frac{A\tau^2}{m^2}[t-\tau(1-\mathrm{e}^{-t/\tau}) - \tfrac{1}{2}\tau(1-\mathrm{e}^{-t/\tau})^2].\end{aligned}$$

Evidently, the motion of a drop for $t \gg \tau$ is diffusive, since $\langle x^2(t)\rangle \sim t$. The corresponding diffusion coefficient is $D = A\tau^2/2m^2$, and the fluctuation–dissipation relation can be expressed as $D = kT/\alpha$.

We now assume that, after an oil drop has fallen through a height h, the uncertainty in its position due to Brownian motion in the vertical direction is negligible, so we can take $\langle z(t)\rangle = h$. With the further assumption that $t \gg \tau$, this implies that the time taken is $t = h/g\tau$ and we obtain

$$\langle x^2\rangle = \frac{A\tau^2}{m^2}\frac{h}{g\tau} = \frac{2kT}{mg}h$$

and

$$r = \sqrt{2\langle x^2\rangle} = 2\sqrt{\frac{kTh}{mg}}.$$

Suppose that the oil drops are spheres of radius a and density ρ. We then obtain $r = \sqrt{3kTh/\pi a^3 \rho g}$. This radius increases as a decreases, so we shall improve our chances of observing the effect by using very small droplets. Taking $h = 1$ m, $T = 300$ K and $\rho = 850$ kg m^{-3}, we find that $r = a$ when $a \simeq 1.4 \times 10^{-5}$ m. Clearly, we need droplets which are considerably smaller than this. While sufficiently small drops might be produced by an atomizer, it would be very difficult to start these droplets falling from a well-defined point, so our problem is somewhat idealized.

• Problem 6.13

(a) The Langevin equation for an ion of charge q and mass m is

$$m\ddot{\boldsymbol{x}}(t) + \alpha\dot{\boldsymbol{x}}(t) = \boldsymbol{\eta}(t) + q\boldsymbol{E}(\boldsymbol{x}(t))$$

where α is a friction coefficient and $\boldsymbol{\eta}(t)$ is a random force. On the assumption that this ion collides principally with neutral molecules which on average are uniformly distributed, this random force is independent of the position of the ion that we are studying and of the other ions in which we are eventually going to be interested. We recall from problem 6.12 that η is taken to have a mean value of zero and an autocorrelation function

$$\langle \eta_i(t)\eta_j(t')\rangle = A\delta_{ij}\delta(t-t').$$

On time scales much greater than the relaxation time $\tau = m/\alpha$, the particle can be considered to reach its terminal velocity almost instantaneously and the acceleration term $m\ddot{\boldsymbol{x}}$ can be neglected in comparison with the viscous drag term $\alpha\dot{\boldsymbol{x}}$. On the assumption that the random part of the motion is in thermal equilibrium with the surrounding fluid that is responsible for $\boldsymbol{\eta}(t)$, the amplitude A is identified as $A = 2\alpha kT$ and the diffusion constant is $D = kT/\alpha = kT\tau/m$.

To obtain an equation for the number density, we first examine the probability density $P(\boldsymbol{x}, t)$ of finding a particular ion at the point $\boldsymbol{x}$ at time t and then reinterpret this as equivalent to the number density of a collection of ions moving independently, but in the same environment. Some care is needed to get the correct equation for P. The first step is to discretize the time evolution into steps of length ε, which will finally be taken to zero. We write $t = n\varepsilon$ and, since an integral $\int \mathrm{d}t$ is represented by the sum $\varepsilon \sum_n$, the delta function $\delta(t-t')$ becomes $\varepsilon^{-1}\delta_{n,n'}$. It is then convenient to define $\boldsymbol{\eta}(n\varepsilon) = \varepsilon^{-1/2}\boldsymbol{\xi}^{(n)}$, so that the correlation function becomes

$$\langle \xi_i^{(n)} \xi_j^{(n')} \rangle = A\delta_{ij}\delta_{n,n'}.$$

For a given value of $\boldsymbol{\xi}^{(n)}$, the Langevin equation now tells us that, if the ion is at the position $\boldsymbol{x}$ at time $t = n\varepsilon$, then its position at time $t' = (n+1)\varepsilon$ is

$$\boldsymbol{x}' = \boldsymbol{x} + \alpha^{-1}\varepsilon^{1/2}\boldsymbol{\xi}^{(n)} + \varepsilon\left(\frac{q}{\alpha}\right)\boldsymbol{E}(\boldsymbol{x}).$$

It follows that, if the ion is in the volume element d^3x containing the point $\boldsymbol{x}$ at time t, it will be found, at time t', in a volume element containing the point $\boldsymbol{x}'$, given by $\mathrm{d}^3x' = J\,\mathrm{d}^3x$, where the Jacobian is

$$J = \det\left|\frac{\partial x_i'}{\partial x_j}\right| = \det\left|\delta_{ij} + \frac{\varepsilon q}{\alpha}\frac{\partial E_i}{\partial x_j}\right| = 1 + \frac{\varepsilon q}{\alpha}\boldsymbol{\nabla}\cdot\boldsymbol{E} + \mathrm{O}(\varepsilon^2).$$

Introducing explicitly the probability density $\mathcal{P}(\boldsymbol{\xi})$ for the random force, we can calculate the probability of finding the ion in the volume element d^3x' at time $t' = t + \varepsilon$ as

$$\begin{aligned}
P(\boldsymbol{x}', t+\varepsilon)\,\mathrm{d}^3x' &= \int_{-\infty}^{\infty} \mathrm{d}^3\xi\, \mathcal{P}(\boldsymbol{\xi}^{(n)}) P(\boldsymbol{x}, t)\,\mathrm{d}^3x \\
&= \int_{-\infty}^{\infty} \mathrm{d}^3\xi\, \mathcal{P}(\boldsymbol{\xi}^{(n)}) P\left(\boldsymbol{x}' - \frac{\varepsilon^{1/2}}{\alpha}\boldsymbol{\xi}^{(n)} - \frac{\varepsilon q}{\alpha}\boldsymbol{E}, t\right)\mathrm{d}^3x'\,J^{-1} \\
&= \Bigg(P(\boldsymbol{x}', t) - \frac{\varepsilon^{1/2}}{\alpha}\langle\boldsymbol{\xi}^{(n)}\rangle\cdot\boldsymbol{\nabla}P(\boldsymbol{x}', t) \\
&\quad + \frac{\varepsilon}{2\alpha^2}\langle(\boldsymbol{\xi}^{(n)}\cdot\boldsymbol{\nabla})^2\rangle P(\boldsymbol{x}', t) - \frac{\varepsilon q}{\alpha}\boldsymbol{E}\cdot\boldsymbol{\nabla}P(\boldsymbol{x}', t) \\
&\quad - \frac{\varepsilon q}{\alpha}\boldsymbol{\nabla}\cdot\boldsymbol{E}P(\boldsymbol{x}', t) + \mathrm{O}(\varepsilon^{3/2})\Bigg)\mathrm{d}^3x' \\
&= \Bigg(P(\boldsymbol{x}', t) + \frac{\varepsilon A}{2\alpha^2}\nabla^2 P(\boldsymbol{x}', t) - \frac{\varepsilon q}{\alpha}\boldsymbol{E}\cdot\boldsymbol{\nabla}P(\boldsymbol{x}', t) \\
&\quad - \frac{\varepsilon q}{\alpha}\boldsymbol{\nabla}\cdot\boldsymbol{E}P(\boldsymbol{x}', t) + \mathrm{O}(\varepsilon^{3/2})\Bigg)\mathrm{d}^3x'.
\end{aligned}$$

The averages here mean $\langle\cdots\rangle = \int \mathcal{P}(\boldsymbol{\xi}^{(n)})\cdots\mathrm{d}^3\xi^{(n)}$. On taking the limit $\varepsilon \to 0$, we obtain

$$\frac{\partial}{\partial t}P(\boldsymbol{x}, t) = \boldsymbol{\nabla}\cdot\left(\frac{A}{2\alpha^2}\boldsymbol{\nabla}P(\boldsymbol{x}, t) - \frac{q}{\alpha}\boldsymbol{E}(\boldsymbol{x})P(\boldsymbol{x}, t)\right).$$

It is worth noting first that the displacement caused by the random force in the time interval ε is of order $\varepsilon^{1/2}$, corresponding to the fact that the average displacement in a random walk is proportional to $t^{1/2}$, and second that the Jacobian associated with the volume element is essential to ensure that $\partial P/\partial t$ is equal to a total divergence. This guarantees (given suitable boundary conditions) that the total probability $\int P(\boldsymbol{x}, t)\, \mathrm{d}^3x$ of finding the particle somewhere is independent of time and can be normalized to unity. The equation we have derived is an example of a *Fokker–Planck equation*, which is essentially equivalent to the original Langevin equation on large time scales. We see that the average motion that it describes is more or less the same as that arising from the random walks studied in earlier problems. The Langevin equation is perhaps more easily recognizable than a hopping probability as originating in the microscopic equation of motion for a particle, although the friction coefficient and the random force are introduced as phenomenological guesses. The analysis exemplified by this example of a single particle can be generalized by replacing $\boldsymbol{x}$ with the point in, say, the $3N$-dimensional configuration space that represents the microscopic state of a large system and $\boldsymbol{E}(\boldsymbol{x})$ with the gradient of the potential energy of this large system. One can also attempt to derive the Langevin equation by coarse-graining the true microscopic equations of motion of a large system, although the friction term and the correlations of the random force usually turn out to be more complicated than those assumed here.

Provided that ions move independently of each other, their number density at a given point is simply proportional to the probability of finding any one ion there. In terms of the diffusion constant identified earlier, we therefore find the pair of equations given in the question:

$$\frac{\partial}{\partial t}[n(\boldsymbol{x}, t)] + \boldsymbol{\nabla}\cdot\boldsymbol{j}(\boldsymbol{x}, t) = 0 \qquad \boldsymbol{j}(\boldsymbol{x}, t) = -D\boldsymbol{\nabla}n(\boldsymbol{x}, t) + \frac{q}{kT}Dn(\boldsymbol{x}, t)\boldsymbol{E}(\boldsymbol{x}).$$

The first of these is the equation of continuity that expresses conservation of the total number of particles, and $\boldsymbol{j}(\boldsymbol{x}, t)$ is their current density. Clearly, we have recovered a familiar equation, describing the diffusion of particles towards regions of lower concentration, together with a drift velocity due to an external force.

(b) It is simple to verify that the stationary distribution

$$n(\boldsymbol{x}, t) = n_0 \exp\left(-\frac{V(\boldsymbol{x})}{kT}\right)$$

is a steady-state solution, with $\boldsymbol{j}(\boldsymbol{x}, t) = 0$, where $V(\boldsymbol{x})$ is the potential energy of an ion in the applied electric field, so that $q\boldsymbol{E}(\boldsymbol{x}) = -\boldsymbol{\nabla}V(\boldsymbol{x})$. This, of course, reproduces the configurational part of the canonical probability density. We see, therefore, that the Langevin equation, together with the fluctuation–dissipation relation discussed in the previous problem, leads to a consistent description of fluctuations about an equilibrium state.

(c) For a fluid containing positive and negative ions, we have a current density for each species:

$$\boldsymbol{j}_+(\boldsymbol{x},t) = -D_+\nabla n_+(\boldsymbol{x},t) + \frac{q}{kT}D_+n_+(\boldsymbol{x},t)\boldsymbol{E}(\boldsymbol{x})$$
$$\boldsymbol{j}_-(\boldsymbol{x},t) = -D_-\nabla n_-(\boldsymbol{x},t) - \frac{q}{kT}D_-n_-(\boldsymbol{x},t)\boldsymbol{E}(\boldsymbol{x}).$$

In general, there will be different diffusion constants, because the two species of ions have different scattering cross sections. The electric current density is $\boldsymbol{j}_\mathrm{e}(\boldsymbol{x},t) = q[\boldsymbol{j}_+(\boldsymbol{x},t) - \boldsymbol{j}_-(\boldsymbol{x},t)]$. It consists of a diffusive part and a drift term representing the response to the applied electric field. Writing the latter as $\boldsymbol{j}_\mathrm{drift} = \sigma\boldsymbol{E}$, which is recognizable as Ohm's law, we identify the conductivity as

$$\sigma(\boldsymbol{x},t) = \frac{q^2}{kT}[D_+n_+(\boldsymbol{x},t) + D_-n_-(\boldsymbol{x},t)] = \frac{q^2\tau_+}{m_+}n_+(\boldsymbol{x},\boldsymbol{t}) + \frac{q^2\tau_-}{m_-}n_-(\boldsymbol{x},\boldsymbol{t})$$

where $\tau_\pm$ and $m_\pm$ are the relaxation times and masses of the two species of ions. Each contribution clearly has the same form as the Drude formula. A simple derivation of this formula supposes that an electron in a metal undergoes an acceleration $-eE/m$ during a time τ_c between collisions with positive ions and that its velocity immediately after a collision is random. This then results in an average drift velocity $v_\mathrm{d} = -eE\tau_\mathrm{c}/m$ and an electric current density $j_\mathrm{e} = -ev_\mathrm{d}n = (e^2\tau_\mathrm{c}n/m)E$. In the Langevin formulation, the relaxation time τ arises from the frictional drag rather than from the random force but, since these are related through the fluctuation–dissipation relation, a loose identification of τ and τ_c is intuitively not unreasonable.

• **Problem 6.14** The temperature at the end of the rod varies with time roughly as a square wave, and a temperature wave passes down the rod. At the point where the temperature is measured, it must be that all the higher Fourier components of the original waveform have been attenuated to a much greater extent than the fundamental component. To investigate this, consider that the variation in temperature along the rod is controlled by the diffusion equation

$$\frac{\partial T}{\partial t} = D\frac{\partial^2 T}{\partial x^2}$$

where $D = \kappa/C\rho = 1.13 \times 10^{-4}\ \mathrm{m}^2\ \mathrm{s}^{-1}$ is the thermal diffusivity. We can find the dispersion relation for sinusoidal waves by substituting a trial solution $T_{\omega,k}(x,t) = \mathrm{e}^{\mathrm{i}(\omega t - kx)}$, and in this way we discover that $k(\omega) = (1-\mathrm{i})\ell(\omega)^{-1}$, where

$$\ell(\omega) = \sqrt{\frac{2D}{\omega}}.$$

The varying component of the temperature at the end of the rod attached to the cup (which we call $x = 0$) is proportional to the function $f(t)$ that has the value 1 for $0 < t < \tau/2$, the value -1 for $\tau/2 < t < \tau$, and so on, where $\tau = 20\,\text{min} = 1200\,\text{s}$. This function has the Fourier representation

$$f(t) = \sum_{n=0}^{\infty} f_n \sin(\omega_n t)$$

with $\omega_n = (2n+1)2\pi/\tau$ and

$$f_n = \frac{2}{\tau}\int_0^{\tau} f(t)\sin(\omega_n t)\,\mathrm{d}t = \frac{4}{(2n+1)\pi}.$$

Thus, the complete solution for the temperature distribution is

$$T(x,t) = T_0 + \frac{4\,\Delta T}{\pi}\sum_{n=0}^{\infty}\frac{1}{2n+1}\,\mathrm{e}^{-x/\ell(\omega_n)}\sin\left(\omega_n t - \frac{x}{\ell(\omega_n)}\right)$$

where $T_0 = 323$ K and $\Delta T = 50$ K. For the fundamental mode, $n = 0$, the attenuation length is $\ell_0 = \sqrt{D\tau/\pi} \simeq 21\,\text{cm}$ and for the higher modes $\ell(\omega_n) = (2n+1)^{-1/2}\ell_0$. Consequently, higher modes are more rapidly attenuated than lower modes. After a distance $x = 50\,\text{cm}$, the mode $n = 1$ has been attenuated relative to the fundamental mode by a factor $\exp[-(\sqrt{3}-1)x/\ell_0] = 0.175$ and the ratio of the two amplitudes is $0.175/3 = 0.06$. By this point, therefore, the temperature is quite well described by just the fundamental mode, whose amplitude is

$$a(x) = \frac{4\,\Delta T}{\pi}\,\mathrm{e}^{-x/\ell_0} \simeq 5.9\,\text{K}$$

for $x = 50\,\text{cm}$. Our solution assumes, of course, that only waves travelling in the positive x direction are important. If the total length of the rod is, say, 1.5 m, then even the fundamental mode has been attenuated by a factor of about 8×10^{-4} in travelling this length, so reflected waves will be negligible.

• Problem 6.15

(a) Let us take $x = 0$ to be the end of the rod whose temperature is raised. The temperature in the rod obeys the heat conduction equation $\partial T/\partial t = D\,\partial^2 T/\partial x^2$, so, by differentiating, we find that $T' \equiv \partial T/\partial x$ obeys the same equation. To solve it, we introduce the Fourier transform

$$T'(x,t) = \int_{-\infty}^{\infty}\frac{\mathrm{d}k}{2\pi}\,\mathrm{e}^{ikx-\alpha(k)t}\,\tilde{T}(k).$$

On substituting this into the conduction equation, we find that it is a solution provided that $\alpha(k) = Dk^2$. To find the function $\tilde{T}(k)$, we use the initial condition that $T'(x,0) = 0$, except at $x = 0$, where the temperature changes abruptly. By taking $\tilde{T}$ to be a constant, we obtain $T'(x,0) = \tilde{T}\delta(x)$, which vanishes for all $x > 0$. We therefore have

$$T'(x,t) = \tilde{T}\int_{-\infty}^{\infty} \frac{\mathrm{d}k}{2\pi}\,\mathrm{e}^{-Dtk^2+\mathrm{i}kx} = \tilde{T}\frac{\mathrm{e}^{-x^2/4Dt}}{2\pi\sqrt{Dt}}\int_{-\infty}^{\infty} \mathrm{d}q\,\mathrm{e}^{-q^2} = \tilde{T}\frac{\mathrm{e}^{-x^2/4Dt}}{\sqrt{4\pi Dt}}$$

where we have used the substitution $k = q/\sqrt{Dt} + \mathrm{i}x/2Dt$. The temperature itself is

$$T(x,t) = T_0 + \Delta T + \int_0^x T'(x',t)\,\mathrm{d}x' = T_0 + \Delta T + \frac{\tilde{T}}{\sqrt{\pi}}\int_0^{x/\sqrt{4Dt}} \mathrm{e}^{-y^2}\,\mathrm{d}y$$

where $y = x'/\sqrt{4Dt}$. When $x \to \infty$, the remaining integral is equal to $\frac{1}{2}\sqrt{\pi}$. In this limit, the temperature must still have its initial value T_0, so we identify $\tilde{T} = -2\,\Delta T$, and the temperature distribution can be expressed as

$$T(x,t) = T_0 + \Delta T\left[1 - \frac{2}{\sqrt{\pi}}\int_0^{x/\sqrt{4Dt}} \mathrm{e}^{-y^2}\,\mathrm{d}y\right] = T_0 + \Delta T\,\mathrm{erfc}\left(\frac{x}{\sqrt{4Dt}}\right)$$

where $\mathrm{erfc}(\cdot)$ is the complementary error function encountered in problem 6.2. This function clearly has the expected behaviour. By construction, after any finite time, the temperature approaches T_0 as $x \to \infty$. On the other hand, at any finite position x, $T(x,t)$ approaches $T_0 + \Delta T$ as $t \to \infty$. At $t = 0$, the temperature is discontinuous at $x = 0$. The value $T_0 + \Delta T$ at the end of the rod is found by setting $x = 0$ and then taking t to zero, but the limit $t \to 0$ for any non-zero value of x yields the correct initial temperature T_0.

(b) Since the temperatures at the ends of the rod are fixed, we can write

$$T(x,t) = T_0 + (T_\ell - T_0)\frac{x}{\ell} + \bar{T}(x,t)$$

where $\bar{T}(x,t)$ is a function that vanishes at $x = 0$ and $x = \ell$. Any such function can be expressed as a Fourier sine series:

$$\bar{T}(x,t) = \sum_{n=1}^{\infty} a_n(t)\sin\left(\frac{n\pi x}{\ell}\right).$$

The heat conduction equation then holds if the coefficients $a_n(t)$ satisfy

$$\dot{a}_n(t) = -\tau_n^{-1}a_n(t) \qquad \tau_n = \frac{\ell^2}{n^2\pi^2 D}$$

so that $a_n(t) = a_n(0)\,\mathrm{e}^{-t/\tau_n}$. Thus, all the Fourier modes relax to zero. The longest relaxation time is τ_1, corresponding to the mode of longest wavelength, so the temperature distribution approaches the steady state, with a uniform temperature gradient, on a time scale no longer than τ_1.

(c) Let the hot end of the pipe be $x = 0$, the cold end be $x = \ell$ and let the position of the interface be $X(t)$. In a time δt, the interface moves a distance $\dot{X}(t)\,\delta t$ and, if the cross-sectional area of the cylinder of ice is A, then a mass $\rho_i A\dot{X}\,\delta t$ has melted. This requires a heat input $\rho_i A\dot{X}L\,\delta t$, which must come from an imbalance between the heat flows in the water and the ice. Let us call the temperature at the hot end of the pipe T_H, the temperature of the cold end of the pipe T_C and the melting temperature T_m. The interface, of course, must be at the temperature T_m; so, assuming uniform temperature gradients, the rate $\dot{Q}_w$ of heat flow into the interface from the water and the rate $\dot{Q}_i$ of heat flow away from the interface through the ice are given by

$$\dot{Q}_w = \kappa_w A \frac{T_H - T_m}{X} \qquad \dot{Q}_i = \kappa_i A \frac{T_m - T_C}{\ell - X}.$$

The difference $\dot{Q}_w - \dot{Q}_i$ is equal to the rate $\rho_i A\dot{X}L$ at which heat is absorbed by the melting ice, so we obtain the equation of motion

$$\frac{dX}{dt} = \frac{a}{X} - \frac{b}{\ell - X} = \frac{a\ell - (a+b)X}{X(\ell - X)}$$

where

$$a = \frac{\kappa_w(T_H - T_m)}{\rho_i L} \simeq 2 \times 10^{-7}\ \mathrm{m^2\ s^{-1}}$$
$$b = \frac{\kappa_i(T_m - T_C)}{\rho_i L} \simeq 3.6 \times 10^{-8}\ \mathrm{m^2\ s^{-1}}.$$

When X is small, $\dot{X}$ is positive and the interface moves as ice melts. However, this process stops when the heat flows balance, and we see that this happens when $X = X_{max} = a\ell/(a+b) \simeq 85$ cm.

Our equation for X is readily integrated, to yield

$$t = \frac{1}{a+b}\left[\frac{1}{2}X^2 - \frac{b\ell}{a+b}X - \frac{ab\ell^2}{(a+b)^2}\ln\left(1 - \frac{X}{X_{max}}\right)\right].$$

Setting $\ell = 1$ m and $X = 0.1$ m, we find that $t \simeq 2.5 \times 10^4$ s $\simeq 7$ h. As the interface approaches X_{max}, the most important term is the logarithm, so we get $X(t) \simeq X_{max}(1 - e^{-t/\bar{t}})$, with the characteristic time $\bar{t} = ab\ell^2/(a+b)^3 \simeq 150$ h.

(d) The water–ice interface is always at the temperature T_m. As it moves, the temperatures at the ends of both the cylinder of water and the cylinder of ice remain fixed, but the lengths of these cylinders change. The analysis of (b) shows that the temperature distributions always relax towards a uniform temperature gradient, and it is intuitively clear that these gradients will always be approximately uniform if the interface moves sufficiently slowly. The question is: how slowly must it move? To estimate this, imagine that it moves in small discrete steps. After each step, the temperature at the end of each cylinder corresponding to the new position of the interface has changed by a small amount. If we imagine this change to be instantaneous, then there is a small discontinuity, affecting an infinitesimal length of the cylinder. To see the immediate effect of this, we might as well consider the cylinder to be infinitely long. Then we can apply the result of (a) to see that the discontinuity changes immediately into a smooth but non-uniform gradient. So we can now apply the result of (b) to investigate how the non-uniformity relaxes. Since the longest relaxation time is that for the Fourier mode of longest wavelength, we need consider only that mode. For the water, the relaxation time is $\tau_\mathrm{w} = X^2/\pi^2 D_\mathrm{w}$ (since the length of the cylinder of water is X) and for the ice it is $\tau_\mathrm{i} = (\ell - X)^2/\pi^2 D_\mathrm{i}$. The relevant thermal diffusivities are

$$D_\mathrm{w} = \frac{\kappa_\mathrm{w}}{\rho_\mathrm{w} C_\mathrm{w}} \simeq 1.4 \times 10^{-7}\ \mathrm{m}^2\ \mathrm{s}^{-1} \qquad D_\mathrm{i} = \frac{\kappa_\mathrm{i}}{\rho_\mathrm{i} C_\mathrm{i}} \simeq 1 \times 10^{-6}\ \mathrm{m}^2\ \mathrm{s}^{-1}.$$

We shall now adopt the criterion that, for each cylinder, the distance $\dot{X}\tau$ through which the interface moves in the time τ should be only a small fraction of the length of the cylinder. This criterion is hard to justify in detail but is intuitively reasonable and, on dimensional grounds, is the only one available. Applying it first to the water, we get $\dot{X}\tau_\mathrm{w} \ll X$ or

$$\frac{a}{\pi^2 D_\mathrm{w}} - \frac{b}{\pi^2 D_\mathrm{w}} \frac{X}{\ell - X} \simeq 0.14 - 0.03 \frac{X}{\ell - X} \ll 1.$$

The left-hand side of this inequality vanishes when $X = X_\mathrm{max}$, corresponding to the fact that $\dot{X} = 0$ there. In fact, we see that this quantity is always less than 0.14, so our criterion is moderately well satisfied for the water. This is true even at the initial time when $X = 0$ and the speed is infinite, because there is only an infinitesimal distance over which the temperature gradient needs to adjust. For the ice, our criterion is $\dot{X}\tau_\mathrm{i} \ll (\ell - X)$ or

$$\frac{a}{\pi^2 D_\mathrm{i}} \frac{\ell - X}{X} - \frac{b}{\pi^2 D_\mathrm{i}} \simeq 0.02 \frac{\ell - X}{X} - 0.004 \ll 1.$$

The left-hand side of this inequality is less than 0.1 when X is greater than about 0.1ℓ. Of course, when X is very small, this quantity is large, and our criterion is not even approximately satisfied. Under these circumstances, however, the first term in the equation for $\dot{X}$, corresponding to heat conduction through the water, is by far the more important, so the error in the second term, due to a non-uniform temperature gradient in the ice, has no serious effect. It seems, then, that our equation of motion for the interface, based on the assumption of uniform temperature gradients, should be a reasonable approximation.

• Problem 6.16

(a) Let us first discuss how the functional $\Phi(\phi)$ can be minimized. Suppose that $\phi(x) = M(x)$ is a minimum. Then, if $\Phi(M + \delta\phi) = \Phi(M) + \delta\Phi$, the change

$\delta\Phi$ must vanish when $\delta\phi$ is sufficiently small. Expanding in powers of $\delta\phi$, we find that

$$\delta\Phi = \int \mathrm{d}x \left(\frac{\mathrm{d}M}{\mathrm{d}x}\frac{\mathrm{d}(\delta\phi)}{\mathrm{d}x} + \left.\frac{\partial V}{\partial \phi}\right|_{\phi=M} \delta\phi + \mathrm{O}(\delta\phi^2) \right) = 0.$$

In particular, we can consider variations $\delta\phi$ which vanish at the spatial boundaries, say, $x = \pm\infty$. Then, on integrating by parts, we get

$$\int \mathrm{d}x \left(-\frac{\mathrm{d}^2 M}{\mathrm{d}x^2} + \left.\frac{\partial V}{\partial \phi}\right|_{\phi=M} \right) \delta\phi = 0$$

and this must hold for any small function $\delta\phi(x)$. For this to be true, the quantity in large parentheses must vanish, and we obtain an equation for $M(x)$, namely

$$\frac{\mathrm{d}^2 M}{\mathrm{d}x^2} - V'(M) = 0$$

where $V'(\phi)$ means $\partial V(\phi)/\partial\phi$. Evidently, this equation must hold for a stationary solution to the time-dependent Ginzburg–Landau equation. The absolute minimum of Φ (which is the most stable thermodynamic state) is the lowest minimum of the potential V, since $(\mathrm{d}\phi/\mathrm{d}x)^2$ contributes a positive term to Φ whenever M is non-uniform. These minima are $M_{\mathrm{eq}} = \pm\sqrt{a/b} + h/2a + \mathrm{O}(h^2)$, the most stable minimum having the same sign as h.

Now let $M(x,t) = M_{\mathrm{eq}} + \mu(x,t)$. On expanding in powers of μ, the time-dependent Ginzburg–Landau equation reads

$$\frac{\partial \mu}{\partial t} = D\left(\frac{\partial^2 \mu}{\partial x^2} - V''(M)\mu + \mathrm{O}(\mu^2) \right).$$

The meaning of this is most easily seen by taking the Fourier transform

$$\mu(x,t) = \int \frac{\mathrm{d}k}{2\pi}\, \mathrm{e}^{\mathrm{i}kx}\, \widetilde{\mu}(k,t)$$

to obtain

$$\frac{\partial \widetilde{\mu}}{\partial t} = -D[k^2 + V''(M)]\widetilde{\mu}.$$

Since M is a minimum of V, we have $V''(M) > 0$, and so each Fourier component $\widetilde{\mu}(k,t)$ decays with time as $\widetilde{\mu}(k,t) \sim \mathrm{e}^{-t/\tau(k)}$, with $\tau(k) = 1/D[k^2 + V''(M)]$. Thus, $\mu(x,t)$ relaxes to zero in a characteristic time set by the modes of longest wavelength, $\tau(0) = 1/DV''(M)$.

(b) For brevity, we set $T(x) = \tanh[q(x - x_0)]$ (this has nothing to do with temperature, to which we make no explicit reference in this problem) and

$M(x) = M_0 T(x)$. We have

$$\begin{aligned}
\frac{\mathrm{d}T(x)}{\mathrm{d}x} &= q\,\mathrm{sech}^2[q(x-x_0)] = q[1-T^2(x)] \\
\frac{\mathrm{d}^2T(x)}{\mathrm{d}x^2} &= -2q^2\,\mathrm{sech}^2[q(x-x_0)]\tanh[q(x-x_0)] \\
&= -2q^2T(x)[1-T^2(x)] \\
V'(M(x)) &= -aM_0T(x) + bM_0^3T^3(x).
\end{aligned}$$

We easily find that this satisfies the equation for a local minimum of Φ, provided that $M_0 = \sqrt{a/b}$ and $q = \sqrt{a/2}$. This solution represents a domain wall, centred at the point x_0, whose width is roughly $1/q$, and which separates two domains in which the magnetization is close to $\pm M_0$. It has a higher free energy than either of the uniform configurations $M(x) = \pm M_0$, but it has the lowest free energy consistent with, say, the boundary conditions $M(-\infty) = -M(+\infty)$. There clearly exists an infinite family of domain wall solutions, corresponding to different positions x_0.

(c) When h is non-zero, the uniform configurations $M = \pm\sqrt{a/b} + h/2a + \mathrm{O}(h^2)$ are the only minima of Φ. As indicated in the question, however, the time-dependent Ginzburg–Landau equation has a solution corresponding to a moving domain wall. If we write a trial solution as $M(z)$, with $z = x - vt$, then the equation becomes

$$\frac{\mathrm{d}^2M}{\mathrm{d}z^2} + \frac{v}{D}\frac{\mathrm{d}M}{\mathrm{d}z} = V'(M)$$

and, on substituting $M(z) = M_1T(z) + m$, we get

$$\begin{aligned}
-2q^2M_1T(z)[1-T^2(z)] &+ \frac{vq}{D}M_1[1-T^2(z)] \\
&= -a[M_1T(z)+m] + b[M_1T(z)+m]^3 - h.
\end{aligned}$$

Comparing coefficients of $T^n(z)$ on each side, we find exactly four equations to be solved for the four constants M_1, m, q and v, so the solution does indeed exist. The expansions for these constants in powers of h are

$$M_1 = \sqrt{\frac{a}{b}}\left(1 - \frac{3b}{8a^3}h^2 + \cdots\right) \qquad m = \frac{1}{2a}h + \cdots$$

$$q = \sqrt{\frac{b}{2}}M_1 = \sqrt{\frac{a}{2}}\left(1 - \frac{3b}{8a^3}h^2 + \cdots\right) \qquad v = -\frac{3\sqrt{2b}D}{2a}h + \cdots.$$

Within our approximation, it seems that the magnetization values $\pm M_1 + m$ in the two domains correspond to the minima of Φ, and this is actually an exact result. Also, the wall velocity v has the opposite sign to h. This means, as might have been expected, that the more stable domain, with M parallel to h, grows, while the other domain shrinks. However, it should be noted that, in real ferromagnetic materials, factors such as magnetostatic energy, anisotropy energy and lattice imperfections have important effects on the equilibrium size of domains and on the structure and motion of the domain walls. None of these factors is taken into account by our simple model.

(d) The stability of the moving domain wall can be ascertained by a generalization of the method used in (a), which is standard in the theory of solitons. We first observe, by differentiating the Ginzburg–Landau equation, that the function $M'(z) = \mathrm{d}M(z)/\mathrm{d}z$ obeys

$$\left(\frac{\mathrm{d}^2}{\mathrm{d}z^2} + \frac{v}{D}\frac{\mathrm{d}}{\mathrm{d}z} - V''(M)\right)M'(z) = 0.$$

Now consider a perturbed configuration, which we write as $M(z) + \mu(z,t)$. Expanding the Ginzburg–Landau equation to first order in μ gives

$$\frac{\partial \mu}{\partial t} = D\left(\frac{\partial^2}{\partial z^2} + \frac{v}{D}\frac{\partial}{\partial z} - V''(M)\right)\mu(z,t).$$

We look for normal-mode solutions of this equation, which vary as $\mu(z,t) = \mu_0(z)\,\mathrm{e}^{-\gamma t}$. If all the possible eigenvalues γ are positive, then any small perturbation will decay to zero. If we write $\mu_0(z) = \mathrm{e}^{-(v/2D)z}\,\mu_1(z)$, then the eigenvalue equation takes the form

$$\left(-\frac{\partial^2}{\partial z^2} + U(z)\right)\mu_1(z) = \frac{\gamma}{D}\mu_1(z)$$

where $U(z) = V''(M) + v^2/4D^2$, and this has the same form as the Schrödinger equation for a particle of mass $m = \frac{1}{2}$ in the potential $U(z)$, if we use units such that $\hbar = 1$. Now, we already know one solution to this equation, namely $\mathrm{e}^{(v/2D)z}\,M'(z) \propto \mathrm{e}^{(v/2D)z}\,\mathrm{sech}^2(qz)$. Its eigenvalue is zero. Moreover, it has no nodes, so the theory of the Schrödinger equations tells us that this is in fact the lowest eigenvalue. A perturbation of this form corresponds just to a small displacement of the moving wall or, equivalently, to a small change in the origin of time. Any other perturbation has a positive eigenvalue γ and will relax to zero. Obviously, this analysis also applies to the static domain wall when $h = 0$.

• **Problem 6.17** When there are no external forces, the full Boltzmann equation is

$$\frac{\partial f}{\partial t} + \frac{1}{m}\boldsymbol{p}\cdot\boldsymbol{\nabla} f = \int \mathrm{d}^3 p_2\,\mathrm{d}^3 p_1'\,\mathrm{d}^3 p_2'\,W_{(\boldsymbol{p}\boldsymbol{p}_2)\to(\boldsymbol{p}_1'\boldsymbol{p}_2')}\,(f_2' f_1' - f_2 f)$$

where for brevity we simply use $\boldsymbol{\nabla}$ to represent $\boldsymbol{\nabla}_{\boldsymbol{r}}$. This is an integro-differential equation that in principle (although with great difficulty in practice) can be solved for the distribution function $f(\boldsymbol{r},\boldsymbol{p},t)$ given some initial distribution $f(\boldsymbol{r}, \boldsymbol{p}, 0)$. Assuming that energy and momentum are conserved in binary collisions, the right-hand side vanishes when $f(\boldsymbol{r}, \boldsymbol{p}, t)$ is equal to the given distribution function $f_0(\boldsymbol{r}, \boldsymbol{p}, t)$, in which $n(\boldsymbol{r}, t)$, $T(\boldsymbol{r}, t)$ and $\boldsymbol{v}(\boldsymbol{r}, t)$ are *arbitrary* functions of position and time. This distribution represents a fluid in *local* thermal equilibrium, which is to say that a small element at the point $\boldsymbol{r}$, which is moving with a velocity $\boldsymbol{v}(\boldsymbol{r}, t)$, has a number density $n(\boldsymbol{r}, t)$ of particles whose momenta relative to its rest frame are distributed according to the Maxwell–Boltzmann distribution corresponding to a temperature $T(\boldsymbol{r}, t)$.

The relaxation time approximation is a rather drastic simplification, in which we suppose that, at a given time t, the true distribution function f is not too different from some state of local equilibrium f_0, and that collisions tend to make f relax towards f_0, with a characteristic relaxation time τ. The approximate Boltzmann equation then is

$$\frac{\partial f(\boldsymbol{r}, \boldsymbol{p}, t)}{\partial t} + \frac{1}{m}\boldsymbol{p} \cdot \boldsymbol{\nabla} f(\boldsymbol{r}, \boldsymbol{p}, t) = -\frac{1}{\tau}[f(\boldsymbol{r}, \boldsymbol{p}, t) - f_0(\boldsymbol{r}, \boldsymbol{p}, t)].$$

Now, it can be shown, by linearizing the full Boltzmann equation, that small perturbations about an equilibrium state do relax back to equilibrium, so the approximation is plausible to this extent. We see, however, that the approximate equation can, in principle, be solved for $f(\boldsymbol{r}, \boldsymbol{p}, t)$, when $f_0(\boldsymbol{r}, \boldsymbol{p}, t)$ is constructed from *any* set of functions $n(\boldsymbol{r}, t)$, $T(\boldsymbol{r}, t)$ and $\boldsymbol{v}(\boldsymbol{r}, t)$: it does not tell us what functions we should choose. Some further considerations are needed if we want a reasonably consistent description of the fluid, and this problem explores one way in which this can be achieved.

(a) The number density in a fluid described by the distribution $f(\boldsymbol{r}, \boldsymbol{p}, t)$ is $n_f(\boldsymbol{r}, t) = \int \mathrm{d}^3p\, f(\boldsymbol{r}, \boldsymbol{p}, t)$. Using $f_0(\boldsymbol{r}, \boldsymbol{p}, t)$, we find that $n_{f_0}(\boldsymbol{r}, t) = n(\boldsymbol{r}, t)$. By integrating the approximate Boltzmann equation, we find that

$$\partial_t n_f(\boldsymbol{r}, t) + \boldsymbol{\nabla} \cdot \boldsymbol{j}(\boldsymbol{r}, t) = -\tau^{-1}[n_f(\boldsymbol{r}, t) - n(\boldsymbol{r}, t)]$$

where ∂_t is an abbreviation for $\partial/\partial t$ and the particle current is

$$\boldsymbol{j}(\boldsymbol{r}, t) = \int \mathrm{d}^3p\, \frac{\boldsymbol{p}}{m} f(\boldsymbol{r}, \boldsymbol{p}, t).$$

However, if no particles are created or destroyed, then n_f must obey the equation of continuity $\partial_t n + \boldsymbol{\nabla} \cdot \boldsymbol{j} = 0$, and therefore $n_f = n_{f_0} = n$.

In a similar way, we can consider the momentum density $\boldsymbol{\pi}(\boldsymbol{r}, t) = \int \mathrm{d}^3p\, \boldsymbol{p} f(\boldsymbol{r}, \boldsymbol{p}, t) = m\boldsymbol{j}(\boldsymbol{r}, t)$. The corresponding current, which tells us the rate at which the ith component of momentum is transported through the fluid in the jth direction by particles with velocity $\boldsymbol{p}/m$, is the stress tensor

$$\theta_{ij}(\boldsymbol{r}, t) = \int \mathrm{d}^3p\, p_i \frac{p_j}{m} f(\boldsymbol{r}, \boldsymbol{p}, t).$$

Conservation of momentum implies the equation of continuity $\partial_t \pi_i(\boldsymbol{r}, t) + \partial_j \theta_{ij}(\boldsymbol{r}, t) = 0$, where ∂_j is an abbreviation for $\partial/\partial x_j$ and we use the summation convention that the repeated index j implies a sum over j. On multiplying the approximate Boltzmann equation by p_i and then integrating over $\boldsymbol{p}$, we discover that

$$\partial_t \pi_i(\boldsymbol{r}, t) + \partial_j \theta_{ij}(\boldsymbol{r}, t) = -\tau^{-1}[\pi_i(\boldsymbol{r}, t) - \pi_i^0(\boldsymbol{r}, t)]$$

where π_i is the momentum density calculated with f, and π_i^0 is that calculated with f_0. Evidently, conservation of momentum requires that these should be equal. In fact, we can now calculate

$$\boldsymbol{j}(\boldsymbol{r}, t) = \frac{1}{m}\boldsymbol{\pi}(\boldsymbol{r}, t) = \int \mathrm{d}^3 p \, \frac{\boldsymbol{p}}{m} f_0(\boldsymbol{r}, \boldsymbol{p}, t) = n(\boldsymbol{r}, t)\boldsymbol{v}(\boldsymbol{r}, t).$$

(This integral is conveniently calculated by making the change of variable $\boldsymbol{p} = \boldsymbol{q} + m\boldsymbol{v}$:

$$\int \mathrm{d}^3 p \, \boldsymbol{p} f_0(\boldsymbol{r}, \boldsymbol{p}, t) = \int \mathrm{d}^3 q \, (\boldsymbol{q} + m\boldsymbol{v}) \frac{n}{(2\pi m k T)^{3/2}} \, \mathrm{e}^{-|\boldsymbol{q}|^2/2mkT} = mn\boldsymbol{v}$$

since the term linear in $\boldsymbol{q}$ integrates to zero. Other similar integrals that we shall need are

$$\int \mathrm{d}^3 q \, |\boldsymbol{q}|^{2s} f_0(\boldsymbol{r}, \boldsymbol{q} + m\boldsymbol{v}, t) = 1 \times 3 \times 5 \times \cdots \times (2s+1) n (mkT)^s$$

$$\int \mathrm{d}^3 q \, q_i q_j |\boldsymbol{q}|^{2s} f_0(\boldsymbol{r}, \boldsymbol{q} + m\boldsymbol{v}, t) = \frac{\delta_{ij}}{3} \int \mathrm{d}^3 q \, |\boldsymbol{q}|^{2(s+1)} f_0(\boldsymbol{r}, \boldsymbol{q} + m\boldsymbol{v}, t)$$

where s is a positive integer.)

The energy density in the fluid is $\varepsilon(\boldsymbol{r}, t) = \int \mathrm{d}^3 p \, (|\boldsymbol{p}|^2/2m) f(\boldsymbol{r}, \boldsymbol{p}, t)$ and there is a corresponding current $\boldsymbol{j}_\varepsilon(\boldsymbol{r}, t) = \int \mathrm{d}^3 p \, (|\boldsymbol{p}|^2/2m)(\boldsymbol{p}/m) f(\boldsymbol{r}, \boldsymbol{p}, t)$. This assumes, of course, that the interparticle potential that causes scattering does not contribute significantly to the internal energy. By repeating the above argument, we find that

$$\varepsilon(\boldsymbol{r}, t) = \varepsilon_0(\boldsymbol{r}, t) = \tfrac{3}{2} n(\boldsymbol{r}, t) k T(\boldsymbol{r}, t) + \tfrac{1}{2} n(\boldsymbol{r}, t) m v^2(\boldsymbol{r}, t).$$

We see that an element of the fluid at the point $\boldsymbol{r}$ whose volume is $\mathrm{d}^3 r$ has an internal energy $u(\boldsymbol{r}, t)\,\mathrm{d}^3 r = n(\boldsymbol{r}, t)\,\mathrm{d}^3 r \, \tfrac{3}{2} k T(\boldsymbol{r}, t)$ and a kinetic energy $n(\boldsymbol{r}, t)\,\mathrm{d}^3 r \, \tfrac{1}{2} m v^2(\boldsymbol{r}, t)$ due to its bulk motion, $n(\boldsymbol{r}, t)\,\mathrm{d}^3 r$ being the number of particles that it contains.

(b) Even our simplified approximate equation is not easy to solve in general. If the relaxation time is sufficiently short, we can attempt to use the further approximation of expanding the solution in powers of τ. Substituting this expansion $f = f_0 + \tau f_1 + \cdots$ on both sides of the equation, we easily find that

$$f_1(\boldsymbol{r}, \boldsymbol{p}, t) = -\partial_t f_0(\boldsymbol{r}, \boldsymbol{p}, t) - \frac{1}{m} \boldsymbol{p} \cdot \boldsymbol{\nabla} f_0(\boldsymbol{r}, \boldsymbol{p}, t).$$

While the derivatives can be performed explicitly, it is generally more convenient to calculate the quantities that we need by integrating over the momentum first, using the integrals given above. Let us see what is implied by the conservation laws considered in (a). The first correction to the number density is

$$\tau n_1(\boldsymbol{r},t) = \tau \int \mathrm{d}^3 p\, f_1(\boldsymbol{r},\boldsymbol{p},t) = -\tau\{\partial_t n(\boldsymbol{r},t) + \boldsymbol{\nabla}\cdot[n(\boldsymbol{r},t)\boldsymbol{v}(\boldsymbol{r},t)]\}.$$

This ought to vanish, and the equation of continuity shows that indeed it does. The first correction to the particle current is

$$\begin{aligned}\tau j_i^1(\boldsymbol{r},t) &= \frac{\tau}{m}\int \mathrm{d}^3 p\, p_i f_1(\boldsymbol{r},\boldsymbol{p},t)\\ &= -\tau\left(\partial_t(nv_i) + \frac{1}{m}\partial_i(nkT) + \partial_j(nv_iv_j)\right).\end{aligned}$$

This quantity does not vanish automatically, so we obtain a constraint on the functions $n(\boldsymbol{r},t)$, $T(\boldsymbol{r},t)$ and $\boldsymbol{v}(\boldsymbol{r},t)$ that can be used in f_0, namely

$$\partial_t(nv_i) + \frac{1}{m}\partial_i(nkT) + \partial_j(nv_iv_j) = \mathrm{O}(\tau).$$

The first correction to the energy density is

$$\begin{aligned}\tau\varepsilon_1(\boldsymbol{r},t) &= -\tau\int \mathrm{d}^3 p\,\frac{|\boldsymbol{p}|^2}{2m}\left(\partial_t f_0(\boldsymbol{r},\boldsymbol{p},t) + \frac{1}{m}\boldsymbol{p}\cdot\boldsymbol{\nabla} f_0(\boldsymbol{r},\boldsymbol{p},t)\right)\\ &= -\tau[\partial_t\varepsilon(\boldsymbol{r},t) + \boldsymbol{\nabla}\cdot\boldsymbol{j}_\varepsilon^0(\boldsymbol{r},t)]\end{aligned}$$

where

$$\boldsymbol{j}_\varepsilon^0(\boldsymbol{r},t) = [\tfrac{5}{2}n(\boldsymbol{r},t)kT(\boldsymbol{r},t) + \tfrac{1}{2}n(\boldsymbol{r},t)mv^2(\boldsymbol{r},t)]\boldsymbol{v}(\boldsymbol{r},t)$$

is the lowest-order contribution to the energy current. This correction vanishes (up to possible contributions of order τ^2) because of the continuity equation for energy, so we obtain no new constraint. The form of $\boldsymbol{j}_\varepsilon^0$ is of some interest, however. It can be written as $\boldsymbol{j}_\varepsilon^0 = (\varepsilon + P)\boldsymbol{v}$, where $P = nkT$ is the pressure, and this indicates two mechanisms for the transport of energy. The term $\varepsilon\boldsymbol{v}$ corresponds to the energy that a moving element of the fluid carries with it, while $P\boldsymbol{v}$ represents the work done by this element on the element just ahead of it.

(c) In general, the heat current is

$$\boldsymbol{q}(\boldsymbol{r},t) = \int \mathrm{d}^3 p\,\frac{|\boldsymbol{p}-m\boldsymbol{v}|^2}{2m}\frac{\boldsymbol{p}-m\boldsymbol{v}}{m} f(\boldsymbol{r},\boldsymbol{p},t).$$

It differs from the energy current defined earlier, in that the particle momenta $\boldsymbol{p}-m\boldsymbol{v}$ being averaged are those measured in the rest frame of an element of the fluid. This means that the quantity whose transport is considered is just the internal energy, and the transport mechanism involved is heat conduction: the bulk motion of the fluid is discounted. When $\boldsymbol{v}$ is non-zero, the expression that we obtain is quite complicated, but we can focus on the heat conduction mechanism by considering a stationary fluid. In that case, $\boldsymbol{q}$ coincides with the energy current. The leading term vanishes when $\boldsymbol{v} = \boldsymbol{0}$ and to order τ we obtain

$$\begin{aligned} \boldsymbol{q}(\boldsymbol{r}, t) &= \int \mathrm{d}^3 p \, \frac{|\boldsymbol{p}|^2}{2m} \frac{\boldsymbol{p}}{m} f(\boldsymbol{r}, \boldsymbol{p}, t) \\ &= -\tau \int \mathrm{d}^3 p \, \frac{p^2}{2m} \frac{\boldsymbol{p}}{m} \left(\partial_t f_0(\boldsymbol{r}, \boldsymbol{p}, t) + \frac{1}{m} \boldsymbol{p} \cdot \boldsymbol{\nabla} f_0(\boldsymbol{r}, \boldsymbol{p}, t) \right) \\ &= -\frac{5\tau}{2m} \boldsymbol{\nabla} \{ n(\boldsymbol{r}, t) [kT(\boldsymbol{r}, t)]^2 \}. \end{aligned}$$

At this point, we must bear in mind the result derived in (b). When $\boldsymbol{v} = \boldsymbol{0}$, this implies that $\boldsymbol{\nabla}(nkT) = \boldsymbol{0}$. Quite reasonably, the pressure gradient must vanish if the fluid is to be stationary. Finally, therefore, we get

$$\boldsymbol{q} = -\kappa \boldsymbol{\nabla} T \qquad \text{with} \qquad \kappa = \frac{5nk^2 T \tau}{2m}.$$

(d) Using the integrals given above, we evaluate the stress tensor to first order in τ as

$$\begin{aligned} \theta_{ij} &= \frac{1}{m} \int \mathrm{d}^3 p \, p_i p_j \left[1 - \tau \left(\partial_t + \frac{1}{m} \boldsymbol{p} \cdot \boldsymbol{\nabla} \right) \right] f_0(\boldsymbol{r}, \boldsymbol{p}, t) \\ &= [nkT - \tau(\partial_t(nkT) + \boldsymbol{\nabla} \cdot (nkT\boldsymbol{v}))]\delta_{ij} \\ &\quad - \tau m \partial_t (n v_i v_j) - \tau[\partial_i(nkT v_j) + \partial_j(nkT v_i)] - \tau m \partial_k (n v_i v_j v_k). \end{aligned}$$

Making use of the equation of continuity for the number of particles and of the result found in (b), this can be simplified to

$$\theta_{ij} = [nkT - \tau n(\partial_t + \boldsymbol{v} \cdot \boldsymbol{\nabla})kT]\delta_{ij} - \tau nkT[\partial_i v_j + \partial_j v_i].$$

The shear viscosity is the coefficient of the velocity gradient terms, namely $\eta = nkT\tau$. For example, if the fluid moves in the x direction, with a speed $v_x(y)$ that depends only on y, then the rate at which p_x is transported in the y direction is $\theta_{xy} = -\eta \, \mathrm{d}v_x/\mathrm{d}y$.

It is simple to verify that

$$\frac{\kappa}{\eta C_V} = \frac{5nk^2 T\tau/2m}{nkT\tau(3k/2m)} = \frac{5}{3}.$$

A variety of approximations yield universal values for this ratio (that is values which do not depend on the particular fluid or on its temperature or pressure), but they do not agree on the actual value. By linearizing the Boltzmann equation, but retaining the correct form of the scattering integral, one obtains instead the value $\frac{5}{2}$. Elementary treatments of transport processes assume, in effect, that each particle carries the average value of the transported quantity and travels with the mean molecular speed for a time τ between collisions. In such treatments, one obtains $\kappa/\eta C_V = 1$, the difference in numerical factors arising essentially from the independent averaging of these various quantities.

• Problem 6.18

(a) In the presence of a force acting independently on each particle, the approximate Boltzmann equation becomes

$$\left(\partial_t + \frac{1}{m}\boldsymbol{p}\cdot\boldsymbol{\nabla}_r + \boldsymbol{F}(\boldsymbol{r})\cdot\boldsymbol{\nabla}_p\right) f(\boldsymbol{r},\boldsymbol{p},t) = -\frac{1}{\tau}[f(\boldsymbol{r},\boldsymbol{p},t) - f_0(\boldsymbol{r},\boldsymbol{p},t)].$$

Because of the new term involving $\boldsymbol{F}$, we shall need to evaluate integrals of the form $\int \mathrm{d}^3 p\, g(\boldsymbol{r},\boldsymbol{p},t)\boldsymbol{F}(\boldsymbol{r})\cdot\boldsymbol{\nabla}_p f(\boldsymbol{r},\boldsymbol{p},t)$, where $g(\boldsymbol{r},\boldsymbol{p},t)$ is a polynomial in $\boldsymbol{p}$. Since $f(\boldsymbol{r},\boldsymbol{p},t)$ goes to zero exponentially as $|\boldsymbol{p}|\to\infty$, this can easily be done using an integration by parts

$$\int \mathrm{d}^3 p\, g(\boldsymbol{r},\boldsymbol{p},t)\boldsymbol{F}(\boldsymbol{r})\cdot\boldsymbol{\nabla}_p f(\boldsymbol{r},\boldsymbol{p},t) = -\boldsymbol{F}(\boldsymbol{r})\cdot\int \mathrm{d}^3 p[\boldsymbol{\nabla}_p g(\boldsymbol{r},\boldsymbol{p},t)]f(\boldsymbol{r},\boldsymbol{p},t)$$

and the integrals listed in the solution to problem 6.17. We first integrate the approximate Boltzmann equation over $\boldsymbol{p}$, to obtain

$$\partial_t n_f(\boldsymbol{r},t) + \boldsymbol{\nabla}_r\cdot\boldsymbol{j}(\boldsymbol{r},t) = -\frac{1}{\tau}[n_f(\boldsymbol{r},t) - n_{f_0}(\boldsymbol{r},t)]$$

where n_f and n_{f_0} are the number densities associated with f and f_0 respectively and $\boldsymbol{j}(\boldsymbol{r},t) = m^{-1}\int \mathrm{d}^3 p\,\boldsymbol{p} f(\boldsymbol{r},\boldsymbol{p},t)$ is the particle current. The term involving $\boldsymbol{F}$ integrates to zero; so, as before, we use the equation of continuity $\partial_t n + \boldsymbol{\nabla}_r\cdot\boldsymbol{j} = 0$ that expresses conservation of particles to conclude that $n_f = n_{f_0} = n$.

Next, multiply the equation by $\boldsymbol{p}$ and integrate. The result is

$$\partial_t \pi_i(\boldsymbol{r},t) + \partial_j\theta_{ij}(\boldsymbol{r},t) - F_i(\boldsymbol{r})n(\boldsymbol{r},t) = -\tau^{-1}[\pi_i(\boldsymbol{r},t) - \pi_i^0(\boldsymbol{r},t)]$$

where $\boldsymbol{\pi} = m\boldsymbol{j}$ and $\boldsymbol{\pi}^0$ are the momentum densities associated with f and f_0 respectively and $\theta_{ij}(\boldsymbol{r},t) = m^{-1}\int \mathrm{d}^3 p\, p_i p_j f(\boldsymbol{r},\boldsymbol{p},t)$ is the stress tensor. Instead of an equation of continuity, the momentum density should now obey

$$\partial_t \pi_i(\boldsymbol{r},t) = -\partial_j\theta_{ij}(\boldsymbol{r},t) + F_i(\boldsymbol{r})n(\boldsymbol{r},t)$$

where the term involving θ_{ij} accounts for transport of momentum, while the term involving $\boldsymbol{F}$ accounts for the acceleration of particles by the external force. Again, we can conclude that $\boldsymbol{\pi}(\boldsymbol{r},t) = \boldsymbol{\pi}^0(\boldsymbol{r},t)$. As in problem 6.17, this implies that $\boldsymbol{j}(\boldsymbol{r},t) = \boldsymbol{j}^0(\boldsymbol{r},t) = n(\boldsymbol{r},t)\boldsymbol{v}(\boldsymbol{r},t)$.

In the presence of the external force, each particle has an energy $|\boldsymbol{p}|^2/2m + \phi(\boldsymbol{r})$. The energy density and current are therefore

$$\varepsilon(\boldsymbol{r},t) = \int \mathrm{d}^3 p \left(\frac{|\boldsymbol{p}|^2}{2m} + \phi(\boldsymbol{r})\right) f(\boldsymbol{r},\boldsymbol{p},t)$$

$$\boldsymbol{j}_\varepsilon(\boldsymbol{r},t) = \int \mathrm{d}^3 p \left(\frac{|\boldsymbol{p}|^2}{2m} + \phi(\boldsymbol{r})\right) \frac{\boldsymbol{p}}{m} f(\boldsymbol{r},\boldsymbol{p},t).$$

They obey

$$\partial_t \varepsilon(\boldsymbol{r},t) = -\boldsymbol{\nabla}_r \boldsymbol{j}_\varepsilon(\boldsymbol{r},t) + \frac{1}{m}\int \mathrm{d}^3 p\, \boldsymbol{F}(\boldsymbol{r})\cdot \boldsymbol{p} f(\boldsymbol{r},\boldsymbol{p},t)$$

where the last term is the rate at which the external force does work. Now multiply the approximate Boltzmann equation by $|\boldsymbol{p}|^2/2m + \phi(\boldsymbol{r})$ and integrate over $\boldsymbol{p}$. We get

$$\partial_t \varepsilon(\boldsymbol{r},t) + \boldsymbol{\nabla}_r \boldsymbol{j}_\varepsilon(\boldsymbol{r},t) - \frac{1}{m}\int \mathrm{d}^3 p\, \boldsymbol{F}(\boldsymbol{r})\cdot \boldsymbol{p} f(\boldsymbol{r},\boldsymbol{p},t) = -\frac{1}{\tau}[\varepsilon(\boldsymbol{r},t) - \varepsilon_0(\boldsymbol{r},t)]$$

where $\varepsilon_0(\boldsymbol{r},t)$ is the energy density calculated with f_0. Clearly, we must have $\varepsilon(\boldsymbol{r},t) = \varepsilon_0(\boldsymbol{r},t)$.

(b) On writing $f(\boldsymbol{r},\boldsymbol{p},t) = f_0(\boldsymbol{r},\boldsymbol{p},t) + \tau f_1(\boldsymbol{r},\boldsymbol{p},t) + \cdots$, we can evaluate the first correction as

$$f_1(\boldsymbol{r},\boldsymbol{p},t) = -\left(\partial_t + \frac{1}{m}\boldsymbol{p}\cdot\boldsymbol{\nabla}_r + \boldsymbol{F}(\boldsymbol{r})\cdot\boldsymbol{\nabla}_p\right) f_0(\boldsymbol{r},\boldsymbol{p},t).$$

The first correction to the particle current can be evaluated as

$$\begin{aligned}\tau j_i^1(\boldsymbol{r},t) &= -\frac{\tau}{m}\int \mathrm{d}^3 p\, p_i \left(\partial_t + \frac{1}{m}\boldsymbol{p}\cdot\boldsymbol{\nabla}_r + \boldsymbol{F}(\boldsymbol{r})\cdot\boldsymbol{\nabla}_p\right) f_0(\boldsymbol{r},\boldsymbol{p},t)\\ &= -\frac{\tau}{m}[m\partial_t(nv_i) + \partial_i(nkT) + m\partial_j(nv_iv_j) - nF_i].\end{aligned}$$

Since the total particle current is the same as that calculated from f_0 alone, we find that

$$\partial_t(nv_i) + \frac{1}{m}\partial_i(nkT) + \partial_j(nv_iv_j) - \frac{1}{m}nF_i = \mathrm{O}(\tau).$$

(c) For a stationary gas, the heat current is the same as the energy current. The lowest-order term vanishes, and to first order of the expansion in τ we find that

$$\begin{aligned}\boldsymbol{q}(\boldsymbol{r},t) &= -\frac{\tau}{m}\int \mathrm{d}^3 p\left(\frac{|\boldsymbol{p}|^2}{2m} + \phi\right)\boldsymbol{p}\left(\partial_t + \frac{1}{m}\boldsymbol{p}\cdot\boldsymbol{\nabla}_r + \boldsymbol{F}\cdot\boldsymbol{\nabla}_p\right) f_0(\boldsymbol{r},\boldsymbol{p},t)\\ &= -\frac{\tau}{m}\{\tfrac{5}{2}\boldsymbol{\nabla}[n(kT)^2] + \phi\boldsymbol{\nabla}(nkT) - (\tfrac{5}{2}nkT + n\phi)\boldsymbol{F}\}\\ &= -\frac{\tau}{m}\left[\frac{5nk^2T}{2}\boldsymbol{\nabla}T + (\tfrac{5}{2}kT + \phi)[\boldsymbol{\nabla}(nkT) + n\,\boldsymbol{\nabla}\phi]\right].\end{aligned}$$

When $\boldsymbol{v} = \mathbf{0}$, the result of (b) is $\nabla(nkT) + n\,\nabla\phi = O(\tau)$. This means that the acceleration of particles by the external force is just balanced by that due to the pressure gradient $\nabla P = \nabla(nkT)$. Consequently, to this order, the heat current is just

$$\boldsymbol{q}(\boldsymbol{r}, t) = -\frac{5nk^2T\tau}{2m}\nabla T.$$

The potential energy is irrelevant, because there is no net transport of particles.

• **Problem 6.19**

(a) If the variation in g is neglected, the gravitational potential energy of a molecule is $\phi(z) = mgz$, where $m = 29 \times 1.66 \times 10^{-27}$ kg $= 4.8 \times 10^{-26}$ kg is the mass of a molecule. For a stationary atmosphere, the result of problem 6.18(b) is $\mathrm{d}P/\mathrm{d}z = -mgn = -(mg/kT)P$. This result can, of course, be obtained by elementary means. The weight of a layer of air of unit area and thickness $\mathrm{d}z$ is $mgn\,\mathrm{d}z$ and this has to be balanced by a pressure difference $\mathrm{d}P$ between its upper and lower surfaces. For an isothermal atmosphere, this equation can be integrated immediately to give

$$P(z) = P_0\,\mathrm{e}^{-mgz/kT}.$$

(b) The mean free path of a molecule in a gas of number density n is $\lambda = 1/n\sigma$, where σ is the scattering cross-section. If the mean molecular speed is $\bar{c}$, then the mean collision time is $\tau = \lambda/\bar{c} = 1/n\bar{c}\sigma$. Using $n = P/kT$ and $\bar{c} \propto \sqrt{T}$, we find that $\tau \propto \sqrt{T}/P$. For future use, we write this as

$$\tau = \tau_0\left(\frac{T}{T_0}\right)^{1/2}\frac{P_0}{P}$$

where τ_0 is the relaxation time in the atmosphere at sea level. (This simple derivation assumes that the distance travelled between collisions and the speed of a molecule can be averaged independently. A more sophisticated treatment will change only a numerical factor, which is absorbed in τ_0.)

(c) In a steady state, and with $\boldsymbol{v} = \mathbf{0}$, the equation for conservation of energy given in problem 6.18(a) gives

$$\nabla\cdot\boldsymbol{q} = \nabla\cdot\boldsymbol{j}_\varepsilon = -\partial_t\varepsilon + \frac{1}{m}\int \mathrm{d}^3p\,\boldsymbol{F}\cdot\boldsymbol{p}f = 0.$$

So in the present case $\mathrm{d}q_z/\mathrm{d}z = 0$, or $q_z =$ constant. Combining this with the results of (b) and of problem 6.18(c), we get

$$q_z = -\frac{5kP\tau}{2m}\frac{\mathrm{d}T}{\mathrm{d}z} = -\frac{5k\tau_0P_0}{2m}\left(\frac{T}{T_0}\right)^{1/2}\frac{\mathrm{d}T}{\mathrm{d}z}$$

showing that $\sqrt{T}\,\mathrm{d}T/\mathrm{d}z$ is a constant, which must be equal $-\sqrt{T_0}\alpha_0$:

$$\sqrt{T}\frac{\mathrm{d}T}{\mathrm{d}z} = -\sqrt{T_0}\alpha_0.$$

The solution to this equation is

$$T(z) = T_0\left(1 - \frac{3\alpha_0}{2T_0}z\right)^{2/3}$$

and we see that this becomes complex when z is greater than $z_{\max} = 2T_0/3\alpha_0 \simeq$ 32 km. When z is smaller than this value, we can integrate the equation

$$\frac{\mathrm{d}P}{\mathrm{d}z} = -\frac{mg}{kT(z)}P$$

to obtain

$$P = P_0 \exp\left\{\frac{2mg}{k\alpha_0}\left[\left(1 - \frac{3\alpha_0}{2T_0}z\right)^{1/3} - 1\right]\right\}.$$

When z is small enough for the temperature not to have changed significantly, this reduces to our previous result.

(d) At $z = z_{\max}$, the temperature goes to zero, and the temperature gradient, being proportional to $T^{-1/2}$, becomes infinite. This is because the thermal conductivity $\kappa \propto \sqrt{T}$ vanishes, while the heat current is constant. The fact that κ becomes very small indicates the breakdown of a crucial assumption, namely that we have a steady state in which the atmosphere is locally in thermal equilibrium, so that $P(z)$ and $T(z)$ are well defined. If heat cannot be transported, then this equilibrium cannot be established. A rough criterion for the existence of approximate local equilibrium is that the ratio $R \equiv (\lambda/T)|\,\mathrm{d}T/\mathrm{d}z|$ should be very small. This means that the fractional change in the local temperature seen by a molecule that travels a distance λ is small. To apply this idea, we need to estimate the value of λ from the information to hand. Since $\lambda = \bar{c}\tau$ and $\kappa = 5kP\tau/2m$, we can eliminate the unknown relaxation time to express λ in terms of κ. Up to a numerical factor of order 1, the mean speed is $\bar{c} = \sqrt{kT/m}$. We also know that κ is proportional to $\sqrt{T}$ and independent of pressure. Using this information, it is straightforward to obtain the estimate

$$\lambda \simeq \sqrt{\frac{m}{kT_0}}\frac{\kappa_0 T_0}{P_0}\frac{P_0}{P}\frac{T}{T_0} \simeq \frac{P_0}{P}\frac{T}{T_0} \times 10^{-7}\ \mathrm{m}$$

where we have again ignored numerical factors of order 1. Then, using our result that $\sqrt{T}\,\mathrm{d}T/\mathrm{d}z = -\sqrt{T_0}\alpha_0$, we estimate the ratio R as

$$R \equiv \frac{\lambda}{T}\left|\frac{\mathrm{d}T}{\mathrm{d}z}\right| = \sqrt{\frac{m}{kT_0}}\frac{\kappa_0\alpha_0}{P_0}\frac{P_0}{P}\left(\frac{T_0}{T}\right)^{1/2} \simeq 5 \times 10^{-12}\frac{P_0}{P}\left(\frac{T_0}{T}\right)^{1/2}.$$

At sea level, the value $R = R_0 \simeq 5 \times 10^{-12}$ is very small, as we hoped it would be. The results of (c) allow us to express the variation with altitude of several interesting quantities as

$$P = P_0 \,\mathrm{e}^{-\gamma(1-x)} \qquad T = T_0 x^2 \qquad n = n_0 \frac{\mathrm{e}^{-\gamma(1-x)}}{x^2}$$

$$\lambda = \lambda_0 x^2 \,\mathrm{e}^{\gamma(1-x)} \qquad \tau = \tau_0 x \,\mathrm{e}^{\gamma(1-x)} \qquad R = R_0 \frac{\mathrm{e}^{\gamma(1-x)}}{x}$$

where $x = (1 - z/z_{\max})^{1/3}$ and $\gamma = 2mg/k\alpha_0 \simeq 11.4$. As z increases towards $z_{\max}$, the variable x decreases to zero, and R becomes infinite. Somewhat surprisingly, we see that the number density n also becomes infinite, with the result that λ and τ actually go to zero. This happens over a rather small distance, however. It is easily found that n decreases, as we would expect, to a minimum value $n_{\min} \simeq 3 \times 10^{-3} n_0$ when $x = 2/\gamma \simeq 0.2$, while λ increases to a maximum value $\lambda_{\max} \simeq 360\lambda_0$ at the same altitude. We might perhaps take this value of x, for which $R \simeq 4.5 \times 10^4 R_0$ is still fairly small, as the limit of validity of our calculations. In terms of altitude, however, we have $z = (1 - x^3) z_{\max}$, and this is close to $z_{\max}$ whenever x is significantly smaller than 1. Consequently, the analysis is reasonably self-consistent up to altitudes not much below $z_{\max}$. Needless to say, however, there are many assumptions (neglect of convective and radiative processes, of the correct chemical composition, of the effects of the sun's radiation, etc) which make this simple model far from realistic.

(e) The gravitational force on a gas particle at a distance r from the centre of the Earth is, of course, proportional to $1/r^2$. It can be written as $F(r) = -mg(r_0/r)^2$, where r_0 is the Earth's radius and g is the acceleration due to gravity at the Earth's surface. The equation obeyed by the pressure in an atmosphere at constant temperature T_0 is therefore $\mathrm{d}P/\mathrm{d}r = -(mgr_0^2/kT_0)r^{-2}P$, and the solution (with $P(r_0) = P_0$) is

$$P(r) = P_0 \exp\left[-\Lambda\left(1 - \frac{r_0}{r}\right)\right]$$

where $\Lambda = mgr_0/kT_0 \simeq 758$. In contrast with the situation considered in (a), we see that the pressure does not vanish when $r \to \infty$. Rather, it approaches the limiting value $P_\infty = P_0 \,\mathrm{e}^{-\Lambda} \simeq 10^{-329} P_0 \simeq 10^{-324}$ Pa. The corresponding number density is $n_\infty = P_\infty/kT_0 \simeq 10^{-303}\,\mathrm{m}^{-3}$. Alternatively, the linear size of a region containing one molecule of this gas is about 10^{101} m. To put this absurdly small density in context, we note that the present age of the Universe is about 10^{10} years $\simeq 10^{17}$ s and the size of the observable Universe is, roughly speaking, the distance that a light ray can travel in this time, namely about 10^{25} m.

Clearly, this limiting pressure is of no practical importance. Indeed, the correction to the simple result found in (a) is completely negligible. In terms of the altitude $z = r - r_0$, our result for the pressure is

$$P(z) = P_0 \exp\left[-\frac{\Lambda}{r_0} z \left(1 + \frac{z}{r_0}\right)^{-1}\right].$$

For $z \ll r_0$, the expression $P(z) \simeq \mathrm{e}^{-\Lambda z/r_0}$ coincides with that found in (a). The characteristic distance over which P falls by a factor of e is $r_0/\Lambda \simeq 8$ km; so, at altitudes of the order of $z \sim r_0$, where the factor $(1 + z/r_0)^{-1}$ begins to differ significantly from unity, the pressure has already become negligibly small.

• **Problem 6.20** From problem 6.17, we know that in a steady state with no convection

$$\boldsymbol{\nabla}(nkT) = \mathbf{0} \qquad \boldsymbol{\nabla} \cdot \boldsymbol{q} = -\boldsymbol{\nabla} \cdot (\kappa \boldsymbol{\nabla} T) = 0$$

where $\boldsymbol{q}$ is the heat current and $\kappa = 5nk^2 T\tau/2m$ is the thermal conductivity. The first relation tells us that the pressure is uniform—any pressure gradient would drive a bulk flow of particles, unless it is balanced by an external force. The second tells us that the net flow of heat into or out of any volume is zero, since no energy is created or destroyed inside the volume and the internal energy is constant in a steady state.

(a) Assuming that κ is constant, we have

$$\nabla^2 T = \frac{1}{r}\frac{\mathrm{d}^2}{\mathrm{d}r^2}(rT) = 0$$

since temperature can vary only in the radial direction. Thus, $\mathrm{d}(rT)/\mathrm{d}r$ is constant, and the solution has the form

$$T(r) = A + \frac{B}{r}$$

where A and B are constants of integration. Using the boundary conditions $T(a) = T_1$ and $T(b) = T_2$, we get

$$T(r) = \frac{1}{b-a}\left[b\left(1 - \frac{a}{r}\right)T_2 - a\left(1 - \frac{b}{r}\right)T_1\right].$$

(b) As discussed in problem 6.19(b), the relaxation time τ that appears in the expression for κ is proportional to $\sqrt{T}/P$. Thus we have $\kappa \propto P\tau \propto \sqrt{T}$ and

$$\boldsymbol{\nabla} \cdot (\sqrt{T}\,\boldsymbol{\nabla} T) = \tfrac{2}{3}\nabla^2\left(T^{3/2}\right) = 0.$$

The solution to this equation can be written down immediately by replacing T in the solution to (a) with $T^{3/2}$. It is

$$T(r) = \left\{\frac{1}{b-a}\left[b\left(1 - \frac{a}{r}\right)T_2^{3/2} - a\left(1 - \frac{b}{r}\right)T_1^{3/2}\right]\right\}^{2/3}.$$

To compare these two solutions, let us define $x = (r - a)/(b - a)$ and, in the case of a constant κ, write

$$T(r) = T_1 + \Delta T\,\theta(x)$$

with $\Delta T = T_2 - T_1$. The function $\theta(x)$ is

$$\theta(x) = \frac{x}{x + (a/b)(1 - x)}.$$

In the case that $\kappa \propto \sqrt{T}$ the solution becomes

$$\begin{aligned} T &= [T_1^{3/2} + (T_2^{3/2} - T_1^{3/2})\theta(x)]^{2/3} \\ &= T_1\left\{1 + \left[\left(1 + \frac{\Delta T}{T_1}\right)^{3/2} - 1\right]\theta(x)\right\}^{2/3} \\ &= T_1\left\{1 + \frac{\Delta T}{T_1}\theta(x) + \mathrm{O}\left[\left(\frac{\Delta T}{T_1}\right)^2\right]\right\} \end{aligned}$$

and we see that this is approximately the same as the first solution when $\Delta T \ll T_1$.

• **Problem 6.21** For orientation, we begin with a brief discussion of the Wigner distribution function in the case of particles which interact with an external potential $\Phi(\boldsymbol{r})$, but not with each other. The state of a system consisting of a single particle can be described by the density matrix

$$\rho(\boldsymbol{r}, \boldsymbol{r}', t) = \sum_{m,n} \rho_{mn} \psi_m(\boldsymbol{r}, t) \psi_n^*(\boldsymbol{r}', t)$$

where the wavefunctions $\psi_n(\boldsymbol{r}, t)$ are a complete set of solutions to the Schrödinger equation

$$\mathrm{i}\hbar \partial_t \psi(\boldsymbol{r}, t) = -\frac{\hbar^2}{2m}\nabla^2 \psi(\boldsymbol{r}, t) + \Phi(\boldsymbol{r})\psi(\boldsymbol{r}, t).$$

The matrix ρ_{mn}, which is Hermitian ($\rho_{mn}^* = \rho_{nm}$) and has unit trace ($\mathrm{Tr}\,\boldsymbol{\rho} = \sum_n \rho_{nn} = 1$) represents our statistical uncertainty about which pure quantum state the particle is actually in. Thus, at any instant of time, the density matrix can be diagonalized, by expressing it in terms of a suitable basis $\{\phi_n(\boldsymbol{r})\}$ as $\rho(\boldsymbol{r}, \boldsymbol{r}', t) = \sum_n P_n(t)\phi_n(\boldsymbol{r})\phi_n^*(\boldsymbol{r}')$, and then the diagonal elements P_n give the statistical probabilities for finding the particle in the quantum state $\phi_n(\boldsymbol{r})$. In terms of the density matrix, the probability density for finding the particle near the point $\boldsymbol{r}$ is $n(\boldsymbol{r}, t) = \rho(\boldsymbol{r}, \boldsymbol{r}, t)$ while the probability density for finding it to have a momentum near $\boldsymbol{p}$ is

$$\tilde{n}(\boldsymbol{p}, t) = \frac{1}{(2\pi\hbar)^3}\int \mathrm{d}^3r \int \mathrm{d}^3r'\, \mathrm{e}^{-\mathrm{i}\boldsymbol{p}\cdot(\boldsymbol{r}-\boldsymbol{r}')/\hbar}\, \rho(\boldsymbol{r}, \boldsymbol{r}', t).$$

These probabilities encode both the indeterminacy inherent in quantum mechanics and the uncertainty arising from our ignorance of the exact microscopic state.

To apply this description to a many-particle system, we reinterpret the probability P_n as counting the average number of particles occupying the nth state, so the normalization is now determined by the condition $\sum_n \rho_{nn} = N$, where N is the total number of particles, and the function $n(\boldsymbol{r}, t)$ is the mean number of particles per unit volume to be found near the point $\boldsymbol{r}$ at time t. Because of the uncertainty principle, it is not possible to define a joint probability density for a particle *both* to be near the point $\boldsymbol{r}$ *and* to have a momentum near $\boldsymbol{p}$. Similarly, we cannot count the number of particles that are near the point $\boldsymbol{r}$ *and* have a momentum near $\boldsymbol{p}$ (which is what the classical distribution function does). Nevertheless, the Wigner distribution function

$$f_{\mathrm{W}}(\boldsymbol{r}, \boldsymbol{p}, t) = \frac{1}{(2\pi\hbar)^3} \int \mathrm{d}^3\xi \, \mathrm{e}^{-\mathrm{i}\boldsymbol{p}\cdot\boldsymbol{\xi}/\hbar} \, \rho(\boldsymbol{r} + \tfrac{1}{2}\boldsymbol{\xi}, \boldsymbol{r} - \tfrac{1}{2}\boldsymbol{\xi}, t)$$

does have the property that

$$\int \mathrm{d}^3 p \, f_{\mathrm{W}}(\boldsymbol{r}, \boldsymbol{p}, t) = n(\boldsymbol{r}, t) \qquad \int \mathrm{d}^3 r \, f_{\mathrm{W}}(\boldsymbol{r}, \boldsymbol{p}, t) = \tilde{n}(\boldsymbol{p}, t)$$

and for many purposes it can be treated in much the same way as the classical distribution function. By using the Schrödinger equation and its complex conjugate, the identity

$$\begin{aligned}[\nabla_r^2 \psi_m(\boldsymbol{r} + \tfrac{1}{2}\boldsymbol{\xi}, t)]\psi_n^*(\boldsymbol{r} - \tfrac{1}{2}\boldsymbol{\xi}, t) - \psi_m(\boldsymbol{r} + \tfrac{1}{2}\boldsymbol{\xi}, t)[\nabla_r^2 \psi_n^*(\boldsymbol{r} - \tfrac{1}{2}\boldsymbol{\xi}, t)]& \\ = 2\boldsymbol{\nabla}_r \cdot \boldsymbol{\nabla}_\xi[\psi_m(\boldsymbol{r} + \tfrac{1}{2}\boldsymbol{\xi}, t)\psi_n^*(\boldsymbol{r} - \tfrac{1}{2}\boldsymbol{\xi}, t)]&\end{aligned}$$

and the formal expression

$$\Phi(\boldsymbol{r} \pm \tfrac{1}{2}\boldsymbol{\xi}) \, \mathrm{e}^{-\mathrm{i}\boldsymbol{p}\cdot\boldsymbol{\xi}/\hbar} = \Phi(\boldsymbol{r} \pm \tfrac{1}{2}\mathrm{i}\hbar\boldsymbol{\nabla}_p) \, \mathrm{e}^{-\mathrm{i}\boldsymbol{p}\cdot\boldsymbol{\xi}/\hbar}$$

the reader should find it straightforward to show that

$$\left(\partial_t + \frac{1}{m}\boldsymbol{p} \cdot \boldsymbol{\nabla}_r + \boldsymbol{F}(\boldsymbol{r}) \cdot \boldsymbol{\nabla}_p\right) f_{\mathrm{W}}(\boldsymbol{r}, \boldsymbol{p}, t) = \mathrm{O}(\hbar)$$

where the external force is $\boldsymbol{F}(\boldsymbol{r}) = -\boldsymbol{\nabla}_r \Phi(\boldsymbol{r})$. When the potential varies only slowly with position, the $\mathrm{O}(\hbar)$ terms are negligible. For the purpose of this and subsequent problems, we assume that the effects of a rapidly varying potential, such as that provided by an ionic lattice, and of scattering of one particle by another can adequately be treated by adding an appropriate collision term to the right-hand side, giving an equation closely analogous to the classical Boltzmann equation. However, more sophisticated methods are needed to investigate this in detail.

In the case of binary collisions, the collision integral should be a difference of gain and loss terms that give the rates at which particles of momentum $\boldsymbol{p}$ are produced and absorbed respectively in collisions. Given a particle with momentum $\boldsymbol{p}$ and another with momentum $\boldsymbol{p}_2$, suppose that the transition probability per unit time to produce a pair of particles with momenta $\boldsymbol{p}_1'$ and $\boldsymbol{p}_2'$ is $W_{(\boldsymbol{p},\boldsymbol{p}_2)\to(\boldsymbol{p}_1',\boldsymbol{p}_2')}$. Then the total rate at which particles with momentum $\boldsymbol{p}$ are lost in such collisions is

$$\left.\frac{\mathrm{d}f_{\mathrm{W}}}{\mathrm{d}t}\right|_{\mathrm{loss}} = \int \mathrm{d}^3p_2\,\mathrm{d}^3p_1'\,\mathrm{d}^3p_2'\,Wf(\boldsymbol{p})f(\boldsymbol{p}_2)[1-f(\boldsymbol{p}_1')][1-f(\boldsymbol{p}_2')]$$

where $f(\boldsymbol{p}_i)$ stands for $cf_{\mathrm{W}}(\boldsymbol{r},\boldsymbol{p}_i,t)$, the factor c being a normalization constant that we shall discuss shortly. The factor $f(\boldsymbol{p})f(\boldsymbol{p}_2)$ gives the probability that the two initial particles are actually present in the appropriate region of space while, for fermions, the factor $[1-f(\boldsymbol{p}_1')][1-f(\boldsymbol{p}_2')]$ accounts for the fact that the collision is forbidden by the Pauli exclusion principle if the final states are already occupied. (For bosons, there is an enhancement factor $[1+f(\boldsymbol{p}_1')][1+f(\boldsymbol{p}_2')]$.) The integrals serve to sum over all possible initial and final states that might be involved. For spin-$\frac{1}{2}$ particles, we must also take account of the two possible spin polarizations. To avoid complicating the notation further, we can take the symbol $\boldsymbol{p}_i$ to stand for $(\boldsymbol{p}_i, s_i)$ and each momentum integral to include the sum $\sum_{s_i}$. If this expression is to make sense, the function $f(\boldsymbol{p})$ must clearly be the occupation probability for a single quantum state. Since the Wigner function was defined, by analogy with its classical counterpart, so that $\int \mathrm{d}^3p\, f_{\mathrm{W}}(\boldsymbol{r},\boldsymbol{p},t) = n(\boldsymbol{r},t)$ is the number density, we must have $f_{\mathrm{W}}(\boldsymbol{r},\boldsymbol{p},t)\,\mathrm{d}^3p = f(\boldsymbol{p})g(\boldsymbol{p})\,\mathrm{d}^3p$, where $g(\boldsymbol{p})\,\mathrm{d}^3p$ is the number of states per unit volume in the momentum range d^3p near $\boldsymbol{p}$. To simplify matters, we assume that the particles move freely between collisions, in which case the density of states (including spin polarizations) is $g(\boldsymbol{p}) = 2(2\pi\hbar)^{-3}$ (see, for example, problem 4.2) and $c = (2\pi\hbar)^3/2$. Correspondingly, the transition probability W must be normalized so that $(\mathrm{d}f_{\mathrm{W}}/\mathrm{d}t)|_{\mathrm{loss}}\,\mathrm{d}^3p$ is the number of collisions per unit time per unit volume made by particles with momentum in the range d^3p.

The gain term of course represents the effect of collisions that produce a particle of momentum $\boldsymbol{p}$ in the final state:

$$\left.\frac{\mathrm{d}f_{\mathrm{W}}}{\mathrm{d}t}\right|_{\mathrm{gain}} = \int \mathrm{d}^3p_2\,\mathrm{d}^3p_1'\,\mathrm{d}^3p_2'\,Wf(\boldsymbol{p}_1')f(\boldsymbol{p}_2')[1-f(\boldsymbol{p})][1-f(\boldsymbol{p}_2)].$$

Provided that the interaction causing the scattering is time reversal invariant (the only known exception is a weak interaction effect in the decay of neutral kaons and perhaps certain other mesons), the transition probabilities W appearing in the loss and gain terms are the same. The net collision integral is therefore (in a further abbreviated notation which should be obvious)

$$\left.\frac{\mathrm{d}f_{\mathrm{W}}}{\mathrm{d}t}\right|_{\mathrm{coll}} = \left.\frac{\mathrm{d}f_{\mathrm{W}}}{\mathrm{d}t}\right|_{\mathrm{gain}} - \left.\frac{\mathrm{d}f_{\mathrm{W}}}{\mathrm{d}t}\right|_{\mathrm{loss}}$$

$$= \int \mathrm{d}p\, W[f_{1'} f_{2'}(1-f)(1-f_2) - f f_2(1-f_{1'})(1-f_{2'})].$$

This collision integral ought to vanish when the system is in a state of local equilibrium. To see the condition for this, let us write the equilibrium distribution function as

$$f_0(\boldsymbol{r}, \boldsymbol{p}, t) = \frac{1}{\exp[\chi(\boldsymbol{r}, \boldsymbol{p}, t)] + 1}.$$

The integrand then contains the factor

$$\exp[\chi(\boldsymbol{r}, \boldsymbol{p}, t) + \chi(\boldsymbol{r}, \boldsymbol{p}_2, t)] - \exp[\chi(\boldsymbol{r}, \boldsymbol{p}_1', t) + \chi(\boldsymbol{r}, \boldsymbol{p}_2', t)]$$

which vanishes if χ is a linear combination of quantities χ_i which are additively conserved in the scattering process:

$$\chi(\boldsymbol{r}, \boldsymbol{p}, t) = \sum_i \alpha_i(\boldsymbol{r}, t)\chi_i(\boldsymbol{p}).$$

To be specific, the transition probability W contains δ functions which vanish unless $\chi(\boldsymbol{p}) + \chi(\boldsymbol{p}_2) = \chi(\boldsymbol{p}_1') + \chi(\boldsymbol{p}_2')$ and so the whole integrand vanishes identically when $f = f_0$. For a system of identical particles, the candidates for the conserved quantities $\chi_i(\boldsymbol{p})$ are a constant (say, the particle number), the momentum and the kinetic energy. The corresponding coefficients $\alpha_i(\boldsymbol{r}, t)$ can be identified in terms of a local temperature, chemical potential and drift velocity, so we finally obtain

$$\chi(\boldsymbol{r}, \boldsymbol{p}, t) = \beta(\boldsymbol{r}, t)\left(\frac{1}{2m}|\boldsymbol{p} - m\boldsymbol{v}(\boldsymbol{r}, t)|^2 - \mu(\boldsymbol{r}, t)\right).$$

The function f_0 that we have constructed makes the collision integral vanish, but it is not necessarily a solution of the Boltzmann equation. If the external forces are time independent, we obtain a steady-state solution when the temperature, chemical potential and drift velocity are also time independent. To obtain an equilibrium solution in the absence of external forces, these quantities must also be independent of position. In that case, we recover the Fermi–Dirac distribution of equilibrium statistical mechanics.

For the important case of electrons moving in an ionic lattice (which is considered in subsequent problems), the most important collisions are not with other electrons but with fixed scattering centres and with phonons. After averaging over the phonon distribution, the collision integral for electrons alone can be taken to have the form

$$\left.\frac{\mathrm{d}f_{\mathrm{W}}}{\mathrm{d}t}\right|_{\mathrm{coll}} = \int \mathrm{d}^3p'\, W\{f(\boldsymbol{p}')[1 - f(\boldsymbol{p})] - f(\boldsymbol{p})[1 - f(\boldsymbol{p}')]\}.$$

We can assume that kinetic energy is conserved, because the typical phonon energy is much smaller than the Fermi energy of an electron, but the momentum carried by the electrons is not conserved, since arbitrary amounts of momentum can be absorbed by the host material. In the function f_0, therefore, the drift velocity (as measured in the rest frame of the host material) must be zero.

• **Problem 6.22** In the relaxation time approximation, the Wigner distribution function for the electron gas obeys the equation

$$\left(\partial_t + \frac{1}{m}\boldsymbol{p}\cdot\boldsymbol{\nabla}_r + \boldsymbol{F}\cdot\boldsymbol{\nabla}_p\right) f_{\mathrm{W}}(\boldsymbol{r},\boldsymbol{p},t) = -\frac{1}{\tau}[f_{\mathrm{W}}(\boldsymbol{r},\boldsymbol{p},t) - f_{\mathrm{W}}^{(0)}(\boldsymbol{r},\boldsymbol{p},t)]$$

where, in SI units, the Lorentz force on electrons of charge $-e$ is

$$\boldsymbol{F} = -e\left(\boldsymbol{E} + \frac{1}{m}\boldsymbol{p}\times\boldsymbol{B}\right).$$

We shall consider the electric field $\boldsymbol{E}$ and the magnetic induction $\boldsymbol{B}$ to be uniform and constant in time. In problem 6.21, we discussed the meaning of the Wigner function and remarked that, for electrons that are scattered principally by impurities and phonons in the host lattice, the momentum of electrons alone is not conserved in collisions. For that reason, the treatment of this system differs in two respects from that of ordinary gases considered in previous problems. The first is that the background distribution function $f_{\mathrm{W}}^{(0)}(\boldsymbol{r},\boldsymbol{p},t)$ cannot contain a drift velocity and depends on momentum only through the kinetic energy $\varepsilon(\boldsymbol{p}) = |\boldsymbol{p}|^2/2m$. Thus, the particle current in this background state is zero. However, the current in the state described by f_{W} is no longer constrained by momentum conservation to be the same as that in the background state and does not necessarily vanish. The second difference is that momentum conservation no longer gives us a relation between the temperature and density gradients. We therefore need an extra assumption (which we shall not be able to justify within our approximation). The simplest possibility is to assume that the chemical potential is constant. For the electrons in a metal, this is probably reasonable, because the chemical potential is, essentially, the Fermi energy, which is much larger than any relevant thermal energy.

To find the electrical conductivity and the Hall coefficient, we look for a spatially uniform steady-state solution to the approximate Boltzmann equation, which then reads

$$-e\left(\boldsymbol{E} + \frac{1}{m}\boldsymbol{p}\times\boldsymbol{B}\right)\cdot\boldsymbol{\nabla}_p f_{\mathrm{W}}(\boldsymbol{p}) = -\frac{1}{\tau}[f_{\mathrm{W}}(\boldsymbol{p}) - f_{\mathrm{W}}^{(0)}(\boldsymbol{p})].$$

We multiply this equation by p_i and integrate over the momentum. On the left-hand side, we can integrate by parts, using the fact that f_{W} vanishes exponentially as $|\boldsymbol{p}| \to \infty$, with the result

$$e\int \mathrm{d}^3p\left(E_i + \frac{1}{m}(\boldsymbol{p}\times\boldsymbol{B})_i\right) f_{\mathrm{W}}(\boldsymbol{p}) = -\frac{1}{\tau}\int \mathrm{d}^3p\, p_i[f_{\mathrm{W}}(\boldsymbol{p}) - f_{\mathrm{W}}^{(0)}(\boldsymbol{p})].$$

We have also used the fact that $\nabla_p \cdot (\boldsymbol{p} \times \boldsymbol{B}) \equiv 0$. On the right-hand side, the term involving $f_{\mathrm{W}}^{(0)}$ vanishes because $f_{\mathrm{W}}^{(0)}$ is an even function of $\boldsymbol{p}$. This result can be written simply as

$$en\boldsymbol{E} + \boldsymbol{B} \times \boldsymbol{j} = \frac{m}{e\tau}\boldsymbol{j}$$

where $n = \int \mathrm{d}^3p\, f_{\mathrm{W}}(\boldsymbol{p})$ is the number density of electrons and $\boldsymbol{j} = (-e/m)\int \mathrm{d}^3p\, \boldsymbol{p} f_{\mathrm{W}}(\boldsymbol{p})$ is the electric current.

In the absence of a magnetic field, we have $\boldsymbol{j} = \sigma \boldsymbol{E}$, with the conductivity given by $\sigma = ne^2\tau/m$, in agreement with the Drude theory and with the result obtained in problem 6.13 on the basis of the Langevin equation. To obtain the Hall coefficient, we consider a rectangular sample, with an electric field applied in the x direction and a magnetic field applied in the z direction. Electrons accelerated by the electric field are deflected in the y direction by the magnetic field, causing electric charge to build up on opposite faces of the sample. In practice, this causes a component of electric field to appear in the y direction, and this will increase in magnitude until the current in the y direction is zero. This effect is not apparent from our equations, which neglect forces acting between electrons. From the point of view of electrons in the bulk of the sample, however, the effect of the surface charge is equivalent to that of an externally applied field. We can therefore simply adjust E_y to the value needed to make j_y vanish. With non-zero field components E_x, E_y and B_z, our result reads

$$enE_x = B_z j_y + \frac{m}{e\tau} j_x \qquad enE_y = -B_z j_x + \frac{m}{e\tau} j_y.$$

If we now require $j_y = 0$, the first equation just reproduces $j_x = \sigma E_x$ and from the second we deduce the Hall coefficient

$$R_{\mathrm{H}} \equiv \frac{E_y}{j_x B_z} = -\frac{1}{ne}.$$

The negative sign corresponds to the fact that electrons are negatively charged. For particles of charge q, the Hall coefficient is $R_{\mathrm{H}} = 1/nq$. Its sign provides a standard method of determining the sign of the predominant charge carriers in a given material.

We obtain expressions for the thermal conductivity and the thermoelectric power (or thermopower for short) by considering the situation in which there is a small temperature gradient, but no electric current flows. Now, a temperature gradient tends to cause a flow of electrons (this is the thermoelectric effect), so we need to apply a small electric field to ensure that no current actually flows. The source of this electric field might be an external apparatus or a charge density formed on the surface of an electrically isolated sample when the temperature gradient is established. In the steady state, the approximate Boltzmann equation reads

$$\left(\frac{1}{m}\boldsymbol{p} \cdot \nabla_r - e\boldsymbol{E} \cdot \nabla_p\right) f_{\mathrm{W}}(\boldsymbol{r}, \boldsymbol{p}) = -\frac{1}{\tau}[f_{\mathrm{W}}(\boldsymbol{r}, \boldsymbol{p}) - f_{\mathrm{W}}^{(0)}(\boldsymbol{r}, \boldsymbol{p})].$$

According to the results of problem 6.21, the background distribution is

$$f_{\mathrm{W}}^{(0)}(\boldsymbol{r},\boldsymbol{p}) = \frac{2}{(2\pi\hbar)^3}\frac{1}{\exp\{[\varepsilon(\boldsymbol{p})-\varepsilon_{\mathrm{F}}]/kT(\boldsymbol{r})\}+1}$$

where $\varepsilon(\boldsymbol{p}) = |\boldsymbol{p}|^2/2m$ and, for a highly degenerate gas with $T \ll T_{\mathrm{F}}$, the chemical potential can be taken as just the Fermi energy. The Fermi energy is given, in a form that will be useful below, by

$$(2m\varepsilon_{\mathrm{F}})^{3/2} = \frac{3(2\pi\hbar)^3 n}{8\pi}$$

(see problem 4.18). Strictly, the number density n will vary slightly with position on account of the temperature gradient but, since we shall work only to lowest order in the gradient, this does not concern us. In this background state, both the electric current and the heat current vanish, so we need to find the first correction, given that the temperature gradient and electric field are small. Formally, this is equivalent to expanding in powers of the relaxation time τ, and we obtain

$$f_{\mathrm{W}}(\boldsymbol{r},\boldsymbol{p}) = f_{\mathrm{W}}^{(0)}(\boldsymbol{r},\boldsymbol{p}) - \tau\left(\frac{1}{m}\boldsymbol{p}\cdot\boldsymbol{\nabla}_r - e\boldsymbol{E}\cdot\boldsymbol{\nabla}_p\right)f_{\mathrm{W}}^{(0)}(\boldsymbol{r},\boldsymbol{p}) + \cdots.$$

The electric current $\boldsymbol{j}(\boldsymbol{r})$ and the heat current $\boldsymbol{q}(\boldsymbol{r})$ are then given by

$$j_i = e\tau\int \mathrm{d}^3p\,\frac{p_i}{m}\left(\frac{1}{m}\boldsymbol{p}\cdot\boldsymbol{\nabla}_r - e\boldsymbol{E}\cdot\boldsymbol{\nabla}_p\right)f_{\mathrm{W}}^{(0)}(\boldsymbol{r},\boldsymbol{p})$$

$$q_i = -\tau\int \mathrm{d}^3p\,\frac{p_i}{m}\frac{|\boldsymbol{p}|^2}{2m}\left(\frac{1}{m}\boldsymbol{p}\cdot\boldsymbol{\nabla}_r - e\boldsymbol{E}\cdot\boldsymbol{\nabla}_p\right)f_{\mathrm{W}}^{(0)}(\boldsymbol{r},\boldsymbol{p}).$$

The terms involving the electric field can again be integrated by parts and, since $f_{\mathrm{W}}^{(0)}$ is an even function of $\boldsymbol{p}$, the quantity $p_i p_j$ inside the integral can be replaced by $\delta_{ij}|\boldsymbol{p}|^2/3$. In this way, we obtain

$$\boldsymbol{j}(\boldsymbol{r}) = e\tau\left(\frac{1}{3m^2}\boldsymbol{\nabla}J_1(\boldsymbol{r}) + \frac{e}{m}\boldsymbol{E}J_0(\boldsymbol{r})\right)$$

$$\boldsymbol{q}(\boldsymbol{r}) = -\tau\left(\frac{1}{6m^3}\boldsymbol{\nabla}J_2(\boldsymbol{r}) + \frac{5e}{6m^2}\boldsymbol{E}J_1(\boldsymbol{r})\right)$$

where the remaining integrals are

$$\begin{aligned}J_n(\boldsymbol{r}) &= \int \mathrm{d}^3p\,|\boldsymbol{p}|^{2n} f_{\mathrm{W}}^{(0)}(\boldsymbol{r},\boldsymbol{p}) \\ &= \frac{4\pi(2m)^{n+3/2}}{(2\pi\hbar)^3}\int_0^\infty \mathrm{d}\varepsilon\,\frac{\varepsilon^{n+1/2}}{\exp[(\varepsilon-\varepsilon_{\mathrm{F}})/kT(\boldsymbol{r})]+1} \\ &= \frac{4\pi(2m\varepsilon_{\mathrm{F}})^{n+3/2}}{(2\pi\hbar)^3(n+3/2)}\left[1 + \frac{(n+\frac{1}{2})(n+\frac{3}{2})\pi^2}{6}\left(\frac{kT(\boldsymbol{r})}{\varepsilon_{\mathrm{F}}}\right)^2 + \cdots\right]\end{aligned}$$

if we keep only the leading terms when $kT \ll \varepsilon_F$. In this approximation, the condition that $\boldsymbol{j} = \mathbf{0}$ gives the field necessary to balance the temperature gradient as $\boldsymbol{E} = S\boldsymbol{\nabla}T$, where

$$S = -\frac{\pi^2 k}{2e}\frac{kT}{\varepsilon_F}$$

is the thermopower. Finally, setting $\boldsymbol{E}$ equal to this value, we can evaluate the heat current as $\boldsymbol{q} = -\kappa\boldsymbol{\nabla}T$, with the thermal conductivity given by

$$\kappa = \frac{\pi^2}{3}\frac{nk^2T\tau}{m}.$$

• **Problem 6.23** In the relaxation time approximation, the Boltzmann equation for a spatially uniform system is

$$(\partial_t - e\boldsymbol{E}\cdot\boldsymbol{\nabla}_p) f_W(\boldsymbol{p}, t) = -\frac{1}{\tau}[f_W(\boldsymbol{p}, t) - f_W^{(0)}(\boldsymbol{p})].$$

We multiply this equation by $(-e/m)\boldsymbol{p}$ and integrate over momentum, to obtain

$$\left(\partial_t + \frac{1}{\tau}\right)\boldsymbol{j}(t) = \frac{ne^2}{m}\boldsymbol{E}(t)$$

where $\boldsymbol{j}(t)$ is the electric current and n is the number density (see problem 6.22 for details). The number density is independent of time. This follows from the equation of continuity for conservation of the number of electrons, $\partial_t n = -\boldsymbol{\nabla}_r \cdot \boldsymbol{j} = 0$ for a spatially uniform system.

Consider an applied electric field of the form $\boldsymbol{E}(t) = \boldsymbol{E}_0\,\mathrm{e}^{\mathrm{i}\omega t}$. We look for a complex solution for the current of the form $\boldsymbol{j}(t) = \sigma(\omega)\boldsymbol{E}_0\,\mathrm{e}^{\mathrm{i}\omega t + \mathrm{i}\phi(\omega)}$ and easily discover that

$$\sigma(\omega) = \frac{\sigma}{\sqrt{1 + (\omega\tau)^2}} \qquad \phi(\omega) = -\tan^{-1}(\omega\tau)$$

where $\sigma = ne^2\tau/m$ is the dc conductivity. Equivalently, we could define a complex conductivity $\tilde{\sigma}(\omega) = \sigma(\omega)\,\mathrm{e}^{\mathrm{i}\phi(\omega)}$ such that $\boldsymbol{j}(t) = \tilde{\sigma}(\omega)\boldsymbol{E}(t)$. Physically, of course, $\boldsymbol{E}(t)$ and $\boldsymbol{j}(t)$ must be real, and the phase angle ϕ just gives the relative phases of the current and field. For example, an electric field $\boldsymbol{E}(t) = \boldsymbol{E}_0\cos(\omega t)$ gives rise to a current density $\boldsymbol{j}(t) = \sigma(\omega)\boldsymbol{E}_0\cos(\omega t + \phi)$. At low frequencies $\omega \ll \tau^{-1}$, we naturally recover the dc conductivity. On the other hand, when $\omega \gg \tau^{-1}$, the conductivity decreases as $\sigma(\omega) \sim 1/\omega$. This can be understood as follows. The conductivity is proportional to the average drift velocity that an electron acquires through being accelerated by the electric field, with $\partial_t \boldsymbol{v} = -(e/m)\boldsymbol{E}$. At low frequencies, this acceleration continues in the same direction until the electron makes a collision; so the velocity acquired is roughly $\boldsymbol{v} = -(e\tau/m)\boldsymbol{E}$. However, if the field changes direction in a time of order ω^{-1} that is much shorter than the collision time, then the maximum velocity that an electron acquires in a given direction is roughly $\boldsymbol{v} = -(e\omega^{-1}/m)\boldsymbol{E}_0$. It should be borne in mind, however, that the typical collision time in a metal is of the order of 10^{-14} s. Consequently, the frequency dependence that we have found will become apparent only at frequencies greater than, say, 10^{13} Hz, corresponding to infrared radiation with wavelengths shorter than about 10^{-5} m. Under these circumstances, it is necessary to consider in some detail the propagation of electromagnetic waves through the conducting medium, and the assumption of a spatially uniform state is unlikely to be valid.

• **Problem 6.24** The strategy used in problem 6.22 to find the electrical conductivity does not work when the relaxation time depends on energy. Instead, we must resort to an expansion of the distribution function in the temperature gradient and the applied field (or, more or less equivalently, in powers of τ) to find all the transport coefficients. Recall that the leading order contribution to both the electric and the heat current vanishes. At the next to leading order, they are given by

$$j_i(\boldsymbol{r}) = \frac{e}{m}\int \mathrm{d}^3p\, p_i\tau(\varepsilon)\left(\frac{1}{m}\boldsymbol{p}\cdot\boldsymbol{\nabla}_r - e\boldsymbol{E}\cdot\boldsymbol{\nabla}_p\right) f_{\mathrm{W}}^{(0)}(\boldsymbol{r},\boldsymbol{p})$$

$$q_i(\boldsymbol{r}) = -\frac{1}{m}\int \mathrm{d}^3p\, p_i\tau(\varepsilon)\varepsilon\left(\frac{1}{m}\boldsymbol{p}\cdot\boldsymbol{\nabla}_r - e\boldsymbol{E}\cdot\boldsymbol{\nabla}_p\right) f_{\mathrm{W}}^{(0)}(\boldsymbol{r},\boldsymbol{p})$$

with $\varepsilon(\boldsymbol{p}) = |\boldsymbol{p}|^2/2m$. The background distribution function is

$$f_{\mathrm{W}}^{(0)}(\boldsymbol{r},\boldsymbol{p}) = \frac{2}{(2\pi\hbar)^3}\frac{1}{\exp\{[\varepsilon(\boldsymbol{p})-\varepsilon_{\mathrm{F}}]/kT(\boldsymbol{r})\}+1}$$

and it is straightforward to see that

$$\boldsymbol{\nabla}_p f_{\mathrm{W}}^{(0)} = \frac{\boldsymbol{p}}{m}\frac{\partial f_{\mathrm{W}}^{(0)}}{\partial\varepsilon} \qquad \boldsymbol{\nabla}_r f_{\mathrm{W}}^{(0)} = -\frac{\boldsymbol{\nabla}T}{T}(\varepsilon-\varepsilon_{\mathrm{F}})\frac{\partial f_{\mathrm{W}}^{(0)}}{\partial\varepsilon}.$$

Thus, the currents can be expressed as

$$j_i(\boldsymbol{r}) = \frac{e}{m}\int \mathrm{d}^3p\, p_i\tau(\varepsilon)\left(-\frac{1}{m}\frac{\boldsymbol{p}\cdot\boldsymbol{\nabla}_r T}{T}(\varepsilon-\varepsilon_{\mathrm{F}}) - \frac{e}{m}\boldsymbol{p}\cdot\boldsymbol{E}\right)\frac{\partial f_{\mathrm{W}}^{(0)}}{\partial\varepsilon}$$

$$q_i(\boldsymbol{r}) = -\frac{1}{m}\int \mathrm{d}^3p\, p_i\tau(\varepsilon)\varepsilon\left(-\frac{1}{m}\frac{\boldsymbol{p}\cdot\boldsymbol{\nabla}_r T}{T}(\varepsilon-\varepsilon_{\mathrm{F}}) - \frac{e}{m}\boldsymbol{p}\cdot\boldsymbol{E}\right)\frac{\partial f_{\mathrm{W}}^{(0)}}{\partial\varepsilon}.$$

Since $f_{\mathrm{W}}^{(0)}$ depends on momentum only through the energy $\varepsilon(\boldsymbol{p})$, we can replace $p_i p_j$ inside the integrals with $|\boldsymbol{p}|^2\delta_{ij}/3 = (2m\varepsilon/3)\delta_{ij}$ and convert the momentum integral to an integral over ε. After an integration by parts, we get

$$\boldsymbol{j}(\boldsymbol{r}) = \frac{2e}{3m}\int_0^\infty \mathrm{d}\varepsilon\,\frac{\partial}{\partial\varepsilon}\left(g(\varepsilon)\tau(\varepsilon)\varepsilon(\varepsilon-\varepsilon_{\mathrm{F}})\frac{\boldsymbol{\nabla}T}{T} + g(\varepsilon)\tau(\varepsilon)\varepsilon e\boldsymbol{E}\right) f_0(\varepsilon)$$

$$\boldsymbol{q}(\boldsymbol{r}) = -\frac{2}{3m}\int_0^\infty \mathrm{d}\varepsilon\,\frac{\partial}{\partial\varepsilon}\left(g(\varepsilon)\tau(\varepsilon)\varepsilon^2(\varepsilon-\varepsilon_{\mathrm{F}})\frac{\boldsymbol{\nabla}T}{T} + g(\varepsilon)\tau(\varepsilon)\varepsilon^2 e\boldsymbol{E}\right) f_0(\varepsilon)$$

where

$$g(\varepsilon)=\frac{4\pi(2m)^{3/2}}{(2\pi\hbar)^3}\varepsilon^{1/2}$$

is the density of states per unit volume and

$$f_0(\varepsilon)=\frac{(2\pi\hbar)^3}{2}f_{\mathrm{W}}^{(0)}(\boldsymbol{p})=\frac{1}{\mathrm{e}^{\beta(\varepsilon-\varepsilon_{\mathrm{F}})}+1}$$

is the Fermi distribution function. We evaluate the integrals using the Sommerfeld expansion

$$\int_0^\infty \mathrm{d}\varepsilon\, F(\varepsilon)f_0(\varepsilon)=\int_0^{\varepsilon_{\mathrm{F}}}F(\varepsilon)\,\mathrm{d}\varepsilon+\frac{\pi^2}{6}\varepsilon_{\mathrm{F}}^2F'(\varepsilon_{\mathrm{F}})\left(\frac{kT}{\varepsilon_{\mathrm{F}}}\right)^2+\mathrm{O}\left[\left(\frac{kT}{\varepsilon_{\mathrm{F}}}\right)^4\right]$$

keeping only the first non-zero term in each case. Thus, the integrals multiplying $\boldsymbol{E}$ acquire only the leading zero-temperature term, while those multiplying $\boldsymbol{\nabla}T/T$, which contain the factor $\varepsilon-\varepsilon_{\mathrm{F}}$, are of order $(kT/\varepsilon_{\mathrm{F}})^2$. We find that

$$\boldsymbol{j}=\frac{ne}{m}\left(\frac{\pi^2k}{2}\frac{kT}{\varepsilon_{\mathrm{F}}}[\tau(\varepsilon_{\mathrm{F}})+\tfrac{2}{3}\varepsilon_{\mathrm{F}}\tau'(\varepsilon_{\mathrm{F}})]\,\boldsymbol{\nabla}T+\tau(\varepsilon_{\mathrm{F}})e\boldsymbol{E}\right)$$

$$\boldsymbol{q}=-\frac{n\varepsilon_{\mathrm{F}}}{m}\left(\frac{5\pi^2k}{6}\frac{kT}{\varepsilon_{\mathrm{F}}}[\tau(\varepsilon_{\mathrm{F}})+\tfrac{2}{5}\varepsilon_{\mathrm{F}}\tau'(\varepsilon_{\mathrm{F}})]\,\boldsymbol{\nabla}T+\tau(\varepsilon_{\mathrm{F}})e\boldsymbol{E}\right)$$

where

$$n=\tfrac{2}{3}g(\varepsilon_{\mathrm{F}})\varepsilon_{\mathrm{F}}=\frac{8\pi}{3}\frac{(2m\varepsilon_{\mathrm{F}})^{3/2}}{(2\pi\hbar)^3}$$

is the number density of electrons. The fact that these results depend only on $\tau(\varepsilon_{\mathrm{F}})$ and $\tau'(\varepsilon_{\mathrm{F}})$ reflects the fact that $f_0(\varepsilon)$ approaches the step function $\theta(\varepsilon_{\mathrm{F}}-\varepsilon)$ at low temperatures. It is now a simple matter to extract the transport coefficients. When there is no temperature gradient, the electric current is $\boldsymbol{j}=\sigma\boldsymbol{E}$, with the thermal conductivity given by

$$\sigma=\frac{ne^2}{m}\tau(\varepsilon_{\mathrm{F}}).$$

When there is a temperature gradient, the electric field needed to make the electric current vanish is $\boldsymbol{E}=S\,\boldsymbol{\nabla}T$, with the thermopower given by

$$S=-\frac{\pi^2k}{2e}\frac{kT}{\varepsilon_{\mathrm{F}}}\left(1+\frac{2}{3}\frac{\varepsilon_{\mathrm{F}}\tau'(\varepsilon_{\mathrm{F}})}{\tau(\varepsilon_{\mathrm{F}})}\right).$$

Finally, using this value of $\boldsymbol{E}$, we find that $\boldsymbol{q}=-\kappa\,\boldsymbol{\nabla}T$, with the thermal conductivity given by

$$\kappa=\frac{\pi^2}{3}\frac{nk^2T}{m}\tau(\varepsilon_{\mathrm{F}}).$$

APPENDIX. USEFUL FORMULAE

Physical constants

$$k = 1.381 \times 10^{-23}\ \mathrm{J\ K^{-1}}$$
$$\frac{\mu_B}{k} = 0.67\ \mathrm{K\ T^{-1}}$$
$$\frac{\hbar c}{k} = 2.29 \times 10^{-3}\ \mathrm{m\ K}$$
$$\frac{m_e c^2}{k} = 5.93 \times 10^{9}\ \mathrm{K}.$$

For interdependent variables x, y and z

$$\left(\frac{\partial x}{\partial y}\right)_z \left(\frac{\partial y}{\partial z}\right)_x \left(\frac{\partial z}{\partial x}\right)_y = -1.$$

The Stirling formula for a large non-negative integer n

$$n! \simeq \sqrt{2\pi n}\, n^n\, \mathrm{e}^{-n}$$
$$\ln(n!) \simeq n \ln n - n.$$

Gamma functions

$$\Gamma(x+1) = x\,\Gamma(x) = x \int_0^\infty u^{x-1}\,\mathrm{e}^{-u}\,\mathrm{d}u \qquad x > 0$$
$$\Gamma(\tfrac{1}{2}) = \sqrt{\pi}$$
$$\Gamma(n+1) = n! \qquad n = 0, 1, 2, \ldots.$$

Volume of a d-dimensional sphere of radius r

$$V_d(r) = \frac{2\pi^{d/2} r^d}{d\,\Gamma(d/2)}.$$

Summation formulae

$$\sum_{i=0}^{n} i = \tfrac{1}{2} n(n+1) \qquad n = 0, 1, 2, \ldots$$
$$\sum_{i=0}^{n} i^2 = \tfrac{1}{6} n(n+1)(2n+1) \qquad n = 0, 1, 2, \ldots$$

$$\sum_{i=0}^{n} q^i = \frac{1-q^{n+1}}{1-q} \qquad q \neq 1 \qquad n = 0, 1, 2, \ldots$$

$$\sum_{i=-n}^{n} q^i = \frac{1-q^{2n+1}}{q^n - q^{n+1}} \qquad q \neq 1 \qquad n = 0, \tfrac{1}{2}, 1, \tfrac{3}{2}, 2, \ldots$$

$$\sum_{i=0}^{n} i^m q^i = \left(q\frac{\mathrm{d}}{\mathrm{d}q}\right)^m \sum_{i=0}^{n} q^i \qquad n, m = 0, 1, 2, \ldots$$

$$\sum_{\{n_i\}} \frac{x_1^{n_1} x_2^{n_2} \ldots x_p^{n_p}}{n_1! n_2! \cdots n_p!} = \frac{1}{m!}(x_1 + x_2 + \cdots x_p)^m \qquad \begin{array}{l} n_i = 0, 1, 2, \ldots \\ n_1 + n_2 + \cdots n_p = m. \end{array}$$

Euler–Maclaurin formula

$$\sum_{k=0}^{\infty} f(k) = \int_0^{\infty} f(u)\,\mathrm{d}u + \tfrac{1}{2} f(0) - \tfrac{1}{12} f'(0) + \tfrac{1}{720} f'''(0) - \cdots.$$

Definite integrals

$$\int_0^1 u^{a-1}(1-u)^{b-1}\,\mathrm{d}u = \frac{\Gamma(a)\Gamma(b)}{\Gamma(a+b)} \qquad a > 0 \qquad b > 0$$

$$\int_{-\infty}^{\infty} \mathrm{e}^{-(au^2+bu)}\,\mathrm{d}u = \sqrt{\frac{\pi}{a}}\,\mathrm{e}^{b^2/4a} \qquad a > 0$$

$$\int_0^{\infty} u^m\,\mathrm{e}^{-au^2}\,\mathrm{d}u = \frac{\Gamma[(m+1)/2]}{2a^{(m+1)/2}} \qquad m > 0 \qquad a > 0$$

$$\int_0^{\infty} \frac{u^m}{\mathrm{e}^u \mp 1}\,\mathrm{d}u = \Gamma(m+1) \sum_{l=1}^{\infty} l^{-m-1} (\pm 1)^{l+1} \qquad m > 0$$

$$\int_0^{\infty} \frac{u^m}{\mathrm{e}^{u-x} + 1}\,\mathrm{d}u = \frac{x^{m+1}}{m+1}\left(1 + \frac{m(m+1)\pi^2}{6} x^{-2} + \cdots\right) x \gg 1.$$

Taylor expansions for $|x| \ll 1$

$$(1+x)^{m/2} = 1 + \frac{mx}{2} + \frac{m(m-2)x^2}{8} + \frac{m(m-2)(m-4)x^3}{48} + \cdots$$

$$\ln(1+x) = x - \frac{x^2}{2} + \frac{x^3}{3} - \frac{x^4}{4} + \cdots$$

$$\tanh x = x - \frac{x^3}{3} + \frac{2x^5}{15} - \cdots$$

$$\tanh^{-1} x = \tfrac{1}{2} \ln\left(\frac{1+x}{1-x}\right) = x + \frac{x^3}{3} + \frac{x^5}{5} + \cdots.$$

Reversion of power series

$$\begin{cases} z = ax + bx^2 + cxy + \text{cubic terms} \\ z = dy + ey^2 + fxy + \text{cubic terms} \end{cases} \rightarrow \begin{cases} x = \dfrac{z}{a} - \left(\dfrac{b}{a^3} + \dfrac{c}{da^2}\right) z^2 + \mathrm{O}(z^3) \\ y = \dfrac{z}{d} - \left(\dfrac{e}{d^3} + \dfrac{f}{ad^2}\right) z^2 + \mathrm{O}(z^3). \end{cases}$$

Error function

$$\operatorname{erf}(x) = \frac{2}{\sqrt{\pi}} \int_0^x \mathrm{e}^{-u^2}\, \mathrm{d}u = \begin{cases} \dfrac{2}{\sqrt{\pi}} x - \dfrac{2}{3\sqrt{\pi}} x^3 + \cdots & x \ll 1 \\ 1 - \dfrac{\mathrm{e}^{-x^2}}{\sqrt{\pi}} \left(\dfrac{1}{x} - \dfrac{1}{2x^3} + \cdots\right) & x \gg 1. \end{cases}$$

INDEX

In this index, references to the introductory material of chapter 1 are given as page numbers, while references to the remaining chapters are identified by problem numbers. Thus, the entry for Black-body radiation indicates that this topic is mentioned on page 9 of chapter 1 and discussed further in problems 2.11, 4.4 and 4.5.